March 24–27, 2013
Stateline, Nevada, USA

Association for Computing Machinery

Advancing Computing as a Science & Profession

I0036702

ISPD'13

Proceedings of the 2013 ACM
International Symposium on Physical Design

Sponsored by:
ACM SIGDA

Technical Co-Sponsor:
IEEE CAS

Supported by:
ATopTech, Cadence, IBM Research, Intel Corporation, Mentor Graphics, Oracle, Synopsys, and TSMC

Association for Computing Machinery

Advancing Computing as a Science & Profession

The Association for Computing Machinery
2 Penn Plaza, Suite 701
New York, New York 10121-0701

Notice to Past Authors of ACM-Published Articles

ISBN: 978-1-4503-1867-9

Additional copies may be ordered prepaid from:

ACM Order Department
PO Box 30777
New York, NY 10087-0777, USA

Phone: 1-800-342-6626 (USA and Canada)
+1-212-626-0500 (Global)
Fax: +1-212-944-1318
E-mail: acmhelp@acm.org
Hours of Operation: 8:30 am – 4:30 pm ET

ACM Order Number: 477135

Printed in the USA

Foreword

On behalf of the organizing committee, we are delighted to welcome you to the 2013 *ACM International Symposium on Physical Design* (ISPD), held at the Ridge Tahoe Resort in Stateline, Nevada. Continuing the great tradition established by its twenty-one predecessors, which includes a series of five ACM/SIGDA Physical Design Workshops held intermittently in 1987-1996 and sixteen editions of ISPD in the current form since 1997, the 2013 ISPD provides a premier forum to present leading-edge research results, exchange ideas, and promote research on critical areas related to the physical design of VLSI and other related systems.

We received 57 submissions from all around the world. After a rigorous, month-long, double-blind review process, the Technical Program Committee (TPC) met face-to-face to select papers to be included in the technical program based on over 250 reviews provided by 21 TPC members and 56 external reviewers. Finally, the committee accepted 17 papers to be presented in the symposium. These papers exhibit latest advancements in floorplanning, clocking, placement as well as routing, address challenges in advanced processes, identify opportunities in emerging technologies such as VeSFET, biochips, and e-beam, and offer new 3D solutions for physical design.

The ISPD 2013 program is complemented by a ISPD keynote speech as well as a TAU/ISPD joint keynote speech, thirteen ISPD invited talks, and five TAU/ISPD joint invited talks, all of which are delivered by distinguished researchers in both industry and academia. Mr. Liam Madden, corporate vice president of FPGA development and silicon technology at Xilinx, will present in the keynote speech future directions of next generation 3D technologies. A commemorative session on Monday afternoon allows us to pay tributes to Professor Yoji Kajitani, one of the most influential pioneers in EDA and a great educator. His collaborators and former students will share with us Professor Kajitani's exceptional works in graph theory and their applications to rectangle packing as well as place-and-route, and his recent research in the area of beyond-die co-design and integration. Professor Kajitani will conclude the commemoration session with a talk about how the concept of permutation links the problems in channel routing, packing, placement, and various other PD areas. Other invited talks will be interspersed with the presentations of accepted papers. The topics covered include the impacts of electromigration on PD, debug, verification and test, lithography advancement and challenges in the advanced technology nodes, new breakthroughs in synchronous designs, system-level PD, and next generation device design and architectures.

This year, we are happy to announce 3 joint sessions with the International Workshop on Timing Issues in the Specification and Synthesis of Digital Systems (TAU). The joint sessions have extended the technical program of ISPD by half a day. The joint TAU/ISPD keynote, given by Dr. Ruchir Puri, IBM fellow and adjunct professor at Columbia University, showcases the challenges and opportunities in EDA for high-performance microprocessors. We are excited to have a joint TAU/ISPD session dedicated to both the ISPD's discrete gate sizing contest and TAU's variability aware timing analysis contest. On Wednesday afternoon, there is a special joint invited session on hierarchical PD and timing.

Another special addition to ISPD's program in 2013 is the Expert Designer/User Session (EDS), which contains six invited presentations followed by a poster session. All EDS speakers are carefully selected to cover a wide range of PD topics. This brings the perspectives from the designer/user community to PD researchers, and encourages discussions and future research collaborations.

Since 2005, ISPD has organized highly competitive contests to promote and advance research in placement, global routing, clock network synthesis, and discrete gate sizing. This year's discrete gate sizing contest, which includes realistic interconnect delay calculation, continues to attract a large number of participants from all over the world. The contest results will be announced by the ISPD Contest Chair on Wednesday morning. Continuing the tradition of all the past contests, a new large-scale real-world benchmark suite for gate sizing with discrete gate library will be released in ISPD website:

http://www.ispd.cc

We would like to take this chance to express our gratitude to the authors, the presenters, the keynote/invited speakers for contributing to the high-quality program, and the session chairs for moderating the sessions. We would like to thank our program committee and external reviewers, who provided insightful constructive comments and detailed reviews to the authors. We greatly appreciate the exceptional set of invited talks put together by the Steering Committee, which is chaired by Jiang Hu. We also thank the Steering Committee for selecting the best paper. Special thanks go to the Publications Chair Azadeh Davoodi and the Publicity Chair Evangeline Young for their tremendous services. We would like to acknowledge the Intel team led by the Contest Chair Mustafa Ozdal for organizing and taking the gate sizing contest to the next level. We are also grateful to our sponsors. The symposium is sponsored by the ACM SIGDA (Special Interest Group on Design Automation) with technical co-sponsorship from the IEEE Circuits and Systems Society. Generous financial contributions have also been provided by (in alphabetical order): ATopTech, Cadence, IBM Research, Intel Corporation, Mentor Graphics, Oracle, Synopsys, and TSMC. We also thank the University of Kitakyushu and Waseda University for their generous sponsorship of the dinner banquet honoring Professor Kajitani. Special thanks should also go to the Technical Committee on VLSI Design Technologies, Engineering Sciences Society, IEICE for making the commemorative session for Professor Kajitani a success. Last but not least, we thank Lisa Tolles and Marie Efinger of Sheridan Communications for their expertise and enormous patience during the production of the proceedings.

The organizing committee hopes that you will enjoy ISPD 2013 and find it useful as well as informative. We look forward to seeing you again in future editions of ISPD.

Cheng-Kok Koh
ISPD 2013 General Chair

Cliff C. N. Sze
Technical Program Chair

Table of Contents

Welcome and Keynote Address
Session Chair: Cheng-Kok Koh *(Purdue University)*

Session 1: 3D Integration and Physical Planning
Session Chair: Markus Olbrich *(University of Hannover)*

Session 2: Validation and Design for Yield
Session Chair: Yao-Wen Chang *(National Taiwan University)*

Session 3: Commemoration for Professor Y. Kajitani
Session Chairs: Yasuhiro Takashima *(University of Kitakyushu)*, Jiang Hu *(Texas A&M)*

Session 4: Advanced Technologies and Design for Manufacturability
Session Chair: Ting-Chi Wang *(National Tsing Hua University)*

Session 5: Routability and Routing
Session Chair: Jackey Yan *(Cadence)*

Session 6: New Frontiers for Physical Design
Session Chair: Ismail Bustany *(Mentor Graphics)*

Session 7: Expert Designer/User Session (EDS)
Session Chairs: Cliff Sze *(IBM)*

Session 8: Logic, Clock Driven PD and Beyond
Laleh Behjat *(University of Calgary)*

Session 9: TAU/ISPD Joint Session on Contests
Session Chair: Charles Liu *(TSMC)*

TAU/ISPD Keynote
Session Chair: Chirayu Amin *(Intel Corporation)*

TAU/ISPD Invited Session: What Will It Take to Tame the Hierarchical Design Trolls?
Session Chair: Tom Spyrou *(Altera)*

ISPD 2013 Symposium Organization

General Chair: Cheng-Kok Koh *(Purdue University)*

Technical Program Chair: Cliff Sze *(IBM Research)*

Past Chair: Jiang Hu *(Texas A&M University)*

Steering Committee Chair: Jiang Hu *(Texas A&M University)*

Steering Committee: Yao-Wen Chang *(National Taiwan University)*
Jason Cong *(UCLA)*
Jiang Hu *(Texas A&M University)*
Gi-Joon Nam *(IBM Research)*
Sachin Sapatnekar *(University of Minnesota)*
Prashant Saxena *(Synopsys)*
P. V. Srinivas *(Mentor Graphics)*

Program Committee: Laleh Behjat *(University of Calgary)*
Ismail Bustany *(Mentor Graphics)*
Steven C. Chan *(GLOBALFOUNDRIES)*
Hung-Ming Chen *(National Chiao Tung University)*
Tung-Chieh Chen *(SpringSoft)*
David Chinnery *(Mentor Graphics)*
Azadeh Davoodi *(University of Wisconsin - Madison)*
Dwight Hill *(Synopsys)*
Shiyan Hu *(Michigan Technological University)*
Charles C. C. Liu *(TSMC)*
Malgorzata Marek-Sadowska *(University of California, Santa Barbara)*
Yu-Yen Mo *(Oracle)*
Markus Olbrich *(University of Hannover)*
Marcelo de Oliveira Johann *(Universidade Federal do Rio Grande do Sul)*
Yiyu Shi *(Missouri University of Science and Technology)*
Bill Swartz *(TimberWolf Systems / University of Texas - Dallas)*
Cliff Sze *(IBM Research)*
Atsushi Takahashi *(Tokyo Institute of Technology)*
Chin-Chi Teng *(Cadence)*
Hailong Yao *(Tsinghua)*
Cheng Zhuo *(Intel)*

Publications Chair: Azadeh Davoodi *(University of Wisconsin - Madison)*

Publicity Chair / Webmaster: Evangeline Young *(Chinese University of Hong Kong)*

Contest Chair: Mustafa Ozdal *(Intel)*

Additional reviewers:

Vahe Arakelyan
Sarvesh Bhardwaj
Jonathan Bishop
Manjit Borah
Ashutosh Chakraborty
Terence Chen
Salim Chowdhury
Cyril Descleves
Duo Ding
Hussein Etawil
Guilherme Flach
Houle Gan
Hui Geng
John Gilchrist
Fang Gong
Rais Huda
Ryoichi Inanami
Rouwaida Kanj
Ivan Kissiov
Yukihide Kohira
Alexander Korshak
C. K. Lam
Lin Liu
Jianchao Lu
Ravi Mamidi
Sravan Marella
Laurent Masse-Navette
Trevor Meyerowitz

Kirill Minkovich
Tarun Mittal
Swamy Muddu
Shigetoshi Nakatake
David Z. Pan
Rajendran Panda
Tamer Ragheb
Logan Rakai
Ahmed Ramadan
Nimish Shah
Joseph Shinnerl
Yasuhiro Takashima
Satoshi Tanaka
Edward Teoh
Yoichi Tomioka
Christophe Vinard
Alex Volkov
Tao Wang
Yue Xu
Jacky Yan
Amir Yazdanbakhsh
Xiaoji Ye
Xi Yin
Evangeline Young
Guo Yu
Min Zhao
Yuchen Zhou
Jianfang Zhu

ISPD 2013 Sponsors & Supporters

Sponsor:

Technical Co-sponsor:

Supporters:

Heterogeneous 3-D Stacking, Can We Have the Best of Both (Technology) Worlds?

Liam Madden
Xilinx, Inc.
2100 Logic Drive
San Jose, CA 95124
liam.madden@xilinx.com

ABSTRACT

Since the advent of integrated circuit technology in 1958, the industry has focused primarily on monolithic integration. Unfortunately, due to physical and economic issues, the vast majority of high performance analog chips, high density memory chips, and high performance digital chips are each built on separate technologies. Therefore, in order to deliver optimum system performance, power and cost, it is desirable to integrate multiple different die, each using its own optimized technology, in a single package.

This paper describes the industry's first heterogeneous Stacked Silicon Interconnect (SSI) FPGA family (3D integration). The heterogeneous IC stack delivers up to 2.78Tb/s transceiver bandwidth. The resulting bandwidth is approximately three times that achievable in a monolithic solution. Mounted on a passive silicon interposer with through-silicon vias (TSVs), the heterogeneous IC stack comprises FPGA ICs with 13.1-Gb/s transceivers and dedicated analog ICs with 28-Gb/s transceivers.

Categories and Subject Descriptors

B.7.1 [**Integrated Circuits**]: General – *advanced.*

General Terms

Design.

Keywords

3D Chip Stacking

1. 3D INTEGRATION

The XC7VH580T [1] is the first commercial FPGA built with heterogeneous SSI. The device consists of a passive silicon interposer and three active die: an 8 x 28Gb/s transceiver IC (GTZ-IC) and two FPGA ICs known as Super Logic Regions (SLR), see Figures 1 and 2.

The enabling technologies which deliver this performance include a silicon interposer with four high-density (~1um pitch) interconnects, through silicon vias (TSVs), and fine-pitch micro-bumps as shown in Figure 3.

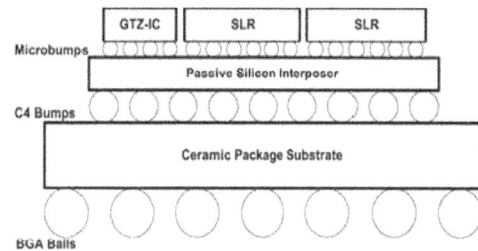

Figure 1. Conceptual cross-sectional view of *XC7VH580T*.

Figure 2. *XC7VH580T* in package substrate. (a) Interposer only; (b) GTZ-IC and two SLRs mounted face-down on interposer.

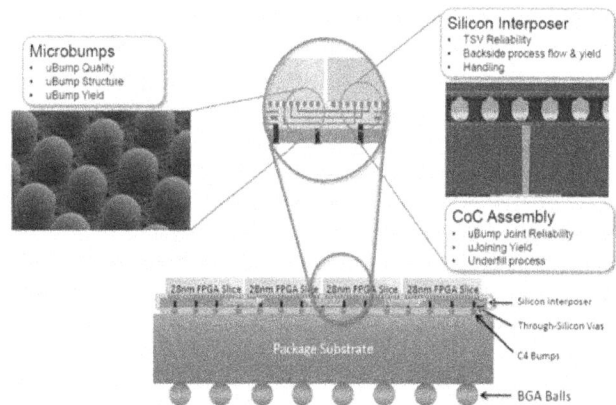

Figure 3. Stacked Silicon Interconnect cross section.

2. INTERPOSER SIGNAL INTEGRITY

Although the silicon interposer is only a hundred microns thick, failure to consider high frequency effects of the TSVs will degrade the rise/fall time, increase crosstalk, increase noise injection, and cause significant performance degradation of the signal transmitted through the high speed channel.

2.1 Silicon Interposer TSV Modeling and Optimization

A full 3D EM field solver is used to model the silicon TSV interposer accurately over a wide frequency range. A broadband S-parameter model is generated with an upper frequency limit of 50GHz. The interposer consists of TSVs, four metal layers for die-to-die connections, micro-bumps for die to interposer connections and C4 bumps for interposer to package connections. A silicon interposer test vehicle is fabricated for measurement and verification. The measured data is compared with simulation results. Figure 4 shows the comparative results of effective capacitances and conductances which were extracted from the test structures. Two different silicon substrates are fabricated to compare the substrate resistivity effects. We observe good agreement between the measured and simulated results. Based on these measurements and simulation results, we conclude that a 20 ohm-cm high resistivity silicon substrate is the best choice for very high speed signaling applications such as the 28Gb/s SerDes.

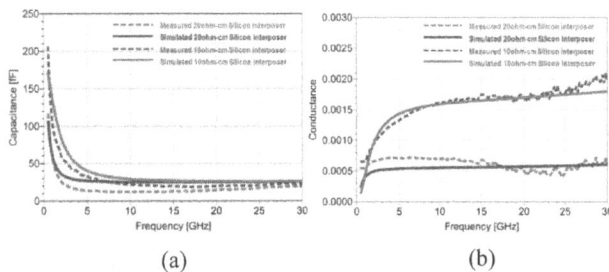

(a) (b)

Figure 4. (a) Effective capacitance correlation[*fF*]; (b) Effective conductance correlation.

2.2 Package Design for Ultra Low Loss

The package substrate can be a major source of losses due to discontinuities, dielectric loss, skin effect and surface roughness loss due to narrower trace widths. A Low Temperature Co-Fired Ceramic (LTCC) package substrate is chosen to minimize loss and discontinuities.

2.3 Embedded Capacitors Design in the Silicon Interposer to Minimize SSO Noise

The ceramic package provides excellent signal integrity in the signal channel but at the same time has a disadvantage in power integrity due to the long vertical power connections. Higher inductance in the package coupled with the on-die capacitance creates higher plane impedance (and therefore noise) at the resonant frequency. To minimize the power plane noise, a finger capacitor constructed from metal layers in the silicon interposer is used. Figure 5 shows one example of worst case simultaneous switching noise simulated at the resonant frequency. Power noise is significantly reduced by the finger capacitor in the silicon interposer as shown.

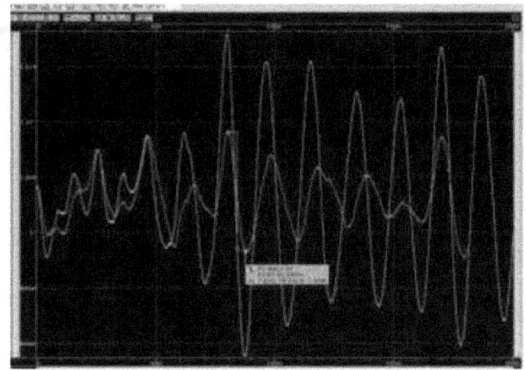

Figure 5. Simultaneous switching noise simulation in system without interposer capacitor (in yellow) and with interposer capacitor (in blue).

2.4 Channel Simulation

Full channel analysis is performed including an optimized silicon interposer model on top of a low loss package substrate and a PCB model. The simulation data is then compared to the measured 28Gb/s eye diagram and shows very good correlation as shown in Figure 6.

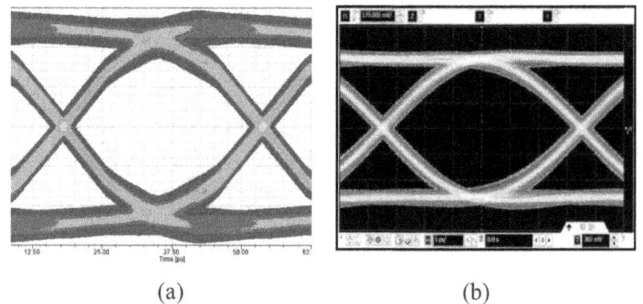

(a) (b)

Figure 6. System level eye diagram comparison: (a) *28Gb/s* simulated eye diagram; (b) *28Gb/s* measured eye diagram.

3. CONCLUSIONS

A heterogeneous FPGA and transceiver solution with up to 2.78Tb/s bandwidth is detailed. The design methodologies to minimize signal insertion loss, reflection loss and power coupling noise in the system are presented. Full channel analysis is performed including an optimized silicon interposer model on top of a low loss package substrate and a PCB model. The simulation data when compared to the measured 28Gb/s eye diagram shows excellent correlation. The proposed optimized channel system enables high performance signaling at 28Gb/s.

4. REFERENCES

[1] Xilinx, Inc., *7 Series FPGAs Overview*, May 5, 2012, http://www.xilinx.com/support/documentation/data_sheets/ds180_7Series_Overview.

Physical-Aware System-Level Design for Tiled Hierarchical Chip Multiprocessors

Jordi Cortadella   Javier de San Pedro   Nikita Nikitin   Jordi Petit [*]
Universitat Politècnica de Catalunya
Barcelona, Spain

ABSTRACT

Tiled hierarchical architectures for Chip Multiprocessors (CMPs) represent a rapid way of building scalable and power-efficient many-core computing systems. At the early stages of the design of a CMP, physical parameters are often ignored and postponed for later design stages. In this work, the importance of physical-aware system-level exploration is investigated, and a strategy for deriving chip floorplans is described. Additionally, wire planning of the on-chip interconnect is performed, as its topology and organization affect the physical layout of the system. Traditional algorithms for floorplanning and wire planning are customized to include physical constraints specific for tiled hierarchical architectures. Over-the-cell routing is used as one of the major area savings strategy. The combination of architectural exploration and physical planning is studied with an example and the impact of the physical aspects on the selection of architectural parameters is evaluated.

Categories and Subject Descriptors: B.7.2 [Integrated circuits]: Design Aids—*placement and routing*

General Terms: Algorithms, Design

Keywords: Network-on-chip, floorplanning, wire planning, chip multiprocessor

1. INTRODUCTION

During the past decade many-core chip multiprocessors have become the major trend in designing scalable computing architectures. Multiple processing units with distributed memory combined with power saving schemes are the platforms used today for exploiting application parallelism while keeping power consumption under control.

Tiled CMP architectures facilitate the design process offering a rapid way of assembling platforms with tens or hun-

---

[*]This work has been supported by a gift from Intel Corp., the Spanish Ministry of Science and Innovation (project FORMALISM, TIN2007-66523) and the Catalan Government (SGR 2009-1137).

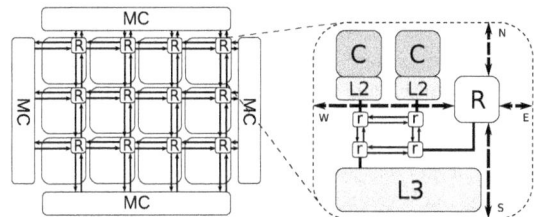

Figure 1: Structural representation of a hierarchical tiled CMP with two-level interconnect: a global mesh and bi-directional rings in clusters.

dreds of units by replicating pre-designed tiles [3, 12, 19]. Nevertheless, large-scale systems obtained by tile replication may suffer from the limited bandwidth of the on-chip interconnect and deliver poor performance because of the communication bottleneck to shared memory. To overcome this problem, *hierarchical* CMP organizations have been proposed to better exploit locality [3, 8].

Figure 1 depicts the block diagram of a tiled hierarchical CMP with 24 cores and distributed L3 cache. The chip is organized as a 4×3 regular grid of tiles (*clusters*), each one including two computing cores (C) with private cache (L2), a distributed shared cache (L3), a router of the global mesh (R) and a local interconnect (bi-directional ring with four routers (r)). The two-level hierarchical interconnect constitutes the backbone of this architecture. The purpose of the global mesh is to provide inter-cluster communication, as well as access to the memory controllers (MC). Intra-cluster communication is supported by low-latency rings that significantly improve the bandwidth of the system given the locality of memory references inherent to the applications.

The problem of *system-level design* for a many-core CMP consists of selecting high-level architectural parameters (e.g., number of cores, size of cache, topology of the interconnect, etc.) so as to maximize system performance for the selected workload and satisfy the design constraints (e.g., area and power). System-level design is performed early in the design cycle. The main complexity of this task is determined by the vast space of potential architectural configurations and the inaccuracy of the models to represent the components of the system and the workload.

To alleviate the problem complexity, most strategies for architectural exploration disregard physical parameters and postpone them to the later stages of design. However, in this paper we show that physical planning has a non-negligible impact on performance and area of a CMP. In this work we propose methods for *floorplanning* and *wire planning* of tiled hierarchical CMPs and show the impact of physical parameters in the configuration of the architecture.

1.1 Networks-on-chip

Physical planning of a CMP is strictly driven by the organization of its on-chip interconnect. In this section we give a brief overview of the interconnect architecture.

Networks-on-chip (*NoCs*) [7] have been firmly established as a the paradigm to implement on-chip communication for many-core CMPs. In a NoC, each link is a short, point-to-point connection that can be shared among different communication flows. NoCs offer several advantages: they enable higher performance, avoid wire underutilization and facilitate reuse and extendability of designs.

The core component of a NoC is the *router* (or *switch*). Each router implements a crossbar switch in order to commute packets from any input port to any output port. Network links connect either different routers, or routers with the endpoint nodes, such as cores and memories. On-chip communication is realized by means of data packet exchange between the nodes. A source node *injects* a packet into the NoC via the attached router, which forwards the packet to other routers, subject to the established routing policy. The destination router consumes the packet from the network and forwards data to the relevant node.

The network *topology* dictates the arrangement of links, routers and nodes. There are several commonly used topologies, such as meshes, rings, butterflies, fat trees, etc [9]. Hierarchical topologies are also used by combining some of the previous topologies.

1.2 Related work

Floorplanning as a part of the VLSI design flow has been extensively studied for decades. Traditional algorithms often try to minimize a linear combination of area and estimated wire length, and leave actual wire planning to posterior stages in the design process. Hierarchical approaches to floorplanning have already been shown to reduce the algorithm runtime. Quite often hierarchical floorplanning is applied to the design of Systems-on-Chip (*SoCs*), for which every component can be considered as a fixed-size block. These blocks can be generated using fixed-outline floorplanners such as [2], while the system-level floorplanning can be solved using the traditional minimal area techniques such as [6]. In this work, we will instead exploit the regularity of tiled CMPs with hierarchical NoCs.

When floorplanning a CMP, it might also be desirable to optimize factors other than area and wire length. Previous approaches exist that evaluate floorplans based on other qualities such as temperature minimization [20, 15] or power consumption [22], using analytical models. For floorplanning at the system-level, [27] proposes a method that creates tile arrangements which minimize the overall wire length for several 3D NoC topologies.

Adding constraints is also considered in modern floorplanning [24]. These constraints usually restrict valid placement of blocks, e.g. adjacency constraints, distance limits between pairs of blocks, and preplaced objects [28]. However, in this work, we want to satisfy constraints imposed by the CMP interconnect nets. In [26] the authors show how over-simplified models for those constraints (e.g., disregarding pin place-

ment) produces sub-optimal floorplans, but only for classic bus-based interconnects.

1.3 Motivation

The problems of physical planning for CMPs are related to traditional problems in VLSI physical design [21]. CMP floorplanning is similar to classical VLSI floorplanning, while wire planning is more common with global routing. However, there are several aspects inherent to tiled hierarchical CMPs which motivate us to extend existing approaches.

As shown in Fig. 1, the tiled organization of CMPs reduces the floorplanning problem from chip to cluster level. However, the cluster floorplan has to satisfy the property of symmetry in the location of the North/South and East/West ports at the boundaries of the tile. This enables the construction of a full chip by replicating and abutting of tiles.

Floorplanning of the local interconnect introduces another complexity into the design. For example, when considering rings, as in Fig. 1, it is required that the links between the ring routers (r) have *balanced* lengths to guarantee similar hop delays. If the link delays are imbalanced the communication through the ring may have a negative impact on performance.

A special type of constraints, such as *adjacency* or *maximum net delay* constraints are required to prevent certain components be placed far from each other. A typical example may be a core and its L2 cache. Placing a cache far from the core may increase its access delay and result into a significant performance penalty. While adjacency of the two components may appear as a too strict constraint, a weaker requirement of the inter-component distance to be less than one hop will be enough to assure no loss of performance.

An important observation is the recent tendency to design CMPs with wide links. Communication links of the on-chip interconnect may incorporate thousands of wires, aiming at transferring a complete cache line in one cycle. Given the ITRS prediction for minimal wire spacing [1], links of a global mesh can have a width of the order of 10^2 μm, occupying a significant amount of chip area.

One of the possible ways to alleviate the area overhead is to benefit from *over-the-component* routing. Some of the CMP components, such as memories, do not use all the metal layers available in the technology and, therefore, these available resources can be used to implement global nets across the chip.

In this scenario, the most complex components using all metals layers may act as blockages for over-the-component routing. Hence, one of the purposes of wire planning is to verify chip *routability*. Another purpose is the estimation of wire length, which is one of the main parameters when evaluating design quality [23].

The main contribution of this paper is the incorporation of physical planning during the architectural exploration of hierarchical CMPs. The algorithms for floorplanning and wire planning are customized to support constraints for tiled configurations. To demonstrate the viability of this methodology, a case study for exploration is presented and the influence of physical planning on the exploration is evaluated.

2. OVERVIEW

This section gives an overview of the main contributions of the paper by using a simple example. The impact of physical planning on architectural exploration will be illustrated.

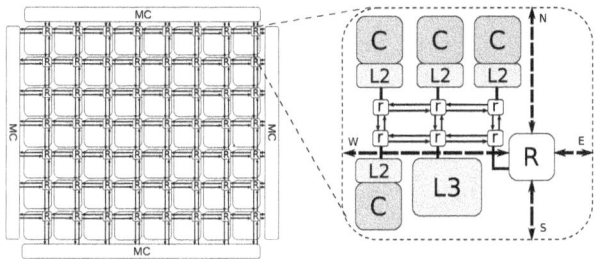

Figure 2: Block diagram of a CMP configuration.

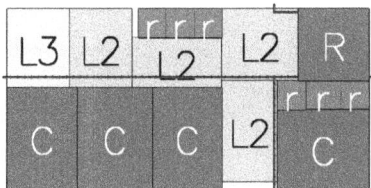

Figure 3: Minimum-area floorplan

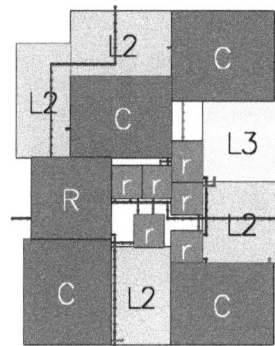

Figure 4: NoC-aware floorplan

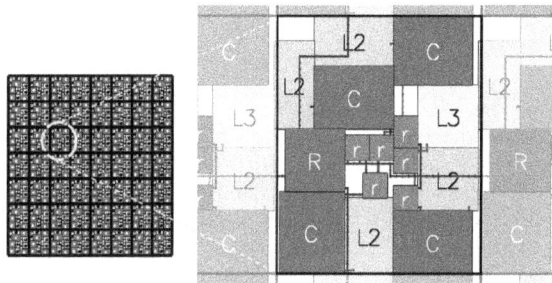

Figure 5: NoC-aware floorplan (full cluster)

Selection of the sample configuration. For this example, let us assume that an architectural exploration tool has generated a configuration such as the one shown in Fig. 2. This configuration has a total of 224 identical cores. The cores are pre-designed and each one occupies an area of 1.2 mm^2. The layout of the cores can be flipped and rotated.

Each core is assumed to include an internal L1 cache and has an associated private L2 cache. We assume that L2 caches take an area of 1 mm^2 and can be implemented as soft blocks, i.e., various layouts with different aspect ratios are allowed. The L3 cache is shared among all cores (a total of 56 MB for the entire chip) and is also a soft block with the same area as the L2 caches.

Conventional floorplanning. When we consider the floorplan the entire system, we face a problem with about 800 components, including cores, L2 and L3 caches and routers. However, in this work we deal with tiled hierarchical CMPs, which have several proven benefits by enabling a divide-and-conquer design strategy. Floorplanning, placement, routing, and timing closure are processes that can be applied to a single tile while guaranteeing correctness for the global system. For this reason, we will center on the floorplanning of a single tile.

The hierarchical CMP in Fig. 2 has 56 identical clusters, interconnected with an 8 × 7 mesh. Each cluster contains 16 components, including cores, caches and routers. The L3 cache is distributed uniformly among all clusters. The L2 and L3 caches and the router are interconnected with a bidirectional ring. The total area of a cluster is 12.52 mm^2.

Figure 3 depicts a minimum-area floorplan that could be obtained by a conventional floorplanner such as CompaSS[6]. From the point of view of a hierarchical CMP, this floorplan has some undesirable problems:

1. Some cores are not adjacent to their private L2 caches, potentially increasing the communication latency between them. Similarly, there are long distances between some caches and the corresponding ring routers.

2. Ring routers for the local interconnect are not evenly separated. In a ring, the wire length of the longest hop dictates the maximum speed for the entire ring. If this distance is too long, some timing constraints might be violated. Therefore, it is desirable to minimize the length of each link hop separately instead of minimizing the total link length.

3. Assuming that the cores (C) and the router (R) use all metal layers, the two rightmost ring routers (r) have no available routing area in their boundaries. Thus, the design cannot be routed without whitespace insertion.

NoC-aware floorplanning. An alternative floorplan is shown in Fig. 4. This floorplan has been generated using all the constraints and enhancements discussed in this work. Since area minimization is no longer the only objective, this floorplan has a 53% area increase (19.12 mm^2). However, all of the cores are now adjacent to their private L2 caches. Additionally, a route can be found between all the ring routers so that the the link length for each hop is always between 0.2 and 0.7 mm, and the distance between a component and its attached ring router is strictly less than 0.4 mm.

As an example, Fig. 5 shows a floorplan for the entire system, including all clusters, based on the cluster floorplan from Fig. 4.

Note that a 53% increase in area may induce an unacceptable overhead in manufacturing cost. This fact may encourage a designer to select an alternative architectural configuration, with a slightly lower performance, although with better floorplan properties.

3. ARCHITECTURAL EXPLORATION

This section overviews the *flow for architectural exploration of CMPs* and introduces the context for physical planning. Consider the problem of maximizing CMP performance (throughput) subject to a resource budget, i.e. constraints on area and power. The given formulation is an example of the architectural exploration problem with the objective of efficiently distributing the chip resources among the components of a multi-core system, e.g. cores, memories and interconnect.

The design space for exploration is specified through a set of models and design constraints. The *models* describe the behavior of individual components. There can be different models for cores characterizing different micro-architectural features that trade-off area, power and performance (in-order/out-of-order execution, multi-threading, etc). The memory models define the size, area and latency of different memory modules. The models for the interconnect topologies define their physical and performance properties (latency, contention, etc).

The expected workload for the CMP requires another type of models that characterize the observable behavior produced by the generated memory patterns (memory locality, burstiness, etc). *Constraints* on power consumption and area are typically defined to confine the design space.

Exploration is a complex optimization problem due to the vast discrete space of architectural variables that determine the configuration of a CMP (e.g. number of cores, cache sizes, interconnect topology, link width). To handle this complexity, in this work we resort to a three-stage divide-and-conquer approach to solve the exploration problem. Figure 6 illustrates our methodology, with the main stages being the *architectural exploration*, *physical planning* and *validation*.

Architectural exploration. During the first stage, analytical models are used to rapidly prune the design space and generate a set of promising configurations in the area/power/performance space. The analytical model from [16] is used to evaluate CMP configurations and discriminate those with poor performance. Static and dynamic power are also evaluated using analytical approximations based on the area and activity of the CMP components. The area is approximated as the sum of the areas of all components on chip.

Analytical models are used as a cost estimator for an iterative metaheuristic-based search to efficiently navigate through the design space. This space is described with a set of architectural variables and a set of *transformations* is defined to explore the neighborhood of any particular configuration. Some examples of transformations include modifying the dimensions of the top-level mesh, the number of cores per cluster or the topology of the local interconnect, among others. *Simulated Annealing* [13] and *Extremal Optimization* [5] are used to explore the design space by probabilistically applying transformations and tracking the best discovered solution.

Physical planning. The objective of this stage is to evaluate wire length and give a more accurate area estimation. The floorplanning and wire planning algorithms at this stage consider physical constraints for individual CMP components, such as the aspect ratio and the number of metal layers. The accuracy in estimation comes at the expense of a higher algorithmic cost, which is however tolerated by

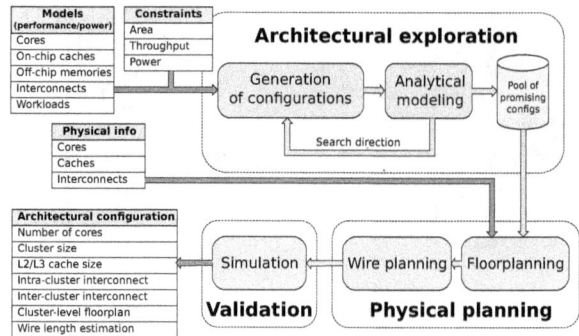

Figure 6: The CMP exploration flow.

performing the planning for a moderate number of configurations, selected during the first stage.

Validation. Finally, the validation phase of the flow is aimed at verifying performance and power, which may differ from the initial analytical estimates. In the current setup we use a cycle-accurate simulation for CMP interconnect, supplied with probabilistic automata models for cores and memories.

This paper focuses on the algorithms for physical planning of hierarchical CMPs. Their objective is to accurately estimate the chip area and wire length, subject to the physical constraints. The methods proposed in this work are applied at the second phase of the described exploration flow.

4. FLOORPLANNING STRATEGY

Floorplanning is the task of defining tentative locations for the blocks of system under certain geometric constraints. The blocks represent pre-designed CMP components such as cores, memories and routers. The blocks can either have a fixed size or accept a set of different *aspect ratios*. The traditional floorplanning problem only considers the minimization of the total area occupied by the components. More advanced floorplanning strategies can also consider the minimization of other metrics such as the estimated wire length.

Because of the complexity of the problem, it is essential to select efficient data structures to represent floorplans. *Slicing trees* [17] are a very popular and compact representation. When combined with compaction, slicing trees are a complete floorplan representation for all non-slicing floorplans [14]. As blocks with multiple aspect ratios are common (e.g., memories), slicing floorplans are very appropriate for CMP floorplanning [29].

In this work, we use Simulated Annealing for the exploration of slicing floorplans similarly as proposed in [25], where the cost function is defined as a linear combination of area and wire length approximated with half-perimeter wire length. In this work we extend this cost function with other components that aim at generating floorplans with some properties and constraints for tiled hierarchical CMPs.

4.1 CMP floorplanning constraints

In Section 1.3 we have discussed some of the requirements for the physical planning of tiled hierarchical CMPs. Next we address them in more detail.

Over-the-cell routing. One important aspect of our approach for floorplanning is that routing is to be done en-

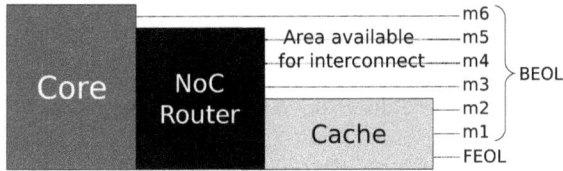

Figure 7: Over-the-cell routing

tirely inside the bounds of the floorplan, using free metal layers on top of placed components (Fig. 7). Because of the prevalence of cache memories in the tiles, we can assume that every configuration can be routed using the available metal layers on top of the components without requiring any extra whitespace. During floorplanning, and as part of the wire length estimation that will be described further in this section, unroutable configurations are discarded.

Abutability. Because only a single tile of a chip is floorplanned, some nets that connect different clusters will have floating terminals that must be placed on one of the boundaries of the tile. However, the placement of this terminal must lie adjacent to the placement of a corresponding terminal on the next cluster. Thus, a special symmetry constraint is created between pairs of nets. All the global interconnect nets have this property.

Wire length constraints. Due to performance reasons, certain critical nets must have a wire length constraint. In case these constraints are violated the floorplan is rejected. This maximum length will depend on the desired interconnect operating frequency, wire sizing and other parameters [1].

Equidistantly-spaced nets. For most interconnect network, the communication delay is determined by the maximum length of a set of links. For example, in a ring, the cycle period must be long enough to allow packets to propagate across the longest of the ring hops. In these cases, it is desirable not to strictly minimize the total wire length, but to balance the individual lengths of the respective links. For this reason, nets that must satisfy this requirements are evaluated differently in the cost function, minimizing the sum of the squares of the lengths instead:

$$WL_{Eq}(\text{Fp}) = \sum_{\forall net \in Ring} WL(net)^2$$

4.2 Cost function

The final cost function used in the search is as follows:

$$\text{Cost}(\text{Fp}) = \alpha Area(\text{Fp}) + \beta WL(\text{Fp}) + \gamma WL_{Eq}(\text{Fp}) + P(\text{Fp})$$

In this expression, Fp is the floorplan being evaluated, *Area* is defined as the effective area of the floorplan, *WL* is the sum of the wire length estimation for each net, and WL_{Eq} is the sum of the squares of the estimated wire lengths for nets in the ring interconnect, if any. The goal of WL_{Eq} is to penalize floorplans where equidistantly-spaced nets have excessively diverging lengths, as mentioned in the previous section.

The last term, $P(\text{Fp})$, aggregates all penalties that are applied when a floorplan does not satisfy one of the constrains detailed in this section. The α, β and γ parameters are weights that a designer can use to guide the search

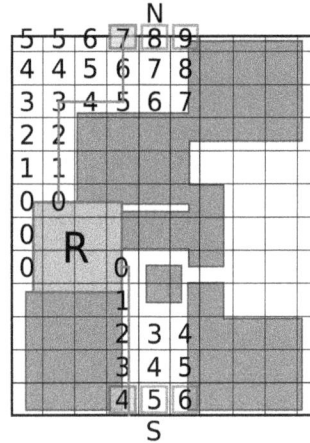

Figure 8: Maze routing a pair of nets with abutability constraint (floorplan from Fig. 4, blockages marked in black).

towards floorplans with smaller area or towards floorplans with smaller wire lengths. An example of this trade-off will be seen in Section 6.3.

4.3 Wire length estimation

A good wire length estimator is important for the evaluation of the cost function. Wire length estimations are used in the $WL(\text{Fp})$ and $WL_{Eq}(\text{Fp})$ terms of the cost function. Additionally, it is used to check satisfiability of some of the constraints, such as abutability and wire length limits.

In over-the-cell routing, the only space considered for routing is the free space over the components that have the top metal layers available. Since cores and routers typically implement a complex internal wiring and thus utilize the highest number of layers, memories are the only components in the entire design that leave some metal layers unused. In fact, the relative area of memories in a cluster is defined by the configuration, but it usually ranges between 50%-60% for the best configurations as seen in our tests.

Thus, the lowest metal layers will typically have no space for routing, while the upper layers will have up to 60% of space available, thereby making over-the-cell routing possible. An example can be seen in Fig. 8, which represents a middle metal layer from the floorplan in Fig. 4, with the area occupied by components marked in a dark color.

The work upon this floorplanning algorithm has been based on, [25], proposes the use of the half-perimeter wire length as an estimator. In this work, we propose the use of Lee's algorithm [18], often known as Maze routing. The algorithm is simplified by a) routing links, not individual wires, b) routing each net independently, and c) working over one metal layer only. Thus, routes might be generated that may be found unfeasible during wire planning. However, for the case of nets with two terminals, we can guarantee that a route found using this method is a valid lower bound. Thus, this information can be used to verify wire length and routability constraints. Because of simplification (a), the size of the routing grid is determined by the minimum link width.

The use of Lee's algorithm also enables checking for violations of the abutability requirement. When planning pairs of nets with such requirement, the algorithm will only accept a path if a matching path has been found on the opposite side for the paired net. The algorithm also will not stop at the first path, but rather collect all paths and select the one where the route is shortest to both opposing extremes of the tile. In Fig. 8, this algorithm is applied to estimate the length of the two vertical mesh links (from the Router to the north side and from the Router to the south). The shortest route for the north net is discarded because at the opposing side of the tile (same column, last row) there has been no path found for the south net.

5. WIRE PLANNING

In order to fully realize the floorplanning estimated in the previous section, we need to establish a wire planning that connects all the required nets between the components and that allows the tiling of the cells. This wire planning must use over-the-cell routing and minimize its wire lengths, while balancing the nets.

This problem corresponds to a routing problem and we solve it in two steps. In the first step, we formulate the routing problem as a Boolean satisfiability problem for which we obtain a feasible solution with a SAT solver. Then, in the second step, we iteratively reduce the wire length of several nets by converting the satisfiability problem to an integer linear programming problem that we solve with an ILP solver. In the following, we describe their essential elements.

Problem formulation. We formulate the routing problem as a Boolean satisfiability problem in the lines of [11], which we extend with some insights that are needed in our context.

The main variables of the SAT problem correspond to the presence (or absence) of a wire segment between two adjacent nodes of the underlying 3D grid. Another set of variables encodes the assignment of wire segments to specific nets. The SAT problem includes several types of constraints:

- Consistency constraints enforce the expected behavior of the variables we have introduced, e.g., if an edge is assigned to a net, then the edge must be occupied by a wire.

- Routability constraints define a legal routing between the components. Basically, these constraints establish that a set of wire segments guarantee the connectivity of all pins of a net. The formulation is similar to the one presented in [11] but extended to handle floating terminals. Our solution is based on the idea that routing must be performed among regions of points that define the endpoints of the nets. These regions are characterized by a set of (not necessarily adjacent nor disjoint) points that may describe the location of a component or the set of all possible locations for a pin. The correctness of our routability constraints is based on Euler's graph theory.

- Abutability constraints ensure the symmetry between the wires that are used to interconnect tiled cells. These constraints assert that if a wire in the North boundary provides a signal for a net that interconnects adjacent cells, another wire for the same cell must be placed in the same position in the South boundary. Similar relations must also occur in the other direction and for East/West boundaries.

- Optionally, constraints for design rules can be requested in order to fulfill fabric requirements or to reduce running time. One of the typical design rules is to assign one direction to each metal layer.

Generation of a feasible solution. As said, solving the previous satisfiability problem provides a first feasible solution for our wire planning problem (or shows the absence of such a solution!). The results presented in this paper use PicoSAT [4] to solve the SAT model.

Reduction of wire length. Once we have a feasible solution for the wire planning problem, we improve it by reducing its wire length while maintaining its feasibility. Our strategy is iterative, where each iteration consists in ripping out a small set of nets from the feasible solution and reroute them, subject to the previously specified constraints and minimizing the total wire length.

To do so, we convert our Boolean satisfiability problem into an integer linear problem: Boolean variables are transformed in 0/1 variables, Boolean constraints are easily converted to linear inequalities and, the linear function that counts the amount of wire is used as the objective function of the ILP.

Since the above process is applied for a small set of nets at each iteration, the resulting problem is tractable and can be solved with efficient solvers in a moderate amount of time. Note that solving the original problem with all the nets and seeking for the absolute minimum is too slow for the sizes of the problems we are faced to.

The currently implemented iterative process proceeds by just ripping out and rerouting one net at a time, with the exception of the set of nets that interconnect tiled cells, which are ripped out and rerouted in one step. This process is repeated while reductions in the wire length are obtained, favoring the reduction of long nets before the reduction of shorter nets.

The results presented in this paper use Gurobi [10] to solve the wire length optimization models.

6. EXPERIMENTAL RESULTS

In this section we demonstrate the impact of using physical planning during system-level exploration, and also show the need for the use of CMP-specific constraints during physical planning for a proper configuration evaluation.

6.1 Exploration setup

All of the experiments from this section use configurations that were obtained using automated system-level exploration. The parameters of this exploration are described in Table 1. We limit the search to tiled hierarchical CMPs using a mesh as the global interconnect, with the second level interconnect being either a bus or a ring (bi-directional or uni-directional). The number of tiles, the number of cores and the distribution of cores among the tiles are exploration variables. We assume that three different models of cores are available (C1, C2 and C3), with different performance and area characteristics obtained by scaling publicly available data of the Intel Core 2 Duo E6400 processor. We also assume that, while cores and interconnect routers occupy all metal layers, cache memories only use two of them. Therefore, routing can be performed over the cache memories. The operating frequency of the interconnect networks has

Parameter	Value		
Maximum chip area	350 mm^2		
Maximum chip power	350 W		
Interconnect frequency	1.6 GHz		
Global interconnect types	Mesh		
Global mesh dimensions	2×2 to 16×16		
Local interconnect types	Bus, Ring		
Local interconnect sizes	Limited by chip area only		
Memory density	$1\text{ mm}^2/\text{MB}$		
Cache latency (per size)	$5.0\cdot CacheSize^{0.5}$ cycles		
Off-chip memory latency	100 cycles		
Interconnect link width	$10\ \mu\text{m}$ (10^3 wires$\times$10 nm)		
Available metal layers	m1, m2, m3, m4		
Used by cores	All		
Used by NoC routers	All		
Used by cache memories	m1, m2		
Core types	*C1*	*C2*	*C3*
Core performance (IPC)	1.75	2	2.5
Core area	1 mm^2	1.25 mm^2	2 mm^2
L1 size	64, 96, 128 KB per core		
L2 size	64 KB to 1 MB per core		
L3 size	Up to 100 MB per chip		

Table 1: Parameters for system-level exploration.

been used to define the constraints on the maximum wire length for the links.

To characterize the memory accesses, a model extracted from the SPEC2006 *soplex* benchmark is used. The exploration generates 200 configurations in around 20 minutes. Each configuration is described as a block diagram of components, such as the one shown in Fig. 2. For example, the best configuration from this exploration has 25 clusters connected with a 5×5 mesh. Each cluster has a bus as local interconnect, two C2 cores and two C3 cores, along with 1 MB of L2 cache per core. The CMP has a total of 50 MB L3 cache distributed across the 25 clusters. It has an estimated throughput of 107.77 IPC.

6.2   Impact of physical planning

In order to prove how the use of physical planning can significantly alter the results of system-level exploration, we applied our physical planning tool to the 200 configurations found by the exploration. This floorplanning process, if run sequentially, takes 5 hours (an average of 90 seconds per configuration). However, on a machine with multiple cores each of the 200 configurations can be run separately.

The results are shown in Fig. 9. For each configuration, *block area* indicates the sum of the areas from all components. The exploration tool, before physical planning, uses this value as estimator for the expected chip area in order to satisfy the maximum area constraint. In this example, no configuration has a block area larger than 350 mm^2. *Conventional floorplan* shows a minimal area floorplan obtained without using any of the constrains described in this work (abutability, link length optimization, etc.). On the other hand, *NoC-aware floorplan* depicts the floorplan with minimal area that satisfies these constraints. A dashed line connects the block area data point with the minimal NoC-aware floorplan area for the same configuration.

Despite the fact that all configurations have a block area lower than the limit, a large number exceeds the area limit once physical planning is taken into account. As an example,

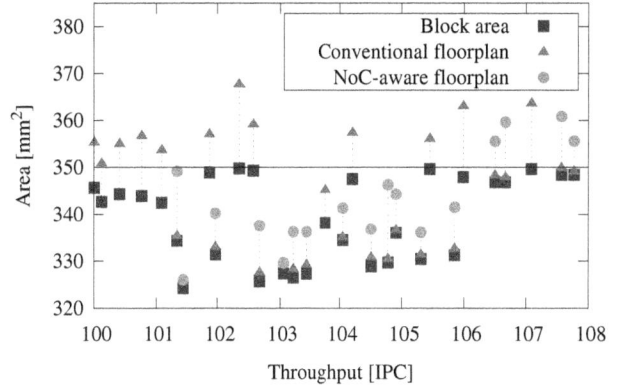

Figure 9: Area for different floorplanning strategies.

the best configuration found by the exploration (rightmost in Fig. 9) has a block area of 348.45 mm^2, which is below the area constraint. A conventional, minimal area floorplan exists with an area of 349.17 mm^2, also below the constraint. However, using the tool presented in this work, we find that the smallest floorplan satisfying all floorplanning constraints has an area of 355.59 mm^2. This violates the area constraint and, therefore, is not actually a valid configuration.

The first viable configuration with area below the limit has a significantly lower performance at 105.85 IPC. Out of the 200 configurations selected during the exploration, 39% of configurations had no floorplan satisfying all the constraints. Even for the configurations for which such a floorplan was found, only 23% satisfy the 350 mm^2 area limit. Configurations using rings as local interconnect, despite their excellent performance characteristics, have much stricter physical constraints and thus often violate design constraints. Without physical planning, those configurations would have been tagged as "promising" and would have been analyzed with more accurate simulation tools.

6.3   Physical planning search space

A single CMP configuration can have a large number of alternative floorplans. Nevertheless, it is desirable to select one or few candidate floorplans. At the same time, we are considering two metrics by which feasible floorplans can be evaluated: area and wire length. Thus, there is a trade-off.

In Section 2 we showed two candidate floorplans where one had much shorter total wire length at the cost of doubling the chip area. Since this trade-off might be inconvenient for some designs, the weights in the cost function (described in Section 4) can be modified to guide the search towards floorplans with better area or towards shorter wire length.

Figure 10 is an example of the available floorplans for a given CMP configuration. In the chart, each point represents a valid floorplan and its position depends on the area and wire length for that floorplan. By changing the weights in the cost function, a designer can decide which points of the Pareto frontier (solid line) are most desirable.

To illustrate, we selected two representative floorplans from the Pareto frontier that we show in Fig. 11. These are, respectively, the floorplan with the minimal area (but satisfying all constraints) and the overall best floorplan assuming we give the same weights to both area and wire length minimization.

9

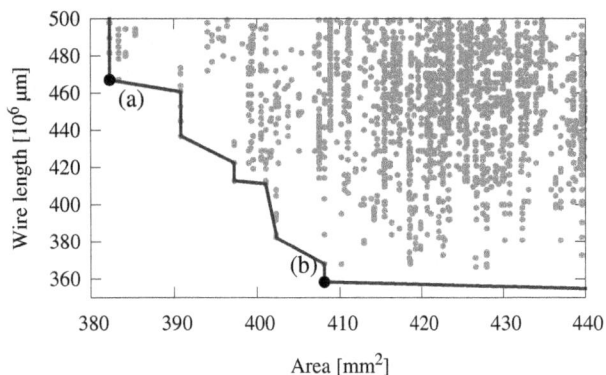

Figure 10: Example of physical planning search space for a single CMP configuration.

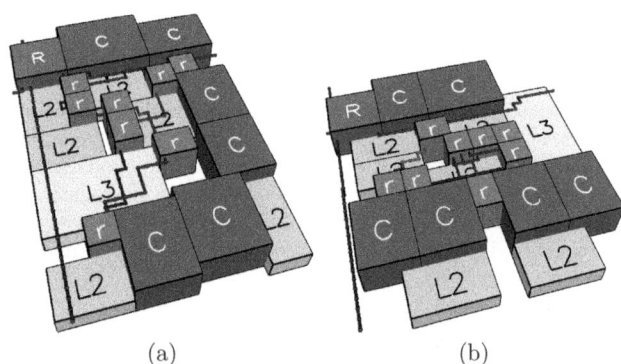

(a)   (b)

Figure 11: Two design points from the exploration space in Fig. 10.

7. CONCLUSIONS

This work has shown the importance of floorplanning and wire planning during the exploration of CMP architectures. Classical approaches for VLSI physical planning have been extended to support constraints specific for tiled CMPs with hierarchical interconnects. The presence of physical constraints has an important impact in deciding the parameters for the design of CMPs and contributes to guide the exploration towards physically-viable architectures. An interesting and important extension of this work would be to incorporate wire sizing as an additional parameter for exploring area/performance trade-offs.

8. REFERENCES

[1] International Technology Roadmap for Semiconductors. http://www.itrs.net/reports.html.
[2] S. Adya and I. Markov. Fixed-outline floorplanning: enabling hierarchical design. *IEEE Transactions on VLSI Systems*, 11(6):1120 –1135, Dec. 2003.
[3] J. Balfour and W. J. Dally. Design tradeoffs for tiled CMP on-chip networks. In *Proc. Intl. Conf. on Supercomputing*, pages 187–198, 2006.
[4] A. Biere. PicoSAT. http://fmv.jku.at/picosat.
[5] S. Boettcher and A. G. Percus. Extremal optimization: Methods derived from co-evolution. In *Proc. of the Genetic and Evolutionary Computation Conf.*, pages 825–832, 1999.
[6] H. H. Chan and I. L. Markov. Practical slicing and non-slicing block-packing without simulated annealing. In *Proc. of the Great Lakes Symposium on VLSI*, pages 282–287, 2004.
[7] W. J. Dally and B. Towles. Route packets, not wires: on-chip inteconnection networks. In *Proc. ACM/IEEE Design Automation Conference*, pages 684–689, 2001.
[8] R. Das, S. Eachempati, A. Mishra, V. Narayanan, and C. Das. Design and evaluation of a hierarchical on-chip interconnect for next-generation CMPs. In *High Performance Comp. Arch.*, pages 175–186, Feb. 2009.
[9] F. Gilabert, F. Silla, M. E. Gomez, M. Lodde, A. Roca, J. Flich, J. Duato, C. Hernández, and S. Rodrigo. *Designing Network On-Chip Architectures in the Nanoscale Era*. CRC Press, 2010.
[10] Gurobi Optimization, Inc. Gurobi Optimizer Reference Manual. http://www.gurobi.com.
[11] W. Hung, X. Song, T. Kam, L. Cheng, and G. Yang. Routability checking for three-dimensional architectures. *IEEE Transactions on VLSI Systems*, 12(12):1371–1374, Dec. 2004.
[12] J. Howard et al. A 48-core IA-32 processor in 45 nm CMOS using on-die message-passing and DVFS for performance and power scaling. *J. Solid-State Circuits*, 46(1):173–183, 2011.
[13] S. Kirkpatrick, C. D. Gelatt, and M. P. Vecchi. Optimization by simulated annealing. *Science*, 220:671–680, 1983.
[14] M. Lai and D. F. Wong. Slicing tree is a complete floorplan representation. In *Proc. Design, Automation and Test in Europe (DATE)*, pages 228–232, 2001.
[15] M. Monchiero, R. Canal, and A. Gonzalez. Power/performance/thermal design-space exploration for multicore architectures. *Parallel and Distributed Systems*, 19(5):666–681, May 2008.
[16] N. Nikitin, J. de San Pedro, J. Carmona, and J. Cortadella. Analytical performance modeling of hierarchical interconnect fabrics. In *Proc. ACM/IEEE International Symposium on Networks-on-Chip (NOCS)*, pages 107–114, May 2012.
[17] R. H. Otten. Automatic floorplan design. In *Proc. ACM/IEEE Design Automation Conference*, pages 261–267, 1982.
[18] F. Rubin. The Lee path connection algorithm. *IEEE Trans. Comput.*, 23(9):907–914, Sept. 1974.
[19] S. Bell et al. TILE64 - processor: A 64-core SoC with mesh interconnect. In *Solid-State Circuits*, pages 88–98, Feb. 2008.
[20] K. Sankaranarayanan, S. Velusamy, C. Stan, C. L, and K. Skadron. A case for thermal-aware floorplanning at the microarchitectural level. *Journal of ILP*, 7, 2005.
[21] N. A. Sherwani. *Algorithms for VLSI Physical Design Automation*. Kluwer Academic Publishers, 1993.
[22] K. Srinivasan and K. S. Chatha. A low complexity heuristic for design of custom network-on-chip architectures. In *Proc. Design, Automation and Test in Europe (DATE)*, pages 130–135, 2006.
[23] X. Tang, R. Tian, and M. D. Wong. Minimizing wire length in floorplanning. *IEEE Transactions on Computer-Aided Design*, 25(9):1744–1753, Sept. 2006.
[24] X. Tang and D. F. Wong. Floorplanning with alignment and performance constraints. In *Proc. ACM/IEEE Design Automation Conference*, pages 848–853, 2002.
[25] D. F. Wong and C. L. Liu. A new algorithm for floorplan design. In *Proc. ACM/IEEE Design Automation Conference*, pages 101–107, 1986.
[26] B.-S. Wu and T.-Y. Ho. Bus-pin-aware bus-driven floorplanning. In *Proc. of the Great Lakes Symposium on VLSI*, pages 27–32, 2010.
[27] T. T. Ye and G. D. Micheli. Physical planning for on-chip multiprocessor networks and switch fabrics. In *Int. Conf. Application-Specific Systems, Architectures, and Processors*, pages 97–107, 2003.
[28] E. F. Y. Young, C. C. N. Chu, and M. L. Ho. Placement constraints in floorplan design. *IEEE Trans. Very Large Scale Integr. Syst.*, 12(7):735–745, July 2004.
[29] F. Y. Young and D. F. Wong. How good are slicing floorplans? In *Proc. Int. Symposium on Physical Design*, pages 144–149, 1997.

Utilizing 2D and 3D Rectilinear Blocks for Efficient IP Reuse and Floorplanning of 3D-Integrated Systems

Robert Fischbach[†*], Johann Knechtel[*], Jens Lienig
Institute of Electromechanical and Electronic Design
Dresden University of Technology, Dresden, Germany
robert.fischbach@eas.iis.fraunhofer.de, johann.knechtel@ifte.de, jens.lienig@ifte.de

ABSTRACT

The reuse of predesigned intellectual property (IP) blocks is critical for the commercial success of three-dimensional (3D) electronic circuits. In practice, IP blocks can be specified as rectangular as well as rectilinear 2D blocks. The 3D equivalent of 2D rectilinear blocks, orthogonal polyhedra, may be utilized for modeling tightly interconnected (sub-)modules placed onto adjacent dies or for design automation of versatile 3D-integrated systems. Such complex block geometries have not been adequately considered until now. We propose a new 3D layout representation that enables native 3D floorplanning of complex-shaped 3D blocks, i.e., orthogonal polyhedra spread onto multiple dies. Furthermore, it can also be applied during 3D floorplanning of both rectangular and rectilinear 2D blocks. In the former case, experiments reveal superior estimated wirelength and packing density compared to previous work.

Categories and Subject Descriptors

B.7.2 [**Integrated Circuits**]: Design Aids—*Layout*

General Terms

Algorithms, Design

Keywords

3D integration, floorplanning, intellectual property blocks reuse, block geometries: orthogonal polyhedra, 3D design, representation

1. INTRODUCTION

Three-dimensional (3D) integration of electronic circuits has recently gained much attention as a promising option to fulfill ever increasing demands on functionality and performance while limiting cost and power consumption [1–5]. In this context, a coarse design style where available 2D intellectual property (IP) blocks are reused is favored in terms of reliability, cost, testability and design effort [6–8]. In practice, hard IP blocks are generally specified as rectangular as well as rectilinear blocks. Thus, the handling of rectilinear 2D blocks or rectilinear 3D block compounds may be required for 3D design, as further motivated in Section 2.1. Design automation of more complex 3D systems such as System-on-Package (SoP), sensor/chip co-design, or micro-electro-mechanical systems (MEMS) stacks [9] may also benefit from the ability to handle rectilinear 3D blocks. However, related approaches require 3D layout representations which are able to efficiently handle such complex-shaped blocks in the 3D solution space.

Prior work on 3D layout representations and floorplanning has overlooked complex-shaped blocks or related block compounds so far. We note that some publications consider (rectangular) block alignment [10–13], which enables simple block compounds. To do so, the related studies account for additional constraints, possibly decreasing efficiency of solution-space exploration. For example, additional edges for constraint graphs have to be determined in an iterative fashion [10]. In any case, most recent work has neglected (to align) rectilinear blocks. Although such blocks have been successfully considered in some 2D representations such as Corner Block List [14], Sequence Pair [15], Bounded Slicing Grid [16] or B*-Tree [17], rarely have they been considered in their counterparts of 3D representations. The sole exception, the (very recent) work by Quiring et al. [18], extended the T-Tree [19] to enable 3D alignment of rectilinear 2D blocks. However, their approach enforces alignment to a common reference point. This means that complex-shaped 3D blocks including offsets between partial (sub-)blocks on several dies cannot be represented.

In the present work, we propose a 3D layout representation called *3D Moving Block Sequence (3D-MBS)*. Our representation enables handling of complex-shaped 3D blocks in a straightforward and efficient manner. However, it is not restricted to such blocks; 3D floorplanning of both rectangular and rectilinear 2D blocks is also possible. Furthermore, as a "real" 3D representation[1], 3D-MBS allows continuous spreading of blocks in the vertical dimension. We believe this approach is useful for next-generation 3D circuits with versatile IP reuse, massive integration densities or for more complex electronic systems like MEMS stacks.

The remainder of this paper is structured as follows. In Section 2, we provide relevant background and motivate the use of complex-shaped 3D blocks. We then discuss our representation in Section 3 in detail. Experimental results are provided in Section 4; we conclude in Section 5 with recommendations on efficient utilization of complex-shaped blocks in 3D electronic systems.

[†]Currently with Fraunhofer Institute for Integrated Circuits, Design Automation Division, Dresden, Germany.
[*]The work of R. Fischbach and J. Knechtel was supported by the German Research Foundation under project 1401/1.

---

[1]It is common to distinguish so-called 2.5D layout representations and inherent or real 3D representations. In the former case, several instances of a 2D representation are used to represent multiple dies. Vertical block relations are often modeled by additional constraints, likely restricting the flexibility of optimization techniques. For real 3D representations, blocks and their relations are modeled in a continuous (possibly larger) 3D solution space.

2. BACKGROUND AND MOTIVATION

2.1 Complex-Shaped Blocks

As mentioned in Section 1, accounting for rectilinear blocks has been acknowledged as a relevant feature in traditional (2D) physical design automation. Besides handling rectilinear hard IP blocks, related techniques may be helpful to enforce abutment of particular rectangular blocks or to consider special layout constraints [20]. Figure 1b illustrates rectilinear design blocks, which result from fundamental L-, Z-, and T-shapes (Fig. 1a) by assigning a specific (active layer) thickness. Complex-shaped blocks (Fig. 1c), possibly spread onto multiple dies, can be efficiently described by *orthogonal polyhedra (OPa)* — an orthogonal polyhedron is the three-dimensional equivalent of a rectilinear block. Polyhedra in general are geometric solids bounded by flat faces and constructed by rectilinear edges. OPa are special polyhedra where faces meet exclusively at right angles. This simplification correlates well with typical design constraints and may reduce algorithmic complexity while efficiently handling corresponding blocks.

Enabling complex-shaped blocks for 3D floorplanning provides several benefits. First, several studies on 3D integration [11, 12, 21] have shown that partitioning design blocks into (few) sub-modules (carefully placed and vertically aligned among adjacent dies) can help to reduce intra-block as well as inter-block wirelength, latency and thermal-related leakage. Given rectilinear 2D (or dedicated 3D) design blocks, such approaches can be facilitated by modeling rectilinear block compounds, i.e., OPa. Second, utilizing OPa may decrease design complexity for large-scale systems containing (rectilinear) blocks and dedicated inter-die interconnect structures [3, 5, 22] — related blocks and interconnect structures can be abstracted as OPa during early design phases. Note that such a modeling approach is not restricted by the integration technology since characteristics of varied interconnects like through-silicon vias, face-to-face bonds or monolithic inter-die vias can be annotated to the modeled 3D blocks for subsequent consideration. Third, custom vertical-alignment constraints on rectilinear blocks can be directly and efficiently enforced by modeling OPa. Fourth, due to utilization of possibly complex-shaped modules, the design of (future) versatile 3D systems may benefit notably from the capability of handling arbitrary rectilinear 3D blocks. Note that the aforementioned modeling of OPa is out of scope for this paper; sophisticated techniques may be required for appropriate modeling.

2.2 Moving Block Sequence

The Moving Block Sequence (MBS) is a 2D layout representation based on a constructive block-insertion process. Unlike other representations where rectilinear blocks are mainly handled by determining rectangular-block subsets and introducing placement constraints (e.g. [20]), the MBS can directly process such blocks. Therefore, it is an interesting candidate for an efficient 3D representation; our related extension is discussed in Section 3. In the following, the basic concept of MBS is explained for readers convenience; further details can be found in [23].

Based on two sequences π and IP, a constructive process transforms a given abstract solution $MBS = (\pi, IP)$ into a physical layout. Sequence $\pi = (\pi_0, \pi_1, \ldots, \pi_{n-1})$ is a permutation of all n blocks and defines the block-insertion order. Sequence $IP = (IP_1, IP_2, \ldots, IP_{n-1})$ defines for each block π_i one out of four possible insertion positions, as illustrated in Fig. 2a. Note that for the first-to-insert block π_0 the position IP_0 is a fixed special case, i.e., the coordinate origin, thus it is not included in IP. For each insertion position, rules for packing the block towards the coordinate origin are applied as follows (Fig. 2). For positions i or iv, packing is only considered downwards or to the left, respectively. For position ii, packing is performed primarily downwards and to

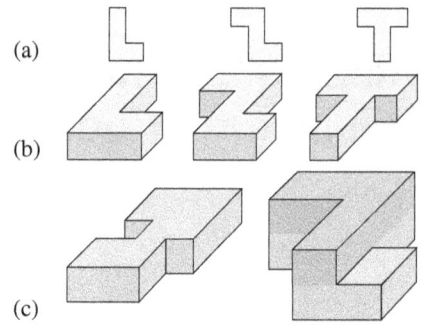

Figure 1: Different types of design blocks. (a) Abstracted blocks in L-, Z-, and T-shaped polygons. (b) Related "real" blocks require a thickness, thus extend the shapes to rectilinear 2D blocks. (c) Complex-shaped blocks, spread onto single or multiple dies, can be described as orthogonal polyhedra (OPa).

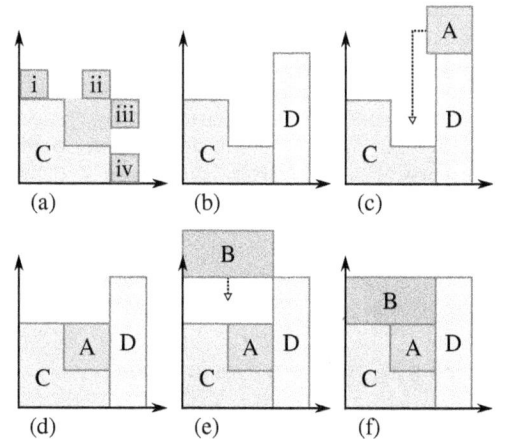

Figure 2: Layout generation of MBS with $\pi = (C, D, A, B)$ and $IP = (iv, ii, i)$. (a) Four different insertion positions are defined. Each position is given specific rules for block packing. (a) to (f) illustrate the stepwise layout generation during constructive block insertion.

the left only if no downward movement is applicable; for position iii, left is the main direction and the downward movement is secondary. Based on the properties of the two sequences, $n! \cdot 4^{(n-1)}$ different abstract solutions exist for n blocks. The layout generation has a runtime complexity of $\mathcal{O}(n^2)$.

Due to the constructive process, no block overlaps occur and only valid solutions are provided. For efficient packing, each block is described by its boundary edges. Given all block edges, considering the edges perpendicular to the packing direction is sufficient for collision detection while inserting a new block. The constructive approach also enables a flexible consideration of additional constraints, such as block symmetries and pre-placed blocks. However, not all possible layouts can be represented. Thus, the MBS is a compacting but not a complete layout representation. Experimental results suggest a similar solution quality compared to common 2D layout representations [23].

3. 3D MOVING BLOCK SEQUENCE

We propose a new real 3D layout representation called *3D Moving Block Sequence (3D-MBS)*, inspired by main features of the MBS (Section 2.2). In particular, we extend the constructive block-insertion process of MBS to the continuous 3D space. Furthermore, the consideration of OPa as well as an efficient collision detection are proposed. Like the original MBS, the 3D-MBS is a non-slicing packing representation.

3.1 Abstract Layout Representation

Similar to the MBS, two sequences defined as follows are used for abstract representation of 3D layouts.

$$3D\text{-}MBS = (\pi, IP') \qquad \begin{array}{l} \pi = (\pi_0, \pi_1, \ldots, \pi_{n-1}) \\ IP' = (IP'_1, IP'_2, \ldots, IP'_{n-1}) \end{array}$$

where n denotes the number of blocks, π defines the insertion order and IP' the insertion positions of blocks. Differing from the MBS, we propose *3D insertion positions* for the 3D-MBS. The positions are denoted as an ordered sequence $IP'_i = C^1 C^2 C^3 \mid C^i \in \{\lambda, X, Y, Z\}$ of coordinates, where λ represents the omission of the particular coordinate. The twelve applied positions are illustrated in Fig. 3a. By proposing specific block-insertion rules (Section 3.2) for various types of positions, we are able to fully and efficiently exploit the 3D space.

The number of possible solutions is defined by $n!$ permutations of π and $12^{(n-1)}$ variations of IP'; the overall number of abstract solutions is $n! \cdot 12^{(n-1)}$. Besides random generation of both sequences, an initial pair can be altered to obtain new solutions (Section 4). Fig. 4 illustrates an exemplary exchange of blocks.

3.2 Layout Generation

Given an abstract solution *3D-MBS*, the constructive process for layout generation stepwise considers pairs (π_i, IP'_i) where $1 \leq i \leq n-1$. Note that block π_0 is always placed in the corner of the coordinate origin, thus no pair (π_0, IP'_0) is required.

During insertion of block π_i, its position IP'_i defines the initial location and allowed shifting directions. To obtain IP'_i, we keep track of the bounding box covering previously inserted blocks. Each outer corner of this box describes a particular insertion position, i.e., initial location (Fig. 3a). The block-shifting directions (towards the coordinate origin, i.e., packing directions) are restricted by the insertion-position types illustrated in Fig. 3a as follows. The red (dark gray) positions provide only one direction (i.e., either along x, y, or z axis), whereas the green (medium gray) positions offer two degrees of freedom. In that case, the shifting process is partitioned into primary and secondary movement, where the first coordinate defines the primary movement direction. A movement along the secondary direction is conducted only if encountering an obstacle during packing along the primary direction. The three yellow (light gray) positions allow movements comprising all three dimensions. That means a ternary shifting direction is considered when both the primary and secondary direction are blocked. Besides previously placed blocks, the coordinate-axis planes (of the first quadrant) are considered as fixed obstacles in any case. Due to the constructive process, additional constraints such as pre-placed blocks can be easily accounted for.

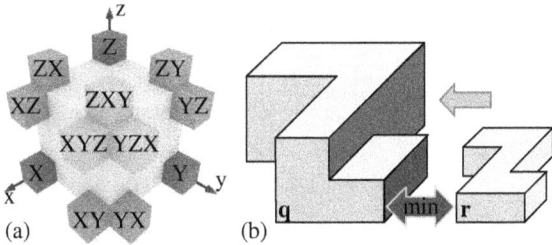

Figure 3: 3D-MBS characteristics. (a) 3D insertion positions. (b) Block shifting and marked faces for collision detection.

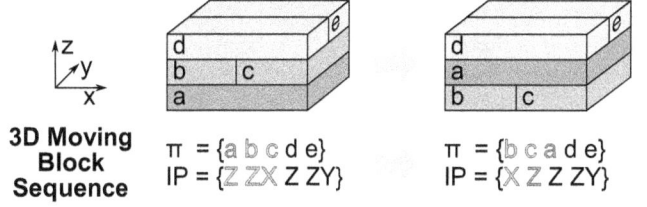

Figure 4: 3D-MBS solutions and corresponding layouts. The exchange of block a with blocks b and c can be accomplished by applying the illustrated modifications to the solution.

The layout-generation process is outlined in Algorithm 1. The obstacle detection, i.e., face-collision detection, is explained in Section 3.3. Note that the runtime complexity of a naive implementation is $\mathcal{O}(c \cdot n^2)$, where c is a complexity factor and collision detection has to consider all previously placed blocks (worst case). The term $c \cdot n$ denotes how many cuboids are required to replace n OPa.

3.3 Collision Detection and Face Segmentation

Block packing is performed in only one direction at any particular moment, even for insertion positions with multiple degrees of freedom (Algorithm 1). Checking for overlaps among a limited subset of block faces is thus sufficient for collision detection, as elaborated in the following. Handling an orthogonal polyhedron is based on partitioning its rectilinear faces into six groups. Figure 5 illustrates such dissection into left (red), right (green), upper (blue), lower (yellow), front (orange), and back (brown) faces. The relevant faces for collision detection are the *boundary faces*, i.e., the inner opposing faces of blocks perpendicular to the packing direc-

---

Algorithm 1: 3D-MBS Layout Generation

Data: number of blocks n; block data B

Input: block-permutation sequence π, insertion-position sequence IP'

Output: layout L

1   Insert π_0 at position $(0, 0, 0)$ into L;
2   **for** $i \leftarrow 1$ **to** $n - 1$ **do**
3   Determine bounding box bb of L;
4   Determine initial position P of π_i based on IP'_i and bb;
5   **while** *shifting towards coordinate origin succeeds* **do**
6   Determine boundary faces bf_p in primary direction;
7   Determine shortest distance d_p of opposing bf_p;
8   Perform primary shift of P until d_p is reached (some bf_p collide);
9   **if** *degrees of freedom for $IP'_i > 1$* **then**
10   Determine boundary faces bf_s in secondary direction;
11   Determine shortest distance d_s of opposing bf_s;
12   Perform secondary shift of P until d_s is reached or no more collision of bf_p exists (primary shift possible);
13   **if** *degrees of freedom for $IP'_i > 2$* **then**
14   Determine boundary faces bf_t in ternary direction;
15   Determine shortest distance d_t of opposing bf_t;
16   Perform ternary shift of P until d_t is reached or no more collision of bf_s or bf_p exists;
17   Insert π_i at position P into L;

13

tion. Figure 3b illustrates the particular boundary faces of block $\mathbf{q}$; assume $\mathbf{q}$ is already placed and $\mathbf{r}$ is to be inserted from the right, i.e. moved along the x-axis. Then, the left faces of $\mathbf{r}$ and the right faces of $\mathbf{q}$ are checked for collision. To do so, these boundary faces are analyzed for any overlap of their rectangular segments while considering the segments' y- and z-coordinates. In case of some overlapping segments, the shifting amount is defined as the minimal distance min between any pair of related segments. In general, if no segments are overlapping (this does not apply to Fig. 3b), the block is to be shifted towards the coordinate-axis plane. Similarly, block packing towards y- or z-direction considers front and back or upper and lower faces as boundary faces, respectively.

As indicated, the collision detection can be further enhanced by *face segmentation*, i.e., rectilinear faces are segmented into (separate) sets of rectangles. Hence, it is sufficient to analyze pairs of rectangles from different sets in order to detect collisions of related faces. Note that performing segmentation is required only once for each block. We predetermined face segments for our experimental validation (Section 4.1). For practical applications, it may be advisable to leverage efficient computational-geometry algorithms for face unfolding and segmentation, e.g., as proposed by Biedl et al. [24] or Keil [25].

4. EXPERIMENTAL INVESTIGATION

We implement the 3D-MBS for experiments using *Python*; implementations including varied optimization heuristics can be retrieved from [26] (see "Evaluation tool").

In Section 4.1 we validate the capability of 3D-MBS to handle OPa properly during 3D floorplanning. We compare with previous work on 3D representations in Section 4.2. This comparison shows that 3D-MBS outperforms other work when applying various optimization heuristics.

4.1 Floorplanning of Complex-Shaped Blocks

In order to enable 3D floorplanning of OPa, we modify the benchmarks *ami33* (MCNC suite) and *n100* (GSRC suite) [27]. In these benchmarks we replace 8 and 10 (randomly selected) blocks with custom blocks, respectively — six rectilinear blocks along with two OPa in *ami33* (Fig. 6) and 6 rectilinear blocks along with four OPa in *n100*. We also predetermine rectangular face segments for simplified collision detection (Section 3.3).

We implement our 3D-MBS representation along with a *simulated annealing (SA)* engine [28]. In this setup, we focus on the packing capability of the 3D-MBS. We therefore define the SA cost function $f_1 = \alpha_1 * A + \beta_1 * WL_w$ where $A = h_d * w_d$ denotes the (common) die footprint and $WL_w = \sum_n w_n * \left(\sum_d \text{HPWL}(bb_n, d) \right)$ a *weighted wirelength estimate* for all nets n. WL_w is determined using the half-perimeter wirelength (HPWL) metric for net bounding boxes bb_n, where net pins are assumed in the block's geometric center. Boxes bb_n are separately determined for each related die d in order to estimate intra-die routing more

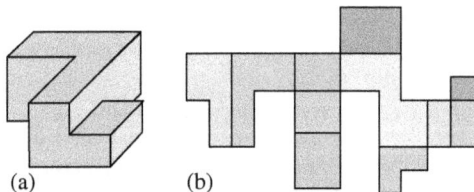

Figure 6: Complex-shaped blocks, for adaption of the benchmark *ami33*. Two new blocks (OPa) spread over multiple dies (and), the others are single-die rectilinear blocks ( ).

accurately. Weight (priority) factors w_n for nets are given in the benchmarks. As SA operations we consider the permutation of π, the variation of randomly selected IP_i', and the rotation of randomly selected blocks i. The SA engine is configured to consider 3 dies with a common outline. Results are presented in Table 1. Generated floorplans are illustrated in Fig. 7. The SA process, i.e., cost reduction, for an particular optimization run is illustrated in Fig. 8.

Considering these results, we make the following observations. First, 3D-MBS can be successfully applied to 3D-floorplanning problems that includes complex-shaped blocks, in particular arbitrarily rectilinear 2D and 3D blocks. Second, the packing capability scales well with the problem size; results for modified *n100* (100 modules) are comparable to results for modified *ami33* (33 modules) in terms of die utilization (reported as *blocks f.p.* in Table 1). Third, applying SA is a simple yet effective approach. Within few (hundred) iteration steps, cost for both wirelength and die footprint can be reduced down to 50% of the initial value.

Since our 3D-MBS is the first real 3D representation to handle OPa directly, we cannot compare this setup to previous work. How-

Table 1: Final results of five subsequent SA runs. *Blocks f.p.* denotes the sum of block footprints in relation to die footprints. Footprints are reported in μm^2, (weighted) wirelength in μm.

	Criteria	1st	2nd	3rd	4th	5th
ami33	*Die footprint*	560011	592735	596573	569504	588464
	Blocks f.p.	43.1 %	45.6 %	45.9 %	43.9 %	45.3 %
	Wirelength	2483694	2394242	2444321	2541327	2395956
n100	*Die footprint*	93009	95718	93796	98356	97940
	Blocks f.p.	43.9 %	45.2 %	44.3 %	46.4 %	46.3 %
	Wirelength	278552	280264	284582	288954	293002

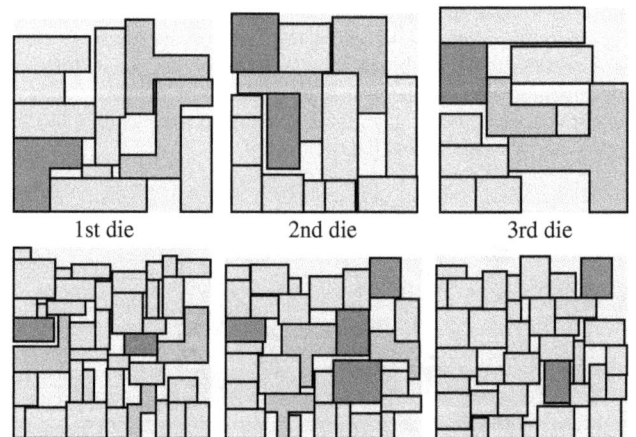

Figure 5: An orthogonal polyhedron (a) can be unfolded into joined rectilinear faces (b). Note that groups of opposite faces cover the respective same area.

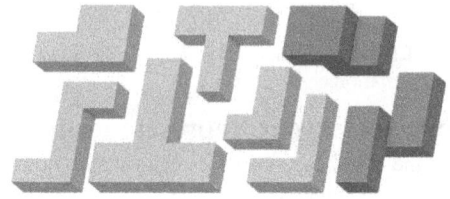

1st die	2nd die	3rd die

Figure 7: Floorplans for adapted *ami33* (top) and *n100* (bottom) obtained by applying SA for 3D-MBS.

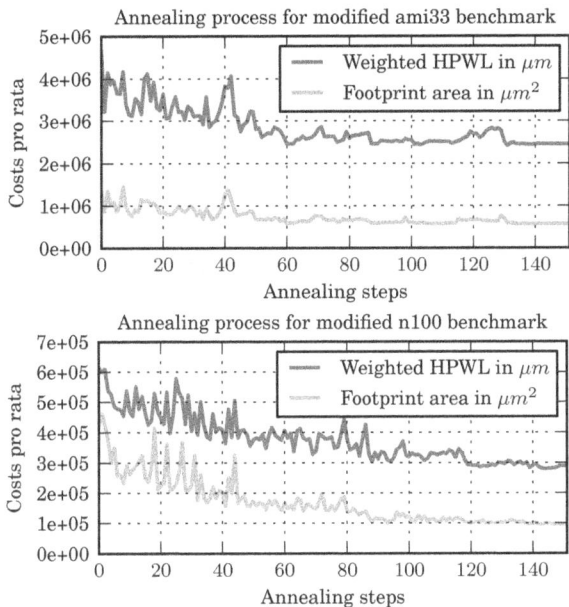

Figure 8: SA process while floorplanning benchmarks including complex-shaped blocks. SA is guided by cost function f_1.

ever, we can apply 3D-MBS also to 3D floorplanning of common rectangular blocks, as discussed in the next subsection.

4.2 Floorplanning of Rectangular Blocks

We apply and evaluate further optimization heuristics besides SA to provide a thorough evaluation of obtainable solution quality. In particular, we leverage concepts for an *evolutionary algorithm (EA)* [29] and the *Great Deluge algorithm (GDA)* [30]. EAs simulate generic mutations and the survival of the fittest in biology. Individuals (i.e., solutions) with favorable features (i.e, low cost) pass their characteristics to the next generation (*selection*). New individuals emerge due to *mutation* or *recombination*. In contrast, the GDA extends the well-known *hill-climbing approach*; this extension is based on a "mountainous" solution space where only "climbers" on mountains above sea level survive. While increasing this threshold value, climbers can also "jump" onto higher mountains, i.e., the search process is not restricted by local maxima. Table 2 provides the main parameters applied in our experiments.

We compare our 3D-MBS to 4 other real 3D layout representations, namely T-Tree [19], 3D Slicing Tree [31], Sequence Triple [32], and Sequence Quintuple [32]. Our implementation of those representations follows the respective publication's description. For the optimization heuristics, only meaningful operations like swapping tree nodes are considered. The applied cost function is defined as $f_2 = \alpha_2 * HESA + \beta_2 * WL_w$ where WL_w denotes the weighted wirelength estimate (Section 4.1) and $HESA = w_s * h_s + w_s * d_s + h_s * d_s$ the *half of enveloping surface area* of the 3D stack where w_s, h_s and d_s denote the die width, die height and stack height, respectively. Thus, $HESA$ measures the 3D packing density. Compared to die footprint or packing volume, we observe that this metric is more efficient in guiding optimization towards a balanced aspect ratio and limited whitespace. We perform 100 full optimization runs for each combination of benchmark and representation, providing detailed cost distributions. (Such distributions are helpful for investigations of the solution quality of 3D layout representations; for further reading refer to [33].) Figure 9 illustrates such distributions for the (original) benchmark *ami33* as box plots; results for further MCNC benchmarks are provided in Table 3. The reported cost f_2' are based on normalized values, i.e.,

Table 2: Applied optimization heuristics parameters

Heuristic	Parameter	Short description	Value
	α_1	Cost factor for die footprint	0.5
	β_1	Cost factor for wirelength	0.5
	α_2	Cost factor for $HESA$	0.5
	β_2	Cost factor for wirelength	0.5
SA	TStart	Initial temperature	1300
	TEnd	Final temperature	100
	TSteps	Temperature-reduction steps	25
	TSamples	Iterations per temperature step	50
EA	NParents	Parents per generation	8
	NChildren	Children per generation	5–15
	NGen	Generations count	30
GDA	LInit	Initial threshold (normalized)	1.0
	NSamples	Iterations per threshold level	2000
	SVapor	Threshold-reduction value	0.005

$f_2' * i = \sum_i \alpha_2/\mathrm{AM}(HESA) * HESA_i + \beta_2/\mathrm{AM}(WL_w) * WL_{w_i}$ for $i = 100$ optimization runs; AM denotes the respective average means, obtained by analyzing a sufficiently large set of previously determined solutions.

We observe that 3D-MBS allows us to *reduce average cost and cost variability for all benchmarks when applying various optimization heuristics*. This emphasizes the efficiency and flexibility of 3D-MBS for 3D floorplanning of rectangular 2D blocks, despite been tailored to handle complex-shaped rectilinear 3D blocks.

5. CONCLUSIONS

The reuse of versatile IP blocks is critical for industry adoption of 3D-integrated circuits. However, such reuse still faces serious challenges. In particular, hard IP blocks may exhibit rectilinear shapes; handling such blocks is not effectively supported by recent 3D layout representations. Furthermore, vertical alignment of (rectilinear) blocks may be required for several 3D integration scenarios, e.g. placement of partitioned design blocks (Section 2.1). However, recent representations support such alignment only by (iteratively) generating and evaluating additional constraints, which likely impedes efficient optimization approaches.

Our proposed layout representation 3D-MBS facilitates the native handling of orthogonal polyhedra (OPa). It allows us to directly and thus efficiently handle designs containing blocks in any rectilinear 2D and/or 3D shape. This also implies that floorplanning of complex-shaped 3D blocks is supported for the first time. The 3D-MBS is based on the classical (2D) MBS; our approach has the advantage of preserving the ease of understanding and flexibility of traditional MBS while supporting OPa and extended collision de-

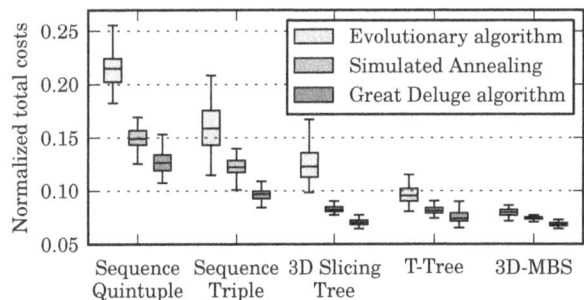

Figure 9: Cost box plots for the benchmark *ami33*. Each box illustrates the interquartile range and the median of normalized cost f_2', obtained by performing optimization 100 times. The whiskers represent the minimal or maximal cost, respectively.

Table 3: Cost comparison for 3D layout representations (REP) applied to floorplanning of MCNC benchmarks. Cost f'_2 are normalized and averaged over 100 optimization runs. SEQ=Sequence Quintuple, SLT=Slicing Tree, SET=Sequence Triple, TTR=T-Tree, Our=3D Moving Block Sequence.

	REP	apte	xerox	hp	M198	ami33	ami49	playout	ø
Simulated Annealing	SEQ	0.077	0.120	0.077	0.148	0.150	0.136	0.087	*0.113*
	SLT	0.042	0.087	0.054	0.082	0.083	0.068	0.047	*0.066*
	SET	0.068	0.112	0.065	0.122	0.123	0.109	0.072	*0.096*
	TTR	0.044	0.099	0.057	0.081	0.082	0.067	0.047	*0.068*
	Our	0.042	0.088	0.055	0.074	0.074	0.060	0.045	*0.062*
Evolutionary Algorithm	SEQ	0.142	0.180	0.138	0.214	0.215	0.204	0.185	*0.183*
	SLT	0.069	0.128	0.089	0.122	0.125	0.112	0.086	*0.105*
	SET	0.117	0.160	0.121	0.153	0.160	0.137	0.126	*0.139*
	TTR	0.081	0.122	0.073	0.095	0.096	0.082	0.062	*0.087*
	Our	0.056	0.109	0.062	0.079	0.080	0.069	0.056	*0.073*
Great Deluge Algorithm	SEQ	0.099	0.138	0.092	0.126	0.127	0.114	0.075	*0.110*
	SLT	0.043	0.101	0.068	0.070	0.071	0.050	0.033	*0.062*
	SET	0.088	0.133	0.085	0.096	0.097	0.080	0.053	*0.090*
	TTR	0.053	0.115	0.065	0.076	0.076	0.054	0.035	*0.068*
	Our	0.041	0.093	0.059	0.069	0.069	0.049	0.038	*0.060*

tection. The proposed techniques may be extended to furthermore facilitate custom design constraints, like preplacement.

The experimental investigation reveals the effectiveness of 3D-MBS for 3D floorplanning of complex-shaped 3D blocks, even along with rectangular and rectilinear 2D blocks. Furthermore, we show that applying 3D-MBS for 3D floorplanning of solely rectangular 2D blocks allows us to reduce cost compared to previous work. Due to the efficiency of 3D-MBS, this holds true for various optimization heuristics and benchmarks. Both observations suggest 3D-MBS as an attractive representation, enabling the reuse of arbitrary rectilinear IP blocks and related 3D-floorplanning tasks.

6. REFERENCES

[1] K. Banerjee *et al.*, "3-D ICs: A novel chip design for improving deep-submicrometer interconnect performance and systems-on-chip integration," *Proc. IEEE*, vol. 89, no. 5, pp. 602–633, 2001. DOI: http://dx.doi.org/10.1109/5.929647

[2] G. H. Loh, Y. Xie, and B. Black, "Processor design in 3D die-stacking technologies," *Micro*, vol. 27, pp. 31–48, 2007. DOI: http://dx.doi.org/10.1109/MM.2007.59

[3] S. Borkar, "3D integration for energy efficient system design," in *Proc. Des. Autom. Conf.*, pp. 214–219, 2011. DOI: http://dx.doi.org/10.1145/2024724.2024774

[4] T. Thorolfsson *et al.*, "Logic-on-logic 3D integration and placement," in *Proc. 3D Sys. Integr. Conf.*, pp. 1–4, 2010. DOI: http://dx.doi.org/10.1109/3DIC.2010.5751451

[5] D. H. Kim *et al.*, "3D-MAPS: 3D massively parallel processor with stacked memory," in *Proc. Int. Solid-State Circ. Conf.*, pp. 188–190, 2012. DOI: http://dx.doi.org/10.1109/ISSCC.2012.6176969

[6] X. Dong, J. Zhao, and Y. Xie, "Fabrication cost analysis and cost-aware design space exploration for 3-D ICs," *Trans. Comput.-Aided Des. Integr. Circuits Sys.*, vol. 29, no. 12, pp. 1959–1972, 2010. DOI: http://dx.doi.org/10.1109/TCAD.2010.2062811

[7] D. H. Kim, R. O. Topaloglu, and S. K. Lim, "Block-level 3D IC design with through-silicon-via planning," in *Proc. Asia South Pacific Des. Autom. Conf.*, pp. 335–340, 2012. DOI: http://dx.doi.org/10.1109/ASPDAC.2012.6164969

[8] J. Knechtel, I. L. Markov, and J. Lienig, "Assembling 2-D blocks into 3-D chips," *Trans. Comput.-Aided Des. Integr. Circuits Sys.*, vol. 31, no. 2, pp. 228–241, 2012. DOI: http://dx.doi.org/10.1109/TCAD.2011.2174640

[9] M. Schuenemann *et al.*, "MEMS modular packaging and interfaces," in *Proc. Elec. Compon. Technol. Conf.*, pp. 681–688, 2000. DOI: http://dx.doi.org/10.1109/ECTC.2000.853232

[10] J. H. Y. Law, E. F. Y. Young, and R. L. S. Ching, "Block alignment in 3D floorplan using layered TCG," in *Proc. Great Lakes Symp. VLSI*, pp. 376–380, 2006. DOI: http://dx.doi.org/10.1145/1127908.1127994

[11] R. K. Nain and M. Chrzanowska-Jeske, "Fast placement-aware 3-D floorplanning using vertical constraints on sequence pairs," *Trans. VLSI Syst.*, vol. 19, no. 9, pp. 1667–1680, 2011. DOI: http://dx.doi.org/10.1109/TVLSI.2010.2055247

[12] Y. Liu *et al.*, "Fine grain 3D integration for microarchitecture design through cube packing exploration," in *Proc. Int. Conf. Comput. Des.*, pp. 259–266, 2007. DOI: http://dx.doi.org/10.1109/ICCD.2007.4601911

[13] X. Li, Y. Ma, and X. Hong, "A novel thermal optimization flow using incremental floorplanning for 3D ICs," in *Proc. Asia South Pacific Des. Autom. Conf.*, pp. 347–352, 2009. DOI: http://dx.doi.org/10.1109/ASPDAC.2009.4796505

[14] Y.-C. Ma *et al.*, "General floorplans with L/T-shaped blocks using corner block list," *J. Comput. Sci. Technol.*, vol. 21, pp. 922–926, 2006. DOI: http://dx.doi.org/10.1007/s11390-006-0922-y

[15] K. Fujiyoshi and H. Murata, "Arbitrary convex and concave rectilinear block packing using sequence-pair," *Trans. Comput.-Aided Des. Integr. Circuits Sys.*, vol. 19, no. 2, pp. 224–233, 2000. DOI: http://dx.doi.org/10.1109/43.828551

[16] M. Kang and W. W.-M. Dai, "General floorplanning with L-shaped, T-shaped and soft blocks based on bounded slicing grid structure," in *Proc. Asia South Pacific Des. Autom. Conf.*, pp. 265–270, 1997. DOI: http://dx.doi.org/10.1109/ASPDAC.1997.600145

[17] G.-M. Wu, Y.-C. Chang, and Y.-W. Chang, "Rectilinear block placement using B*-trees," in *Proc. Int. Conf. Comput. Des.*, pp. 351–356, 2000. DOI: http://dx.doi.org/10.1109/ICCD.2000.878307

[18] A. Quiring *et al.*, "3D floorplanning considering vertically aligned rectilinear modules using T*-tree," in *Proc. 3D Sys. Integr. Conf.*, pp. 1–5, 2012. DOI: http://dx.doi.org/10.1109/3DIC.2012.6263030

[19] P.-H. Yuh, C.-L. Yang, and Y.-W. Chang, "Temporal floorplanning using the T-tree formulation," in *Proc. Int. Conf. Comput.-Aided Des.*, pp. 300–305, 2004. DOI: http://dx.doi.org/10.1109/ICCAD.2004.1382590

[20] Z. Liu *et al.*, "VLSI fast initial placement with abutment constraints and L-shaped/T-shaped blocks based on less flexibility first principles," in *Proc. Int. Conf. Comm. Circ. Sys.*, vol. 2, pp. 1228–1232, 2004. DOI: http://dx.doi.org/10.1109/ICCCAS.2004.1346396

[21] M. Healy *et al.*, "Multiobjective microarchitectural floorplanning for 2-D and 3-D ICs," *Trans. Comput.-Aided Des. Integr. Circuits Sys.*, vol. 26, no. 1, pp. 38–52, 2007. DOI: http://dx.doi.org/10.1109/TCAD.2006.883925

[22] F. Li *et al.*, "Design and management of 3D chip multiprocessors using network-in-memory," in *Proc. Int. Symp. Comput. Archit.*, pp. 130–141, 2006. DOI: http://dx.doi.org/10.1109/ISCA.2006.18

[23] J. Liu *et al.*, "Moving block sequence and organizational evolutionary algorithm for general floorplanning with arbitrarily shaped rectilinear blocks," *Trans. Evol. Computation*, vol. 12, no. 5, pp. 630–646, 2008. DOI: http://dx.doi.org/10.1109/TEVC.2008.920679.

[24] T. Biedl *et al.*, "Unfolding some classes of orthogonal polyhedra," in *Proc. Canadian Conf. Comput. Geom.*, pp. 70–71, 1998.

[25] J. M. Keil, "Polygon decomposition," *Handbook of Computational Geometry*, vol. 2, pp. 491–518, 2000. DOI: http://dx.doi.org/10.1016/B978-044482537-7/50012-7

[26] http://www.ifte.de/english/research/3d-design/index.html

[27] http://vlsicad.cs.binghamton.edu/benchmarks.html

[28] S. Kirkpatrick, C. D. Gelatt, and M. P. Vecchi, "Optimization by simulated annealing," *Science*, vol. 220, no. 4598, pp. 671–680, 1983. DOI: http://dx.doi.org/10.1126/science.220.4598.671

[29] T. F. Gonzalez, *Handbook of Approximation Algorithms and Metaheuristics*. CRC Taylor & Francis, 2007. DOI: http://dx.doi.org/10.1201/9781420010749

[30] G. Dueck, "New optimization heuristics: The great deluge algorithm and the record-to-record travel," *J. Comput. Phys.*, vol. 104, no. 1, pp. 86–92, 1993. DOI: http://dx.doi.org/10.1006/jcph.1993.1010

[31] L. Cheng, L. Deng, and M. D. F. Wong, "Floorplanning for 3-D VLSI design," in *Proc. Asia South Pacific Des. Autom. Conf.*, pp. 405–411, 2005. DOI: http://dx.doi.org/10.1145/1120725.1120899

[32] H. Yamazaki *et al.*, "The 3D-packing by meta data structure and packing heuristics," *IEICE Trans. Fundamentals Elec. Comm. Comput. Scie.*, vol. 83, no. 4, pp. 639–645, 2000. [Online]. Available: http://ci.nii.ac.jp/naid/110003208571/en/

[33] R. Fischbach, J. Lienig, and J. Knechtel, "Investigating modern layout representations for improved 3D design automation," in *Proc. Great Lakes Symp. VLSI*, pp. 337–342, 2011. DOI: http://dx.doi.org/10.1145/1973009.1973076

Benchmarking for Research in Power Delivery Networks of Three-Dimensional Integrated Circuits

Pei-Wen Luo*, Chun Zhang+, Yung-Tai Chang*, Liang-Chia Cheng*, Hung-Hsie Lee*,
Bih-Lan Sheu*, Yu-Shih Su*, Ding-Ming Kwai* and Yiyu Shi+

*Industrial Technology Research Institute
Hsin-Chu, 31040, Taiwan, R.O.C.
{peiwen, yungtai, aga, sheu, dmkwai}@itri.org.tw

+ECE Dept., Missouri S&T
Rolla, MO, 65409, U.S.A.
{zhanchun, yshi}@mst.edu

ABSTRACT

Power integrity is generally considered to be one of the major bottlenecks hindering the prevalence of three-dimensional integrated circuits (3D ICs). The higher integration density and smaller footprint result in significantly increased power density, which threatens the system reliability. In view of this, there has been groundswell of interest in academia to model, design or optimize the power delivery networks (PDNs) in 3D ICs. Unfortunately, while several PDN benchmarks exist for 2D PDNs, none is available in the context of 3D. As a consequence, most existing literature resorts to ad-hoc designs by artificially stacking 2D PDNs for experiments, rendering the results less convincing. In this paper, we put forward a set of ten PDN benchmarks that are extracted from industrial 3D designs. These designs are carefully selected such that they cover a wide range of functionality, size, TSV number, tier number and packaging style. We hope that the released benchmarks can facilitate and promote research in 3D PDNs.

Categories and Subject Descriptors

B.7.2 [Integrated Circuits]: Design Aids - Simulation

General Terms: Design, Standardization

Keywords

3D ICs, benchmarks, power delivery networks.

1. INTRODUCTION

Three-dimensional (3D) integration, which stacks multiple dies together vertically with through-silicon-vias (TSVs), has become one of the most promising solutions in the More-than-Moore regime. However, the threat in power integrity has become one of the major hindrances to the prevalence of 3D ICs.

On one hand, more power needs to be delivered to the chip with the increased number of tiers integrated; on the other hand, the chip footprint is reduced due to the newly introduced vertical dimension. As such, the power density of 3D ICs increases significantly, which leads to severe power integrity and the system reliability problems. For example, the study in [1] has shown much larger IR drop and Ldi/dt variance in 3D power delivery networks (PDNs) compared to the 2D cases. The situation gets even worse when the noise margin becomes more stringent with the supply voltage scaling.

As a result of its importance, 3D PDN has become an active research area in recent years. For instance, Xu et.al. [2] propose SPICE-level 3D power network modelling for accurate electrical performance simulation. The IR drop simulation is performed in [3] based on finite volume method. In [4], the impact of TSVs on IR drop are analysed with varying TSV density and dimension, and a non-uniform power TSV placement method is proposed to reduce the total number of power TSV needed under certain noise margin. The design and optimization of 3D PDN has also been explored [5]. The traditional methods for 2D PDN design, such as decoupling capacitor insertion has been extended to 3D PDN by formulating and solving a linear programming problem [9]. The problem of optimal TSV placement is also studied in [10] for both power and thermal integrity improvement.

Benchmarks have played a key role in promoting the research and development of new techniques. 2D PDN benchmarks were first made available by IBM, where a set of six cases were initially released for DC and transient analysis [11]. A few more cases were added later for power grid simulation contest [12], which stimulates the advancement of fast and accurate PDN solvers[13]-[17]. However, there still lacks a set of high quality industrial-level 3D PDN benchmarks in the literature, in spite of all the dedicated research works in 3D PDN modelling, analysis and optimization.

The lack of real 3D PDN benchmarks brings several disadvantages. First, researchers have to build ad-hoc 3D PDNs by directly stacking 2D designs with proper TSV placement. This would take a considerable amount of time, which distracts the researchers from focusing on solving the problem itself. Second, many realistic design constraints cannot be easily considered without the context of die functionalities, such as JEDEC wide-IO logic-memory interface definitions [18], as will be demonstrated in our 3D-PAC benchmark design. Third, it makes comparison between different techniques difficult in the absence of common benchmarks. Fourth, to evaluate the practicability of a newly proposed technique, benchmarks with industrial-level complexities have to be applied. For example, PDN models and solvers need to be tested against real benchmarks to justify their effectiveness. Finally, real 3D PDN benchmarks provide accurate information such as current loads at different nodes, which can be helpful in design optimization algorithms like decoupling capacitance insertion.

To further facilitate and promote the research in 3D PDNs, based on a collaborative effort between academia and industry, we propose for the first time a set of ten 3D PDN benchmark designs extracted from real industry chips, five for DC analysis and five for transient analysis. We also plan to release more designs in the near future. The major characteristics of the proposed benchmarks are:

- All the benchmark circuits are extracted from real 3D designs of industrial complexities. As such, it exposes researchers to more practical real-world problems.

- The benchmark set covers a wide range of functionality (multi-core, DSP, encryption, etc.), size, TSV number, tier number and packaging style. Specifically, the gate-count size of benchmarks varies by 51.3 times; and the number of TSVs varies by 280 times. The largest 3D-PAC design reaches the scale of nine million nodes and over two thousand P/G TSVs. We also include both two-tier and three-tier designs with both wire-bonding and flip-chip packaging techniques. Such a wide coverage can help to expose any potential limitation of a technique.

- Flexible interfaces are provided such that the benchmark can also be utilized for various PDN optimization methods such as decoupling capacitor insertion, PDN wire sizing, power/ground TSV placement, etc.

All benchmark designs and the related information can be downloaded at [19].

The remainder of the paper is organized as follows. In Section II, the design and extraction process of the proposed benchmark set is introduced. Section III elaborates the TSV model utilized in the benchmark. Then each design is described in detail in Section IV, with the application notes briefed in Section V. We finally conclude this paper in Section VI.

2. BENCHMARK DESIGN AND EXTRACTION FLOWS

In this section, we briefly review the flow used to design the PDNs in 3D ICs, as well as how the netlists and related information are extracted for benchmarking.

Our overall in-house design implementation flow is illustrated in Figure 1(a), which employs the Cadence Encounter [20] design tools. During the floorplanning stage, the number of power/ground (P/G) TSVs needed is estimated based on the total power consumption at each tier. Then, we insert these TSVs into the layout as an initial solution. If specific constraints such as wide IO are present, they will be honored. Otherwise, we simply adopt a uniform TSV placement since no detailed power map is available before placement. However, if certain macro-

architecture level power profile is known in advance, a non-uniform P/G TSV allocation will also be set up. In addition, an initial PDN with estimated metal width/pitch is placed at this stage. During the placement stage, the P/G TSV locations, PDN metal width and pitch are iteratively refined based on the noise distribution. For instance, we will try to move P/G TSVs closer to the area with more noise.

The overall netlist extraction flow is illustrated in Figure 1(b), which employs the Cadence QRC [21] and PowerMeter [22] tools. Two types of netlists are extracted: one for DC analysis (resistance only) and the other for transient analysis (resistance and capacitance). The detail process is discussed below.

Step 1. PDN parasitic extraction: To obtain the parasitic information for PDNs, we use Cadence QRC with "R-only" and "RC-decoupled" options for DC benchmarks and transient benchmarks, respectively. The power mesh and ground mesh in each tier are extracted separately (with power pads and ground pads included). As a result, the netlists in DSPF format (Detailed Standard Parasitic Format) can be obtained for each tier.

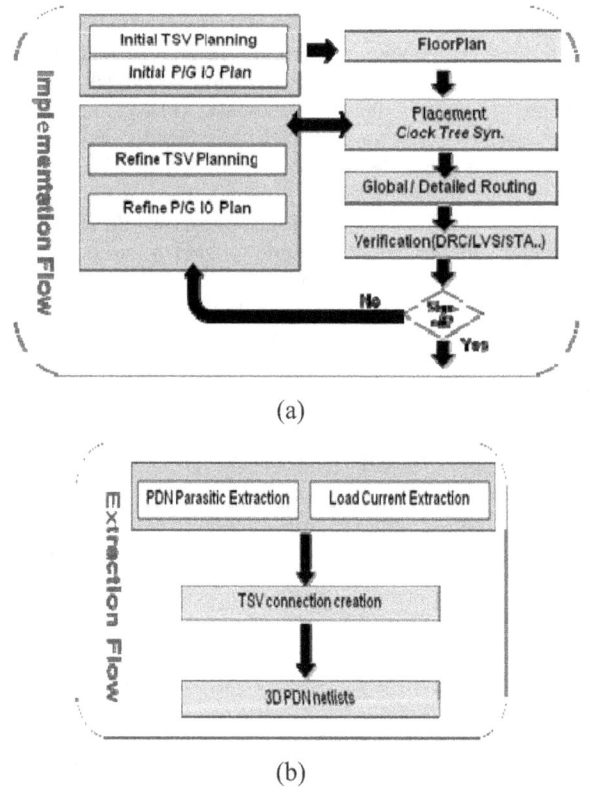

(a)

(b)

Figure 1. Benchmark generation flows: (a) design implementation flow and (b) PDN extraction flow

Step 2. Load current extraction: To obtain accurate load current information, complete signal and power traces in the entire chip need to be considered. Accordingly, we perform parasitic extraction for the entire chip (excluding global nets). The resulted netlists are in SPEF (Standard Parasitic Extraction Format). We then use the vectorless analysis engine in PowerMeter along with the properly specified signal activity rate and timing windows file (TWF) from Cadence Encounter to get the load current values for each instance (power map). Besides, we split each load current into two components, i.e., from VDD to ideal ground and from

VSS to ideal ground. Finally, the length of the time window used for transient current is different in each design, depending on data complexity.

Step 3. TSV connection creation: In this step, we connect the extracted PDN netlists in each tier by stamping the TSV SPICE model. This can be easily done by identifying the two nodes each TSV are connecting to in the layout. The detailed TSV model we use is described in Section III.

Finally, to provide reference noise for each benchmark design, we use UltraSim [23] to simulate the obtained DC and transient netlists with ".usim_pn" option. For very large designs, we further speed-up the simulation speed with reduced accuracy through "sim_mode" and "speed" options. Thus, it is important to note that the noise information provided in the benchmark can only be used as a reference and is not guaranteed to be accurate.

3. TSV MODELS

In this paper, we adopted the lumped circuit model proposed in [24] for TSV parasitics. Here the silicon substrate was treated as a lossy dielectric material connected with ground, so that the shunt path between the TSV and the ground can be modeled using the circuit components in Figure 2, where RT models the TSV resistance. The capacitance C1 models the thin SiO2 insulator layer between the TSV and the silicon substrate. Meanwhile, the silicon substrate between the TSV and the ground can be modeled as a shunt conductance G0 (due to the conductive loss of silicon) in parallel to C2 (due to the displacement current through the silicon substrate). Note that for DC analysis, only the TSV resistance RT is kept.

To get the exact values for the RC elements, a full-wave simulation using Ansys HFSS [25] was performed using our TSV geometry (height = 50 um, diameter = 10um, SiO2 thickness = 0.5um). The obtained results were then used to fit the C1, C2 and G0 values. For more details on the boundary conditions used in full wave simulation and the extraction process, interested readers are referred to [26].

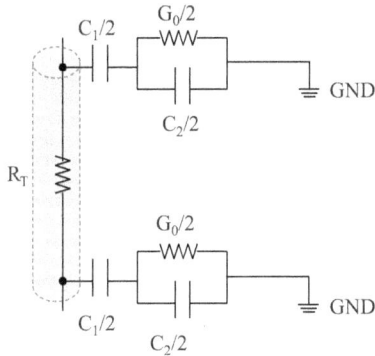

Figure 2. Lumped TSV model.

4. BENCHMARK DESCRIPTION

In this section, details about five 3D designs are given for better understanding of their characteristics. Note that each design corresponds to two benchmark circuits, one for DC analysis and the other for transient analysis. As a summary, Table I (page 7) shows the key information for each design. All designs use face-to-back integration technology, in which the top metal layer in tier n is bonded with the substrate of tier-(n+1). Tier-0 denotes the tier closest to the package.

4.1 3D-μ P (Multi-Processor)

3D-μ P is a 2-tier multi-processor design with 1.8V power supply, as illustrated in Figure 3. Flip-chip packaging is used. The bottom tier (i.e., tier-0) is the logic layer which contains five 32-bit microprocessors, as well as components like AMBA High-performance Bus (AHB). The top tier (i.e., tier-1) is dedicated to the distributed SRAM banks. Each tier has an area of 4,832 x 4,832 um^2.

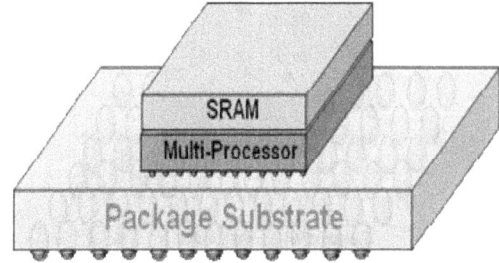

Figure 3. 3D-μP stacking illustration

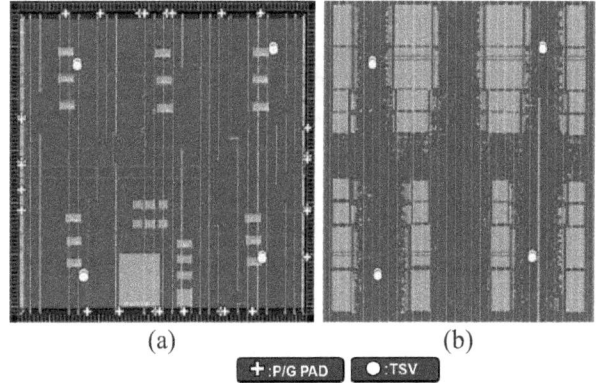

Figure 4. PDN designs of 3D-μP: (a) tier-0; (b) tier-1

The corresponding PDN designs are shown in Figure 4. The TSVs and the power/ground pads are marked. The blue areas are filled with rows of follow pins. For the PDN, a total of eight P/G TSVs are inserted, four for VDD and four for VSS. Due to the regularity of the design, we choose to uniformly distribute the P/G TSVs in the chip area.

We perform extraction to get the complete netlist with load currents. Figure 5 (next page) depicts the power maps of 3D-μP. The 8k bytes on-chip memory in tier-0 and clock I/O have higher power consumption compared with the other parts and thus can be clearly identified. Based on extracted netlist, we obtain that the maximum DC noise is 61 mV and the average noise is 37 mV.

4.2 3D-μ PD (Multi-Processor with Extra Decaps)

It is sometimes necessary for researchers to consider the impact of decoupling capacitor cells (decaps). Towards this, we release a second version of the 3D-uP, namely 3D-uPD, with increased PDN density and additional decoupling capacitors. Everything else is kept the same. Such design pairs are especially useful for exploring 3D PDN optimization related problems

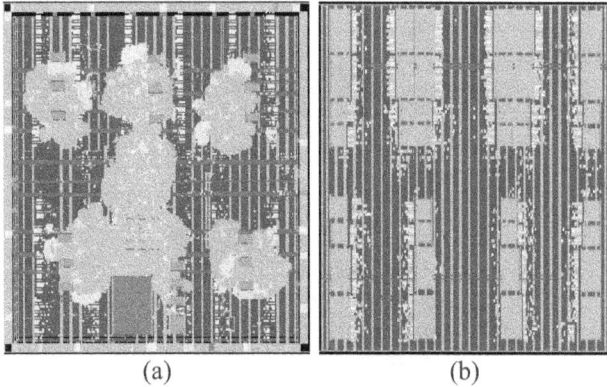

Range(mW)	Cell Count	%
14.516~28.0321	2	0.00%
1~14.516	21	0.02%
0.001~1	24.771K	19.76%
0.0001~0.001	71.546K	57.07%
1e-05~0.0001	24.916K	19.88%
1e-06~1e-05	2.731K	2.18%
1e-07~1e-06	737	0.59%
0~1e-07	639	0.51%

Figure 5. Power maps of 3D-μP: (a) tier-0; (b) tier-1

The resulting PDN designs are shown in Figure 6. The blue rectangle illustrates the area with increased PDN density. A total number of 39,443 decap cells are inserted in whitespace to suppress the noise bound. We use the what-if function of commercial IR drop analysis tool to calculate the required number and location of decap cells, as shown in Figure 7. The yellow regions show the appropriate locations to place decap cells. Due to decap insertion, the leakage power will increase slightly while dynamic noise can be reduced effectively. Simulation results indicate that the maximum and average DC noise of 3D-μPD are 44mV and 33mV, respectively, a 17 mV and 4 mV reduction from 3D-uP. In addition, maximum dynamic noise can be reduced by 120 mV.

4.3 3D-TxRx

3D-TxRx is a hardware cryptography design with three stacked tiers and 1.8V power supply. Specifically, tier-0 contains an AES module for hardware-accelerated AES encryption. Other auxiliary logics such as clock generator, test wrapper, glue logic, etc. reside in the second (i.e., middle) tier. The top tier is in charge of communication with modules like baseband, MAC, DEU and TX/RX to send and receive encrypted information. As shown in Figure 8, flip-chip technology is used for packaging. The area of each tier is 1,900 x 1,900 um^2.

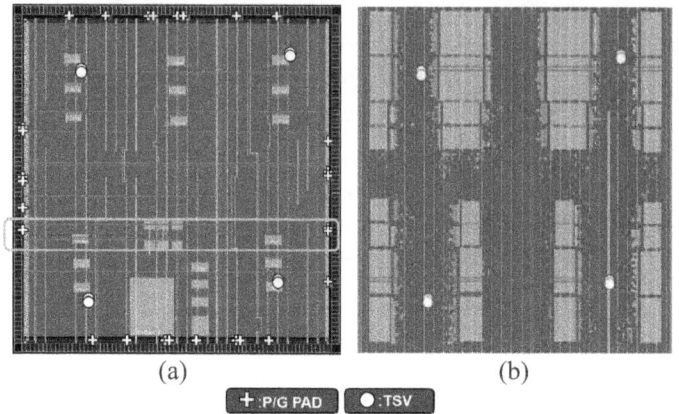

Figure 6. PDN designs of 3D-μPD: (a) tier-0; (b) tier-1

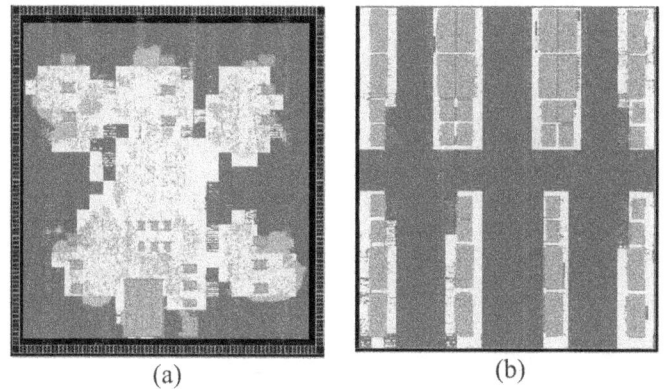

```
Data filtering results for decoupling cap requested for IR fix:
     Overall data minimum:          0F
     Overall data average: 3.15623e-14F
     Overall data maximum: 6.70113e-12F

Filter 1: 595483 of 595483 data values fell into this filter.
     filtered data range:      0F - 6.70113e-12F
     filtered data average: 3.15623e-14F
          8 values were in range 1:      5e-12F -    1e-09F
        111 values were in range 2:      3e-12F -    5e-12F
        276 values were in range 3:      2e-12F -    3e-12F
       1705 values were in range 4:      1e-12F -    2e-12F
       6087 values were in range 5:      5e-13F -    1e-12F
      16548 values were in range 6:      2e-13F -    5e-13F
       8526 values were in range 7:      1e-13F -    2e-13F
     562222 values were in range 8:         0F -    1e-13F
```
(c)

Figure 7. Decap requested maps of 3D-μPD: (a) tier-0; (b) tier-1; (c) requested number of decaps

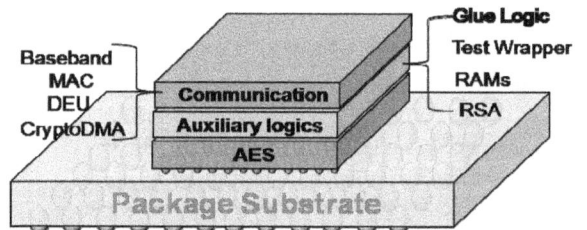

Figure 8. 3D-TxRx stacking illustration

The PDN design of each tier is shown in Figure 9. In order to test each tier before die stacking, test-only pads for pre-bond testing are placed on the lower left corner of tier-1 and tier-2. Since the power consumption of chip is relatively low, we adopt a simple PDN structure which is composed of power ring around tier-0, vertical power strips and horizontal follow pins. A total of 16 P/G TSVs are inserted, 8 for VDD and 8 for VSS. Due to the regularity of the design, we choose to distribute the P/G TSVs in the chip area as uniformly as possible.

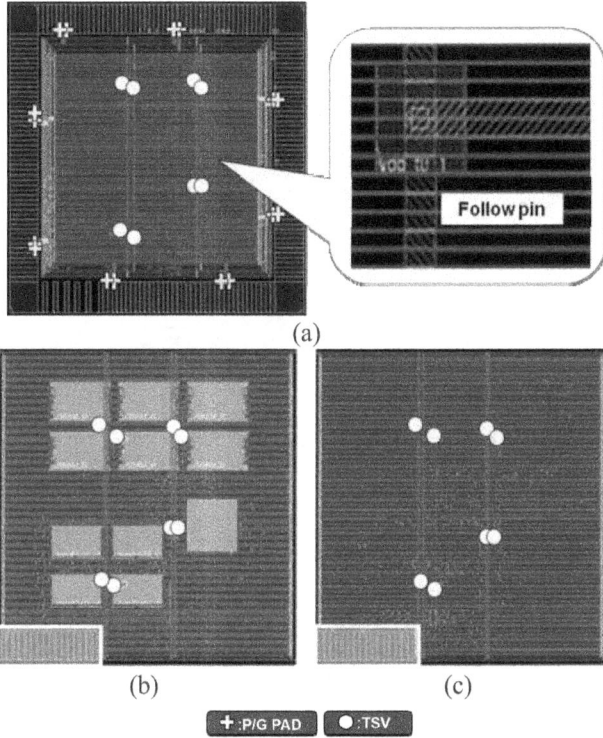

(a)

(b)   (c)

+ :P/G PAD   ● :TSV

Figure 9. PDN designs of 3D-TxRx: (a) tier-0; (b) tier-1; (c) tier-2

Figure 10 shows the power maps of 3D-TxRx at different tiers. A maximum of 19 mV and an average of 10 mV DC noise are observed.

4.4 3D-SAP(Sensor Application Processor)

3D-SAP is an image sensor application processor (ISP) with an array of 128 stacked analog-to-digital converters (ADC) and 1.8V power supply as shown in Figure 11. Wire-bonding technology is used for packaging. Since the ISP has high-speed data streaming, we employ eight processing cores to implement the parallel computation function. The ADC array is placed on tier-1 for data conversion. The areas of tier-0 and tier-1 are 9,000 x 7,000 um^2 and 7,280 x 5,390 um^2 separately.

For mixed-signal design, multiple power domains exist, and they share the same ground plane to reduce parasitic effects. To reduce the benchmark complexity, we only extract one power domain. The corresponding PDN designs are shown in Figure 12, and a total of 268 P/G TSVs are used, 55 for VDD and 213 for VSS. Again, this asymmetry is due to the extraction of a single power domain. Since tier-1 is an ADC design in compact array structure to capture the

pixel signals of the image senor, P/G TSVs have to be limited to the two sides of the die to avoid disturbing the pixel array.

The extracted power maps in Figure 13 (next page) clearly show that memory cells for instruction and data have higher power consumption than the other parts. Simulation results indicate that the maximum and average DC noise of 3D-SAP as 58 mV and 27 mV, respectively.

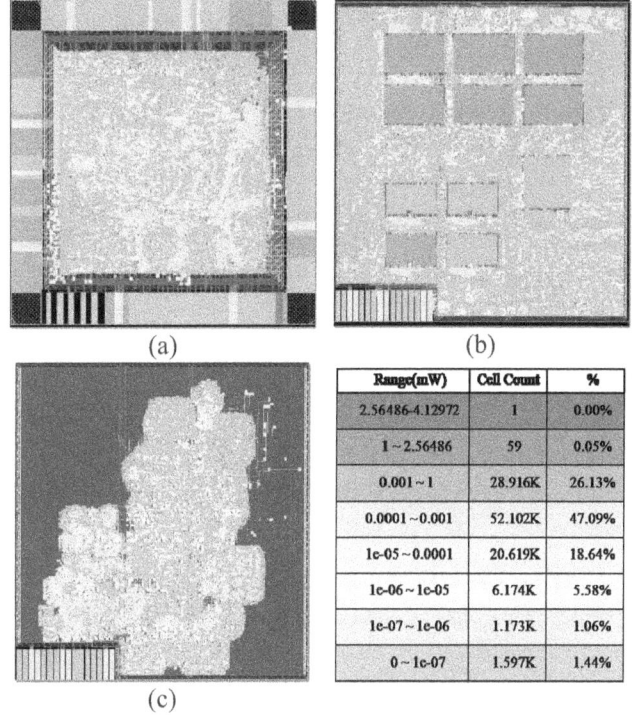

(a)   (b)

(c)

Range(mW)	Cell Count	%
2.56486-4.12972	1	0.00%
1 ~ 2.56486	59	0.05%
0.001 ~ 1	28.916K	26.13%
0.0001 ~ 0.001	52.102K	47.09%
1e-05 ~ 0.0001	20.619K	18.64%
1e-06 ~ 1e-05	6.174K	5.58%
1e-07 ~ 1e-06	1.173K	1.06%
0 ~ 1e-07	1.597K	1.44%

Figure 10. Power maps of 3D-TxRx: (a) tier-0; (b) tier-1; (c) tier-2

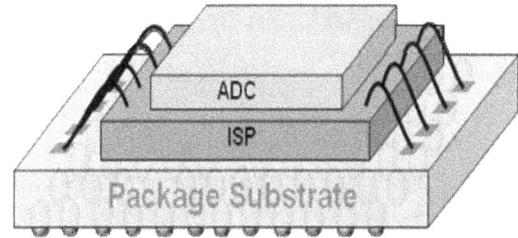

Figure 11. 3D-SAP stacking illustration

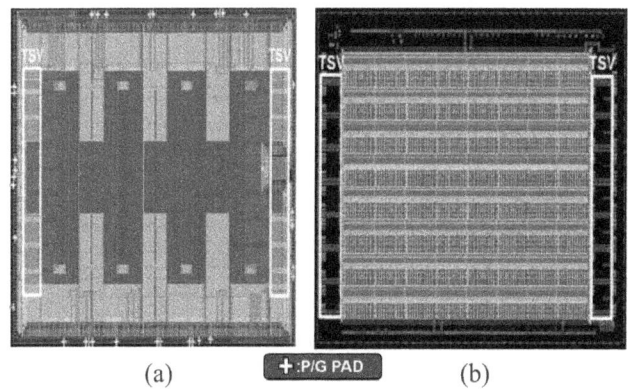

(a)   + :P/G PAD   (b)

Figure 12. PDN designs of 3D-SAP: (a) tier-0; (b) tier-1

Range(mW)	Cell Count	%
10.38-19.76	34	0.00%
1~10.38	177	0.02%
0.001~1	213.939K	23.58%
0.0001~0.001	546.776K	60.27%
1e-05~0.0001	94.202K	10.38%
1e-06~1e-05	12.957K	1.43%
1e-07~1e-06	15.265K	1.68%
0~1e-07	23.813K	2.62%

(a)

Range(mw)	Cell Count	%
Off		
Off		
0.001~1	95.143K	36.29%
0.0001~0.001	148.185K	56.53%
1e-05~0.0001	18.688K	7.13%
1e-06~1e-05	64	0.02%
1e-07~1e-06	0	0.0%
0~1e-07	64	0.02%

(b)

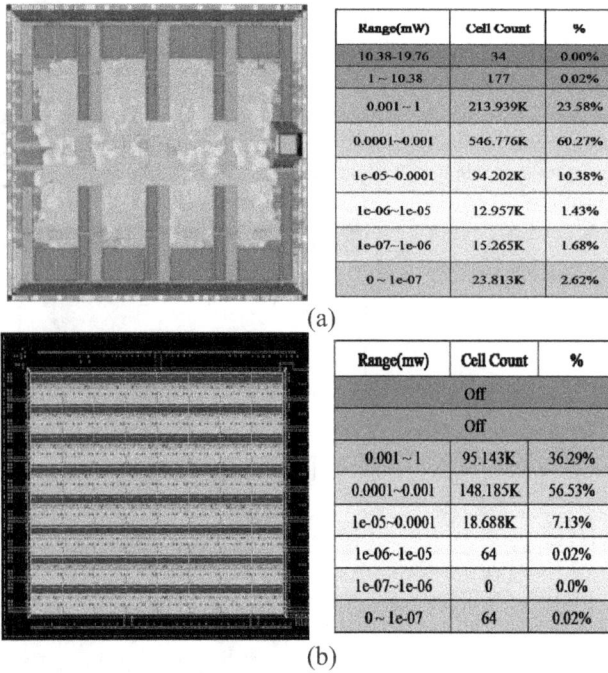

Figure 13. Power maps of 3D-SAP: (a) tier-0; (b) tier-1

4.5 3D-PAC (Parallel Architecture Core)

3D-PAC is a multi-DSP design with stacked SRAM banks and 1.0V power supply as shown in Figure 14. Wire-bonding technology is used for packaging. It is designed for the next generation high-quality multimedia application which is low-power system-on-chip (SOC). Specifically, the architecture is composed of an ARM9 core, two PAC digital signal processor (DSP) cores, EMDMA (Enhanced multi-media DMA), DDR2 controllers and AMBA AXI/AHB. Different from 3D-μ P, tier-0 is dedicated to two DSP cores with one ARM core as co-processor, and tier-1 is consisted of SRAM banks. Since 3D-PAC adopts DVFS (dynamic voltage /frequency scaling) technology, it contains 16X more instances and consumes 1.62X more die area than 3D-μ P. The areas of tier-0 and tier-1 are 7,800 x 7,800 um^2 and 3,880 x 3,880 um^2 separately.

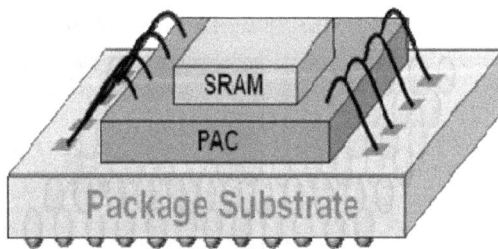

Figure 14. 3D-PAC stacking illustration

The corresponding PDN designs are shown in Figure 15, which contains two channels memory design and a total of 2,234 P/G TSVs are used, 1,114 for VDD and 1,120 for VSS. According to JEDEC wide-IO logic-memory interface definitions [27], these TSVs need to be clustered. Different from 3D-SAP, we place them in the centre of the memory die to allow for signal synchronization. This design, as well as the previous 3D-SAP, are

good examples to indicate why 3D PDN benchmarks are necessary: those special P/G TSV placement constraints only emerge in real 3D designs, and their locations would apparently affect the subsequent PDN sizing. Existing 2D benchmarks do not consider those effects for sizing, and as a result, simply stacking them results in less realistic designs. Note that for tier-0, the two DSP cores in the upper left and upper right corners as well as the PLL in the lower right corner are in different power domains. Accordingly, they are not extracted for benchmarking (masked in the figure).

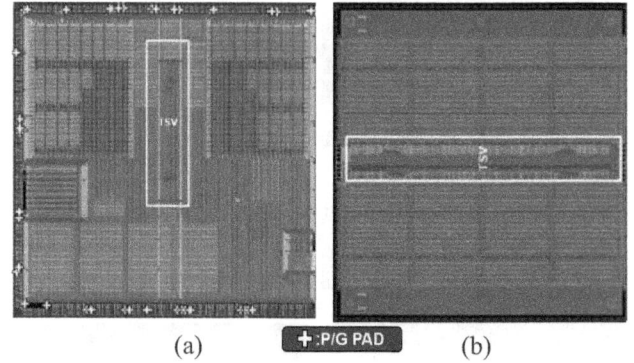

(a)   **+ :P/G PAD**   (b)

Figure 15. PDN designs of 3D-PAC: (a) tier-0; (b) tier-1

The extracted power maps in Figure 16 clearly show that memory cells of AMBA AXU/AHB have higher power consumption than the other parts. Simulation results indicate that the maximum and average DC noise of 3D-PAC as 36 mV and 25 mV, respectively.

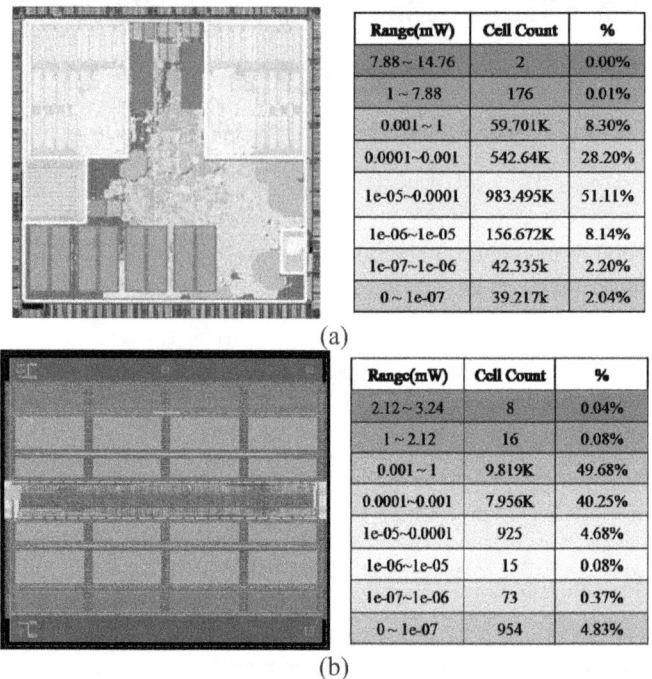

Range(mW)	Cell Count	%
7.88~14.76	2	0.00%
1~7.88	176	0.01%
0.001~1	59.701K	8.30%
0.0001~0.001	542.64K	28.20%
1e-05~0.0001	983.495K	51.11%
1e-06~1e-05	156.672K	8.14%
1e-07~1e-06	42.335k	2.20%
0~1e-07	39.217k	2.04%

(a)

Range(mW)	Cell Count	%
2.12~3.24	8	0.04%
1~2.12	16	0.08%
0.001~1	9.819K	49.68%
0.0001~0.001	7.956K	40.25%
1e-05~0.0001	925	4.68%
1e-06~1e-05	15	0.08%
1e-07~1e-06	73	0.37%
0~1e-07	954	4.83%

(b)

Figure 16. Power maps of 3D-PAC: (a) tier-0; (b) tier-1

Table I. Information of the 3D designs used to extract PDN benchmarks

Chip name			3D-μ P	3D-μ PD	3D-TxRx	3D-SAP	3D-PAC
Chip information		VDD	1.8V	1.8V	1.8V	1.8V	1.0V
		VSS	0.0V	0.0V	0.0V	0.0V	0.0V
		Tier #	2	2	3	2	2
		P/G TSV #	8	8	16	268	2,234
		Tier-0 die area (um^2)	4,832x4,832	4,832x4,832	1,900x1,950	9,000x7,000	7,880x7,880
		Tier-1 die area (um^2)	4,832x4,832	4,832x4,832	1,900x1,950	7,280x5,390	3,880x3,880
		Tier-2 die area (um^2)	--	--	1900x1950	--	--
		Instance #	125,388	551,889	110,657	1,032,275	2,044,743
		Pad #	304	304	232	412	500
PDN Information	General	Metal layers #	7	7	7	7	10
		Current source #	239,530	239,530	218,066	2,902,474	3,364,314
		Voltage source #	28	28	16	36	51
	DC only	Node #	826,524	1,021,185	244,757	4,588,540	8,983,606
		Resistor #	913,154	1,178,081	313,414	7,638,905	13,234,026
	Tran. only	Node #	826,542	1,021,203	244,789	4,589,076	8,988,071
		Resistor #	913,170	1,178,097	313,446	7,639,442	13,238,494
		Capacitor #	762,239	967,057	276,223	4,147,603	6,067,910
		Current source step #[1]	200	140	140	160	100
Static noise (DC only)		Max(mV)	61	44	12	58	36
		Average(mV)	37	33	4	27	25

5. APPLICATION NOTES

Each benchmark design is composed of three files as follows:

1) DESIGN_NAME_dc.sp (DESIGN_NAME_tran.sp), the extracted netlist for DC (transient) simulation. The netlists for DC analysis only contain parasitic resistance, while those for transient analysis contain both parasitic resistance and capacitance. It is possible that we will also include inductance in the future.

2) DESIGN_NAME_info.txt, the information sheet for the benchmark design. It contains related information such as the total number of nodes, branches, TSVs, current sources, and voltage sources. It also provides the detailed locations (in coordinates) of each TSV.

3) DESIGN_NAME.sol, which provides a reference solution. For DC analysis, it contains the static voltage at each node. For transient analysis, dynamic voltages at selected nodes are provided.

File #2 is pretty self-explanatory. Most of the information there is also summarized in Table I (except for the TSV locations). Files #3 comes from the standard output file of Ultrasim. Below we will briefly explain the DC/ transient netlists (File #1). The general structure of the netlist for DC analysis is illustrated in Figure 17 and Figure 18 is for transient analysis. In the beginning, a number

of comment lines provide the x-y coordinate information of each node. For example, the following comment vivfom

*|I (t0_PGTSV_vss_0:VSS_t0 t0_PGTSV_vss_0 VSS_t0 905.97 3955.25)

indicates that the node "t0_PGTSV_vss_0:VSS_t0" belongs to the instance "t0_PGTSV_vss_0", and is connected to the ground mesh "VSS_t0". The node has the x-y coordinate "905.97 3955.25". Note that node name follows the <name:VSS/VDD_tier> style, where name could be an arbitrary string, VSS/VDD stands for power or ground, and an optional tier indicates the tier the node resides in such as t0, t1, etc. Under such naming convention, both power and ground networks, as well as elements in different tiers, are easily described in the same netlist.

To distinguish different elements, all voltage source names start with "V", while all current source names start with "IVDD" or "IVSS", depending on whether they are connected to power mesh or ground mesh. In the netlists for static analysis, DC current sources are used, while in those for transient analysis, piecewise linear (PWL) current sources are used. For relatively small benchmarks such as 3D-μP and 3D-TxRx, complete dynamic waveforms are used. For large benchmarks such as 3D-SAP and 3D-PAC, we only extract the waveform in selected windows. Furthermore, all TSV resistance names start with "Rtsv" to distinguish them from other resistance names, which start with "R".

---

[1] This is the number of steps used in transient analysis.

```
*|I (node_name instance_name p/g_mesh_name X Y)

*parasitic R for VSS/VDD

R<number>      <node1>        <node2>      value

*current source for VSS/VDD

IVSS<number>    0             <node>       value

IVDD<number>  <node>          0            value

*TSV parasitic R

Rtsv<number>  <node1>         <node2>      value

*voltage source

V<number>      <node>         0            value
```

Figure 17. Structure of the netlist for DC analysis.

```
*|I (node_name instance_name p/g_mesh_name X Y)

*parasitic RC for VSS/VDD

R<number>    <node1>   <node2>       value

C<number>    <node1>   <node2>       value

*current source for VSS/VDD

IVSS<number>    0            <node>

                 PWL t1 value1 t2 v2 .....

IVDD<number>  <node>          0

                 PWL t1 value1 t2 v2 .....

*TSV parasitic RC

Rtsv<number>  <node1>   <node2>     value

Ctsv<number>  <node1>   <node2>     value

*voltage source
```

Figure 18. Structure of the netlist for transient analysis.

6. CONCLUSIONS

In this paper, we put forward a set of PDN benchmarks that are extracted from industrial 3D designs. These designs are carefully selected such that they cover a wide range of functionality, process, size, TSV number, tier number and packaging style. We hope that the released benchmarks can facilitate and promote the research in 3D PDNs.

7. REFERENCES

[1] N.H. Khan, S.M Alam and S Hassoun. 2009. System-level comparison of power delivery design for 2D and 3D ICs. *Proc. Intl. Conf. on 3D System Integration*. 1-7. DOI= http://dx.doi.org/10.1109/3DIC.2009.5306539.

[2] Z. Xu, J. Lu, B.C. Webb and J.U. Knickerbocker. 2011. Electromagnetic-SPICE modeling and analysis of 3D power network. *Proc. Electronic Components and Technology Conf.*. 2171-2178. DOC= http://dx.doi.org/10.1109/ECTC.2011.5898820.

[3] J. Xie and M. Swaminathan. 2010. DC IR drop solver for large scale 3D power delivery networks. *Proc. Intl. Conf. on Electrical Performance of Electronic Packaging and Systems*. 217-220. DOI= http://dx.doi.org/10.1109/EPEPS.2010.5642582

[4] M. Jung and S.K. Lim. 2010. A Study of IR-drop Noise Issues in 3D ICs with Through-Silicon-Vias. *Proc. Intl. Conf. on 3D System Integration*. 1-7. DOI=http://dx.doi.org/10.1109/3DIC.2010.5751462

[5] P. Singh, R. Sankar, X. Hu, W. Xie, A. Sarkar and T. Thomas. 2010. Power delivery network design and optimization for 3D stacked die designs. *Proc. Intl. Conf. on 3D System Integration*. 1-6. DOI= http://dx.doi.org/10.1109/3DIC.2010.5751475

[6] H. Chen, H. Lin, Z. Wang and T. Hwang. 2011. A new architecture for power network in 3D IC. *Proc. Design Automation & Test in Europe Conf.*. 1-6.

[7] Z. Li, Y. Ma, Q. Zhou, Y. Cai, Y. Wang, T. Huang and Y. Xie. 2012. Thermal-aware power network design for IR drop reduction in 3D ICs. *Proc. Asia and south Pacific Design Automation Conf.*. 47-52. DOI= http://dx.doi.org/10.1109/ASPDAC.2012.6164995

[8] P. Falkerstern, Y. Xie, Y. W. Chang and Y. Wang. 2010. Three dimensional integrated circuit (3D IC) floorplan and power/ground network co-synthesis. *Proc. Asia and south Pacific Design Automation Conf.*. 169-174. DOI= http://dx.doi.org/10.1109/ASPDAC.2010.5419899

[9] P. Zhou, K. Sridharan and S.S. Sapatnekar. 2009. Optimizing decoupling capacitors in 3D circuits for power grid integrity. *IEEE Design & Test of Computers*, 26, 5. 15-25. DOI= http://dx.doi.org/10.1109/MDT.2009.120

[10] H. Yu, J. Ho and L. He. 2006. Simultaneous power and thermal integrity driven via stapling in 3D ICs. *Proc. Intl. Conf. on Computer-Aided Design*. 802-808. DOI=http://dx.doi.org/10.1109/ICCAD.2006.320123

[11] S. Nassif. 2008. Power grid analysis benchmarks. *Proc. Asia and South Pacific Design Automation Conf.*. 376-381. DOI= http://dx.doi.org/10.1109/ASPDAC.2008.4483978

[12] Z. Li, R. Balasubramanian, F. Liu and S. Nassif. 2011. 2011 TAU power grid simulation contest: benchmark suite and results. *Proc. Intl. Conf. on Computer-Aided Design*. 478-481. DOI=http://dx.doi.org/10.1109/ICCAD.2011.6105371

[13] Z. Zeng, T. Xu, Z. Feng and P. Li. 2011. Fast static analysis of power grids: algorithms and implementations. *Proc. Intl. Conf. on Computer-Aided Design*. 488-493. DOI= http://dx.doi.org/10.1109/ICCAD.2011.6105373

[14] C. H. Chou, N. Y. Tsai, H. Yu, C. Lee, Y. Shi and S. Chang. 2011. On the preconditioner of conjugate gradient method – a power grid Simulation perspective. *Proc. Intl. Conf. on Computer-Aided Design*. 488-493. DOI= http://dx.doi.org/10.1109/ICCAD.2011.6105374

[15] T. Yu, Martin D.F. Wong. 2012. PGT_SOLVER: An Efficient Solver for Power Grid Transient Analysis. *IEEE Intl. Conf. on Computer-Aided Design*. 647-652.

[16] J. Yang, Z. Li, Y. Cai and Q. Zhou. 2012. PowerRush: efficient transient simulation for power grid analysis. *Proc. Intl. Conf. on Computer-Aided Design*. 653-659.

[17] X. Xiong and J. Wang. 2012. Parallel forward and back substitution for efficient power grid simulation. *Proc. Intl. Conf. on Computer-Aided Design*. 660-663.

[18] J. Kim, et al. 2011. A 1.2V 12.8 GB/s 2Gb mobile wide-I/O DRAM with 4x128 I/Os using TSV-based stacking. *Proc. Intl. Solid-State Circuits Conf.*. 496-498. DOI= http://dx.doi.org/10.1109/ISSCC.2011.5746413

[19] http://web.mst.edu/~yshi/3D-PDN.html

[20] Cadence, "Encounter™ user guide".

[21] Cadence, "QRC extraction user manual".

[22] Cadence, "VoltageStorm® PowerMeter manual".

[23] Cadence, "VirtuosoR UltraSim simulator user guide".

[24] H. Wang, J. Kim, Y. Shi and J. Fan. 2011. The effects of substrate doping density on the electrical performance of through-silicon-via. *Proc. of Asia-Pacific EMC Symposium*.

[25] HFSS website, "http://www.ansys.com/"

[26] T. Wang, P.W. Luo, Y. S. Su, L. C. Cheng, D. M. Kwai and Y. Shi. 2012. Capturing the phantom of the power grid - on the runtime adaptive techniques for noise reduction. *Proc. Asia and South Pacific Design Automation Conference*. 640-645. DOI= http://dx.doi.org/10.1109/ASPDAC.2012.6165035

[27] "Wide-IO memories," http://www.gsaglobal.org/3dic/docs/20111216_Wide_I-O_Std_JC_42.6.pdf

High Performance and Low Power Design Techniques for ASIC and Custom in Nanometer Technologies

David Chinnery

Mentor Graphics, 46871 Bayside Parkway, Fremont, CA, USA

david_chinnery@mentor.com

ABSTRACT

Traditionally, synthesized application-specific integrated circuits (ASICs) have been slower and higher power than custom integrated circuits due to a variety of factors. This paper details how this gap has decreased in the past few years. ASICs have adopted higher performance and lower power design techniques with the aid of better CAD tool support. To improve productivity, many full custom designs have migrated to a semi-custom design methodology that is more amenable to the use of standard CAD tools and makes greater use of synthesis.

Categories and Subject Descriptors

B.7.0 [**Integrated Circuits**]: General

General Terms

Design, Performance

Keywords

ASIC, CAD, clock, comparison, custom, EDA, energy, frequency, gap, leakage, methodology, power, speed, standard cell, synthesis

1. INTRODUCTION

Historically, custom designed circuits were faster, lower power, and smaller area than synthesized ASICs. A variety of factors contribute to this: microarchitecture, placement quality, clocking methodology, logic style, voltage scaling, and process technology [8][9]. These advantages have decreased in today's technologies with improvements in automated design techniques.

Herein, ASIC refers to digital circuits constructed of standard cells by automatic synthesis, place, and route (SPR) from a register transfer level (RTL) description [39]. A high productivity electronic design automation (EDA) flow is necessary in today's competitive marketplace where time to market is important for product success. The resulting designs may not have optimal performance or the lowest power for the given design constraints.

Custom design refers to full custom and semi-custom design techniques. A semi-custom designer selects cells from a standard cell library that may have custom cells for the project, entering in a schematic their implementation of RTL. The circuit netlist is then typically placed manually in standard cell rows, either by layout entry or by text commands specifying cell placement. A full custom designer creates additional cells specifically for their particular design, and the placement is not necessarily restricted to standard cell rows. High speed dynamic logic and other techniques that are not amenable to a standard cell flow may be used to achieve higher performance.

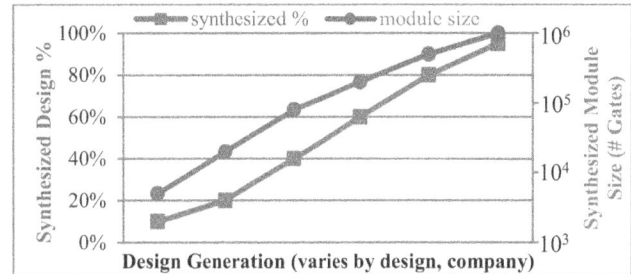

Figure 1. Approximate trends in synthesized portion of design and synthesized module size for traditionally custom designs.

Custom design uses CAD tools to aid construction, analysis, and verification. A vendor router can be used on a manually placed semi-custom design with some pre-routed nets, for example [11]. Many of these are in-house tools with no EDA vendor equivalent.

Custom circuits can be 3 to 8× higher performance than ASICs [8]. ASICs may consume 2.6 to 7× the power of a custom design when the same performance constraint can be met [9]. Related research has shown that FPGA implementations have an order of magnitude higher power and silicon area compared to ASICs, and are also 3 to 5× slower [25]. An overview of contributing factors to the gap between ASIC and custom is given in Table 1.

This paper focuses on digital circuit design techniques. It does not consider on-chip communication architecture, memory hierarchy, or higher system and software level factors that also have a major impact on performance and power consumption. Also, parallel datapaths are not considered as the benefits depend on application.

Trends toward greater use of synthesis are discussed in Section 2. Contributing factors to the performance and power gaps between ASIC and custom in newer technologies are detailed in Section 3. Additional EDA tool improvements are suggested in Section 4. The impact of design methodology choices on the EDA flow are considered in Section 5. Then Section 6 provides a summary.

2. CUSTOM DESIGN TRENDS

In custom designs in earlier technologies, synthesis was reserved for small control logic sub-modules that were not timing critical, as shown in Figure 1. In newer technologies, the majority of the design is synthesized to improve productivity and process portability. Custom design is only used for more timing critical datapaths and memories, such as register banks. Switching to use primarily SPR can improve productivity by a factor of four or more compared to a mix of custom design and SPR [2][42].

Semi-custom design was used for memory and datapath blocks in a 45nm DSP core [11]. In the memory block, manually tiling 14% of the cells and pre-routing 3% of the wires through non-standard pitch tracks to fix routing congestion reduced area by 10% and wire length by 13% versus using only SPR. The memory arrays were full custom, the rest was SPR. Manually tiling 77% of the cells in the 32 bit datapath reduced area by 34% and wire length by 44%. A full custom 16KB data cache was replaced by a semi-custom design with the same performance and area, reducing flow time from a few weeks to a week from RTL to routed design.

Table 1. Factors contributing to ASICs having lower performance & higher power than custom, at a tight performance constraint. Some factors from previous analysis [8][9] shown in red are updated in blue from analysis in this paper. "Excellent" is what ASICs may achieve with excellent low power and high performance techniques. "Power" is total power when a circuit is active – power gating and other standby leakage reduction techniques are omitted. Custom power savings include using timing slack from high speed techniques to reduce power. The combined impact of these factors is not multiplicative. In particular, the microarchitectural power factor accounts for reduction in IPC and increase in registers per stage for deeper pipelines, timing overhead per pipeline stage, power for clock distribution and registers, and use of timing slack for gate downsizing and voltage scaling, using a detailed pipeline performance and power model [9].

Contributing Factor	ASIC Slower vs. Custom		ASIC Power vs. Custom		Major Impacts of Factor
	Typical	Excellent	Typical	Excellent	
microarchitecture	1.80× → 2.10×	1.30× → 1.00×	3.5× → 3.7×	1.9× → 2.0×	# of pipeline stages, IPC
clock distribution, clock gating, registers	1.45× → 1.60×	1.10× → 1.15×	1.6× → 1.8×	1.0× → 1.1×	timing overhead per pipeline stage
logic style	1.40× → 1.20×	1.40× → 1.20×	2.0× → 1.5×	2.0× → 1.5×	combinational delay & power
logic design	1.30×	1.00×	1.2×	1.0×	combinational delay & power
technology mapping	1.00×	1.00×	1.4×	1.0×	combinational delay & power
floorplanning and placement	1.40×	1.00×	1.5×	1.1×	combinational delay & power
cell design, cell sizing, wire sizing	1.45×	1.10×	1.6×	1.1×	combinational delay & power
voltage scaling	1.00× → 1.10×	1.00×	4.0× → 2.0×	1.0×	overall delay & power
process technology, process variation	2.00×	1.20×	2.6×	1.3×	overall delay & power, conservatism

Intel's 45nm Atom processor was 9% full custom, 45% semi-custom, and 46% synthesized [14]. AMD's 40nm Bobcat processor was synthesized from RTL at the full core level. It comprised 1.1 million gates constructed by a traditional SPR flow with 35 instances of 7 custom memory macros [13].

Module size encompasses both flat and hierarchical modules. Larger modules are usually hierarchical and are very useful for initial design exploration However, higher quality results are achieved with sub-modules in the 50,000 to 100,000 gate range. These intermediate block sizes aid floor-planning, specification of timing constraints, and minimizing use of global wiring resources. Also, different voltage regions are usually separate sub-modules.

Various design techniques that were historically primarily custom, such as adding customized standard cells to the library, are used throughout the design flow to improve performance and reduce power consumption. Such approaches have been adopted more widely in the industry for use in synthesized designs.

Toshiba's synthesized streaming processing unit (SPU) had clock frequency of 4GHz at 1.4V in 65nm [42]. The SPU is a major part of the synergistic processor elements (SPEs) in the Cell processor. The 90nm SPE was mostly full custom with blocks of several tens of thousands of transistors, 19% dynamic logic, and synthesized control logic blocks of only a few thousand gates each [44]. The synthesized SPU had 30% lower area, was 12.5% faster, and productivity was 4× with 10 to 100× larger blocks, versus a 65nm full custom version [42]. The SPU used a clock mesh to limit total skew to 20ps, and increased speed by 15% with a height optimized standard cell library and by 10% with wire sizing.

3. FACTORS CONTRIBUTING TO THE GAP BETWEEN ASIC AND CUSTOM

3.1 Microarchitecture

Microarchitectural techniques with poor energy efficiency, such as speculative execution, are less viable now as power is a primary design constraint that limits clock frequency [16]. Maintaining high instructions per cycle (IPC) with accurate branch prediction for deeper pipelines and other techniques is important for energy efficiency [2][3][27].

Bobcat had 13 integer pipeline stages [2], which increased to 14 in AMD's next generation 28nm Jaguar core providing 10% higher clock frequency, with an extra cycle for a mispredicted branch. IPC was improved from 0.95 to 1.10 [37]. Power is less at the same performance, as the headroom allows lower supply voltage. Both of these designs were mostly synthesized.

The fanout-of-4 (FO4) delay metric is used to analyze pipeline logic depth and timing overhead. An FO4 delay is the delay of an inverter driving a load capacitance that has 4× the inverter's input capacitance. This metric is not substantially changed by process technology or operating conditions. Other static and domino logic fanout-of-four gates have at most 30% range in FO4 delay [15].

Deep pipelines have a logic depth of as low as 8 FO4 per stage for optimal performance [17]. Critical stages in ARM's A15 were limited to 15 to 16 gates [27]. AMD's 32nm Bulldozer core reduced FO4 delays per clock cycle by 20%, limiting it to 12 to 14 gates between flops [31]. Along with IPC improvements and lower power, this increased AMD's Trinity APU performance by 25% at 35W TDP, versus AMD's older Llano 32nm APU [32].

ASIC and custom designers have similar architectural choices today. The high performance synthesizable ARM Cortex A15 has a three-way out-of-order pipeline with 15 integer stages and up to 24 floating point stages [22], which is similar in pipeline depth to modern, high performance AMD and Intel processors [2][12].

Reduction in supply voltage and increase in clock frequency have slowed with smaller process technologies, but transistor density is still increasing rapidly with Moore's Law [1][19]. Thus, silicon area has become cheaper. The performance improvements in newer technologies can no longer come from scaling up clock frequency which is power limited. Multi-core architectures are now being used to boost performance by parallel computation.

ARM's big.LITTLE architecture pairs a lower power Cortex A7 processor with the A15. Switching between them in response to performance needs takes 30,000 to 50,000 cycles, and is fast as they have the same instruction set architecture (ISA). The A7 has a two-way in-order 8 to 10 stage pipeline that increases energy efficiency at the cost of lower performance. The pairing enables 50% energy reduction for web browsing with audio playback in the background [22]. Microarchitectural simulations show that fine-grained switching can reduce energy by 18% with only a 5% performance penalty. The lower power core has 8% higher power due to shared hardware that enables faster switching [30].

ASICs with excellent microarchitecture, e.g. the A15, have closed this factor's performance gap. Performance from deep pipelining a typical ASIC is calculated assuming: 180 FO4 total unpipelined combinational delay [9]; 10 FO4 timing overhead per stage (see

Table 2. ASIC and custom timing overheads in FO4 delays.

Flop Type	Typical ASIC	Excellent ASIC	Custom
	low power mux-D	fast mux-D	fast LSSD
Clock Type	CTS	MSCTS	mesh
Clock-to-Q Delay	2.0	1.4	1.6
Setup Time	1.1	0.9	0.1
Clock Skew	4.3	1.3	0.5
Clock Jitter	2.6	1.3	0.3
Total	10.0	4.9	2.5

Table 3. Area and timing for 28nm mux-D and LSSD scan flops [7]. Times are percentage of high frequency clock period.

	Mux-D Flip-Flop		LSSD Flip-Flop	
	Fast	Low Power	Fast	Low Power
Relative Area	1.18	1.00	1.97	1.97
Hold Time	-4.3%	-6.6%	15.0%	6.8%
Clock-to-Q Delay	13.2%	19.0%	14.6%	15.7%
Setup Time	8.5%	10.0%	1.3%	10.7%
C-to-Q + Setup	21.7%	29.0%	15.9%	26.4%

Table 2); and microarchitectural techniques can maintain high IPC, as in Jaguar. A typical ASIC may have only five stages [8] and thus is (10+180/5)/(10+180/15)=2.1× slower versus a 15 stage pipeline. This factor is higher as today's pipelines are deeper.

A detailed model of pipeline performance and power can analyze trade-offs for deeper pipelining, timing slack for gate downsizing and voltage scaling, et al. [9]. The decrease in timing overhead per stage for custom designs from 3 FO4 to 2.5 FO4 results in a small increase in the power gap listed for the microarchitecture factor in Table 1. (The typical ASIC microarchitecture factor of 5.1× in [9] accounted for pipeline stage delay imbalance of 10 FO4 extra, which has been corrected to 10% extra stage delay.)

3.2 Clock Tree/Mesh, Clock Gating, Registers

The timing overhead per pipeline stage for clock skew and jitter, register setup time and clock-to-Q delay limits the effectiveness of deeper pipelining. These pipeline timing overheads are analyzed below to determine their impact. Then, clock power is discussed.

3.2.1 Register Type

ASICs typically use mux-D scan flops that are a good choice for lower power in slower designs. Mux-D scan flops are slower than level sensitive scan design (LSSD) flops because of the input multiplexer for the scan and data inputs. LSSD flops are larger and higher power. Vendor tools provide better support for mux-D scan flops but have sufficient support for LSSD scan flops [7].

Recent analysis of 28nm high speed single-clock soft edge flops (SSEFs) shows clock-to-Q delay and setup timing overhead of 15.9% of the clock period for a high frequency microprocessor (see Table 3) [7]. In comparison, high speed mux-D scan flops add 6% to the clock period which is costly for a high speed design. SSEFs are LSSD flops with two non-overlapping scan clocks to avoid scan datapath hold issues. Similar to latches, the high speed soft edge flops are transparent for 10% of the clock period to provide some immunity to clock skew and jitter, and late signal arrivals via time borrowing, but have higher hold time [31].

The high performance standard cell libraries available to ASICs have improved, and ASIC designers may also develop custom cells to address library limitations if necessary. The high speed mux-D scan flip-flop in Table 3 has a clock-to-Q delay and setup timing overhead of 21.7% of the high frequency clock period. This is about 2.3 FO4, compared to 4.0 FO4 for flops or pairs of latches in a good ASIC in older technologies [8].

3.2.2 Clock Distribution, Skew, and Jitter

Clock skew differs with the clock distribution methodology. Typical ASIGs use clock tree synthesis (CTS) from a single clock source, achieving clock skew in the range of 70 to 100ps in 32nm technology, down to 50ps with significant manual work. CTS produces deep clock trees with varying buffering depth, e.g. 15 to 17 levels in a GPU [6]. Deeper clock trees and varying depth both exacerbate clock skew and the impact of process variation [42].

Multi-source CTS (MSCTS) reduces the clock tree depth to reduce the clock skew. ASICs can achieve 30 to 50ps skew with MSCTS in current technologies. The 40nm Bobcat core was able to achieve 38ps skew by over-constraining CTS to limit it to 6 to 8 levels driven by a top level mesh [13]. This required a several thousand line TCL script and had difficulties driving the clock around macros. The clock jitter was 5% for the synthesized clocks. AMD's 28nm Jaguar restricted the clock tree to three levels to reduce skew: a root inverter buffer at each source grid point, another level of inverters, then integrated clock gaters (ICGs) driving the flops [6]. Inverter buffers, instead of non-inverting, reduce clock insertion delay and thus reduce skew.

Bulldozer used only a single level of clock gaters driving from the clock mesh to the flops and had clock skew of 12ps [31]. Intel's 32nm Xeon with Sandy Bridge cores had skew of 14ps [28]. Both skews are about 5% of the clock period. Intel's 45nm Nehalem had peak-to-peak jitter of 2.4% of the clock period [26]. These percentages and that for the SSEF custom flop overhead are very similar to earlier custom designs back to 0.35um [8].

The mesh reduces clock skew in a similar manner to cross-links in clock trees. The mesh capacitance reduces IR drop on the clock signal. Clock meshes consume significantly more power than CTS clock trees. Placing gaters near the clock spines reduces clock power by enabling trimming of the mesh to reduce wire capacitance and reduce use of higher metal layers [14][31].

Vendor tools now support clock mesh placement restrictions when cloning gaters and enable latches [6]. The clock tree depth can be restricted to a fixed number of levels to minimize clock skew by only cloning clock buffers and gaters in the input netlist to CTS. The problem for vendors is that very few design companies use clock meshes, and there are major differences in implementations.

A clock mesh with a single level of gaters driving the flops can only have two levels of gating: coarse gating at the block level and finer grained at the gater. The coarse and fine gating and enable signals were specified in the RTL for Bulldozer [6]. A design with more clock tree levels could utilize vendor tools' capability to determine appropriate enable signals to insert ICGs.

Reducing the RC delay from gater to flops reduces clock skew by decreasing the difference in clock arrival between flops that are close to and far from the gater. Gater-to-flop RC delay contributed only 2ps of Bulldozer's skew [6]. Custom pre-routes in a fishbone pattern and balanced routing in vendor tools help reduce RC delay at the price of higher wire capacitance.

3.2.3 Pipelining Timing Overhead

Timing overheads for ASIC and custom designs are summarized in Table 2. These values are similar to previous analysis [8][9]. Significant improvements are that ASICs can now use high speed flops which have better support in vendor tools than latches, and use of MSCTS reduces clock skew and jitter from 4.0 to 2.6 FO4.

There is also an overhead for unbalanced combinational delays between pipeline stages. After careful rebalancing of pipeline stages with RTL changes, this imbalance can still increase clock period by about 10% in a typical ASIC [8]. To address this, SPR tools support register retiming and useful skew to add delay to some clock paths to provide slack to timing critical paths [10]. These approaches to pipeline rebalancing are significantly simpler techniques for SPR than using latches to allow time borrowing.

ASICs with deep pipelines for high performance, like the 2.5GHz A15 [27], may achieve a clock period of about 17 FO4 without other high speed techniques. With timing overheads per Table 2 and 12 FO4 for combinational logic, a typical ASIC is slower than custom by (10+12)/(2.5+12)=1.5×, and (4.9+12)/(2.5+12)=1.15× is the gap for an excellent ASIC. 10% pipeline stage imbalance increases this to 1.6× for the typical ASIC.

Comparing to previous values in Table 1, the performance gap from pipeline timing overhead for clock skew, registers, et al. has become more significant because ASICs are using deeper pipelines. The contribution for this factor is less significant at lower performance requirements for shallower pipelines.

3.2.4 Clock Power and Clock Gating

The clock tree and registers can contribute 20% to 40% of the total power. Reducing clock load allows downsizing of buffers in the clock tree or driving the clock tree – power savings by reducing clock load increase by about 1.5× accounting for this [6].

Clock gating disables the clock to idle circuitry to prevent unnecessary switching activity. A typical ASIC may not make sufficient use of clock gating, resulting in 1.6× higher power [9]. SPR tools can identify unnecessary switching activity, and automatically insert ICGs. In Jaguar, 92% of the circuit is clock gated when active, and 99% when idle [37]. Clock gating increased by 3% and 7% respectively compared to Bobcat.

A significant limitation of SPR tools is that each ICG has an enable latch to prevent glitching of the enable signal and a gater. The additional enable latches contribute unnecessary clock power, as a separate enable latch can drive multiple gaters. The enable signal from the enable latch can be buffered to further increase fanout if the same enable is used for many flops. SPR tools need to better support separate enable latches and gaters. They need to trade off more enable buffering to reduce clock load, versus cloning more enable latches for timing critical enables.

Bulldozer had two to three clock gaters per enable latch [6]. Designers had little time to manually fix timing critical enables that had too many buffer levels, so more enable latches were cloned with limited enable buffering by default. Bulldozer required at least 8 flops per gater to reduce clock load [31]. One enable latch per ten gaters might be possible by requiring at least 80 flops per unique enable. Note that too fine-grained clock gating is sub-optimal for power as more enable latches are required.

SPR tools do not cluster registers to reduce wire load from gaters to the flops. TCL scripts for tightly clustering flops together can reduce this wire load by 30% [6]. It can be reduced more by relative placement constraints to abut flops in columns or rows, with flop cells designed to align the input clock pins. Multiple flops can be combined into a single cell to share clock circuitry to reduce power. Automated mapping of nearby flops with the same enable signal to such cells can save 27% clock power [40].

Clock power is worse because vendor CTS tools do not optimize wire width and spacing, only a single non-default rule (NDR) may be used. Clock nets have high activity, so wider wires are used for nets with high load capacitance, and hence high currents, to avoid electromigration. Wider wires and spacing also help reduce RC delay and thus reduce skew. Designers often use wires with double minimum width and double minimum spacing in the clock tree. However, this is not needed everywhere as some nets are shorter or have smaller loads. TCL scripts can apply different NDRs based on load after CTS, but this is sub-optimal.

The Bulldozer core family uses varying NDRs for the gater-to-flop nets to reduce wire load and thus reduce clock power. The in-house tool to clone gaters supports choosing NDR based on load, performing gater and wire sizing in the inner routine that assigns flops to gaters [6]. Inaccuracies in wire load from not considering routing congestion motivated a collaboration to enhance a vendor CTS tool to support clock mesh placement restrictions to replace the in-house tool. Use of the enhanced vendor tool reduced clock mesh load by 15% and reduced clock skew by 1ps.

Additional clock power due to the above CTS tool limitations can add 1.1× to 1.2× total power versus custom, as noted for an excellent ASIC and typical ASIC respectively in Table 1.

3.3 Logic Style

ASICs use complementary static CMOS logic because it is more robust to noise and supply voltage variation. Unlike static CMOS logic, pass transistor logic (PTL) and dynamic domino logic require additional verification to ensure that such logic cells have been composed correctly in the custom designs. An open pass gate passes charge in both directions, so a standard cell library for synthesis may not have cells with bare pass gates at the inputs or outputs. Domino logic is pre-charged and evaluates only once on a cycle, so there must be no glitches. Domino logic cannot use low threshold voltage transistors because of leakage [31].

Fast pulsed static CMOS (PSCMOS) logic skews pull-up/pull-down drive strengths to optimize the critical timing edge through the logic. PSCMOS is nearly as fast as domino in 32nm [31]. Intel's 45nm Nehalem CPU core replaced domino with static CMOS logic with no reduction in clock frequency reduction [38]. Only custom designs can use PSCMOS as it must be glitch free.

Domino logic can only be used in full custom macros in SPR. Semi-custom designs may use PSCMOS in SPR with size only or don't modify attributes to prevent incorrect remapping. Semi-custom design with PSCMOS is significantly higher productivity than full custom design with domino [31].

Vendor tools have little ability to check connectivity of PTL and domino logic, as the market for this is very limited. At companies using custom design, in-house CAD tools check such logic by classifying charge-coupled transistors to validate against allowed constructs. Custom cells can also be identified as particular logic cells or flops to enable logical equivalency checking versus RTL.

Static CMOS combinational logic is lower power than domino [46], but 50% slower [4]. In Table 1, the delay factor for logic style is calculated assuming 8 FO4 for combinational domino logic in a high performance pipeline, and 8×1.5=12 FO4 for static CMOS combinational logic. Assuming for both 2 FO4 pipelining timing overhead for clock skew, jitter, and register delay, this gives a factor for logic style of (12+2)/(8+2)=1.4× [8].

As custom designs now mostly avoid domino logic, ASICs lag custom due to logic style by only the 25% slower delay for regular static CMOS logic versus PSCMOS and PTL [4]. This gives a delay factor for logic style of (10+2)/(8+2)=1.2×, calculated from 8 FO4 for combinational PSCMOS versus 8×1.25=10 FO4 for static CMOS, with 2 FO4 pipelining timing overhead for both.

Correspondingly, the power gap due to use of timing slack from high speed logic on critical paths decreases to 1.5× at a tight delay constraint. This is calculated from the pipeline performance and power model developed in [9] that accounts for timing slack used for voltage scaling and gate downsizing. The performance and power gap from logic style is less at relaxed delay constraints.

3.4 Logic Design

The logic structure and topology used to implement functional datapaths is referred to as logic design. Whether in a domino or static logic style, there are many different implementations for adders: ripple-carry, carry-lookahead, choice of radix (the number of carries merged in each step of a carry tree), etc. These have significant differences in speed and power trade-offs, attributes that have changed with technology and with more significant wire loads, requiring careful analysis to determine the best choice [46].

Specifying the logic design may require careful RTL coding, or a gate level netlist, to ensure a good synthesis implementation. Module hierarchy or size only attributes may be needed to prevent remapping in optimization. Synthesis tools also support compiling to arithmetic modules that achieve similar performance and power versus well structured RTL. Thus ASIC designers can achieve logic design comparable to custom techniques [9].

3.5 Technology Mapping

Compared to schematic entry, synthesis tools achieve relatively good timing and area when mapping RTL to logic gates from a standard cell library for a given process technology. Technology mapping minimizes delay with area as a secondary objective, or minimizes area subject to a delay constraint. With better knowledge of the circuit implementation later in the flow, only small groups of cells on timing critical paths are remapped to better optimize them, e.g. with inversion "bubble pushing" or De Morgan's transform. The initial mapping has a major impact on the final circuitry produced by the SPR flow.

SPR tools should minimize power when technology mapping. Switching activity can be determined from RTL simulation with appropriate benchmarks. Leakage power should also be included. Minimizing cell area does roughly correspond to minimizing power as gate capacitance depends on transistor area, but activity must also be considered. Different combinations of cells can implement the same logic with different switching activity. Thus ASICs may be 1.4× higher power than custom due to sub-optimal technology mapping per previous analysis [9].

3.6 Floorplanning and Placement

Significant time is spent creating a good floorplan to drive SPR to a good final placement. A square floorplan, instead of a tall one, improved placement of Toshiba's SPU, decreasing area by 30% and wire length by 28%, and improved timing by 10% [42]. Updating a traditional top-down floorplan can take 1.5 weeks for a large design. A mix of bottom-up and top-down floorplanning reduced floorplan updates to 2 days after delivery of synthesized netlists for Intel's multi-core Nehalem system-on-chip [29].

Historically, floorplanning was primarily a manual task. Macros are typically placed around the edge of a block to reduce congestion. Creating a good floorplan can take a few iterations through the flow. The design may initially be synthesized at the top level, or hierarchically in parallel via sub-modules to reduce runtime. Based on placement congestion and timing issues, the floorplan is then updated, pins moved, and timing constraints changed in negotiation with other block owners. Automated hierarchical floorplanning with macro and I/O pin placement is now supported in EDA tools. A 90nm quad core ARM11 took 18 hours from RTL through a hierarchical flow with automated floorplanning and SPR to GDS, compared to a flat SPR runtime of 44 hours excluding floorplanning [34].

SPR placement quality and capacity have improved. In modern process technologies with complicated design rules, SPR tools can achieve better area density for random logic than custom design. This reduces wire length and wire capacitance, thus reducing power and increasing speed. To improve correlation and final design quality, physical synthesis performs initial placement with global routing congestion estimates, then this is provided as a starting point to the automated placer. In recent competitions, there has been significant improvement in automated placement and routing, which will further improve such tools [24].

Automated datapath placement is difficult. Wire lengths from automatic placement can be 3× the length of custom datapaths [5]. It is difficult for tools to recognize bit slices within a datapath that is only semi-regular in structure to place them well. Recent work has reduced datapath Steiner wire length by 30% [43]. This coupled with improvements in automatic recognition of datapaths [45] will help improve EDA tool capabilities. In contrast, memory macros have high layout regularity and are usually generated from in-house or vendor automated memory generators.

As detailed previously for floorplanning and placement, typical ASICs lag custom by 1.4× delay and 1.5× power [8][9]. Datapath bit slicing and placement restrictions in an excellent ASIC reduce this to a 1.1× power gap versus manual placement.

3.7 Cell Design, Cell Sizing, and Wire Sizing

Cell design and cell size granularity significantly impact area, delay, and power. SPR tools must consider all cells in the netlist when optimizing cell size and threshold voltage (V_t) to reduce area and power. They must run quickly on large designs, preferably with linear or sub-linear runtime growth with circuit size, so optimality is sacrificed to reduce runtime. Wires can also be sized to reduce RC delay, but this is not included in optimization in today's SPR tools.

Recent research in global circuit optimization has demonstrated sub-optimality for power in existing iterative greedy sizing and V_t assignment in vendor tools. Compared to commercial software, gate sizing with linear programming reduced total power by 16% on average [9]; gate sizing and V_t assignment with Lagrangian relaxation reduced leakage power by 29% on average and also reduced total negative slack [33]. New global optimization approaches are fast enough to use on large blocks, e.g. 13 hours runtime on a design with 361,000 gates [33].

SPR tools support only a fixed non-default rule (NDR) for wire width and wire spacing during optimization which is sub-optimal, as discussed for CTS in Section 3.2.4. Wider wires have lower resistance but slightly higher capacitance [42]. Wider spacing reduces cross-coupling capacitance with neighbors. Wider wires are less subject to process variation. NDRs should be used only where needed as there are limited routing resources. TCL scripts can assign NDRs based on appropriate criteria, such as load capacitance to limit electromigration. Optimal selection of minimum and double wire width to halve RC delay reduced the worst path delays by 10% in Toshiba's SPU [42].

Taller standard cells are faster particularly for datapaths, but they have larger area, increasing wire length and wire capacitance. In Toshiba's SPU, a grid height of 12 was 15% faster than a grid height of 9, which to meet the same speed would have required 10% higher power by increasing supply voltage [42]. A shorter standard cell height reduces power for lower performance designs.

Fine granularity in cell sizes in the standard cell library helps avoid oversized cells, which reduces gate capacitance, delay, and power. Some library cells may be sub-optimal, e.g. several of the gaters in a 28nm library were too slow or too fast which increased the clock skew contribution of the gaters from 2ps to 10ps [6]. Such cells with significant influence on delay or power should be analyzed carefully to ensure that they have good characteristics. Fixing a library can take three months to lay out new library cells and characterize them, then they need to be double-checked.

While standard cell library quality has improved, significant delay and power issues remain as discussed above, so a typical ASIC can be significantly slower and higher power than custom. Excellent ASICs may use a cell library optimized for the design project and TCL scripts to perform wire sizing to reduce this gap. However, custom designs can still use some cell designs that are not safe for automated tools to utilize, e.g. bare pass gates at the inputs and outputs when connecting to nearby gates. Thus the gap for this factor remains similar to earlier analysis.

There are some additional issues below that should be considered for wire load and cell size to improve design convergence.

3.7.1 Wire Load Estimation
Transistor sizes and thus gate capacitances have decreased as circuit technologies have shrunk but wires have remained taller to limit the wire resistance, so there is relatively higher wire capacitance. For clock gater-to-flop nets, wire capacitance increased from an average of 40% of the load in 45nm to 50% in 32nm technology [6]. It is exacerbated at lower supply voltages where both gate drive strength and gate capacitance reduce [18], and wire loads contribute a greater portion of the total load capacitance.

Previously, wire load models were used pre-route to calculate the wire capacitance and resistance from the number of fanouts and the Steiner tree wire length estimate. Accuracy has improved in modern EDA tools with global routing and congestion estimates. However, there are still discrepancies between post-route wire loads and those estimated pre-route. Derating factors may be used to penalize longer nets or those with more fanouts to be more conservative when estimating wire load pre-route. This improves timing correlation, reducing the number of design flow iterations.

If it turns out that cells are oversized post-route, they can be downsized with minimum perturbation if there are smaller size cells that have the same footprint, and are pin compatible – i.e. the pins are on the same track so only a minor ECO reroute is needed. This technique was used to fix the 60% of gaters that were oversized for post-route loads in a 32nm CPU core [6]. The problem was conservative wire load estimates by an in-house gater cloning tool that was not cognizant of routing congestion.

3.7.2 Cell Delay Predictability
Small drive strength cells have significantly worse delay if the load capacitance is higher than expected due to pre-route versus post-route wire load miscorrelation. Small cells can be disallowed during synthesis and automated placement to mitigate this. They may be downsized safely later if post-route loads are small. It is also preferable to avoid minimum width transistors in cells as they are more subject to process variation and thus have poorer yield.

Tighter load capacitance constraints can be specified on low drive cells to add some conservatism during synthesis and automated placement. This approach can also be used to reduce clock skew by tightly controlling delay when cloning clock gaters.

Custom designers generally don't need to take these precautions with small cell size as manual placement has better wire length predictability. However, cells may be sized incorrectly requiring timing analysis and other design checks to resize where necessary.

3.8 Voltage Scaling, Process Technology, and Process Variation
Supply voltage, transistor channel length and threshold voltage greatly impact delay and power and their ranges are determined by the choice of process technology.

3.8.1 Channel Length and Threshold Voltage
ASICs have a good choice of supply voltage and V_t for low power and high performance. Foundries offer four supply voltages from 0.85V to 1.05V and five V_t in 28nm [41].

As an example of leakage and dynamic power trade-offs with V_t, it may be preferable to use regular or lower threshold voltage for the buffers and first level of clock gaters to reduce clock power as the clock is high activity up to the clock gater. Leakage will increase but dynamic power is more significant, and power is reduced overall as smaller cell sizes can be used. A higher threshold voltage may be used to reduce power where the switching activity is significantly lower. Avoid mixing threshold voltages at the same level within the clock tree, as it is bad for clock skew due to process variation differences.

Additional V_t requires additional expensive process masks, which changing channel length does not require. Both can be used to increase the variety of trade-offs for delay, dynamic power, and leakage power. For example, the 40nm Bobcat core and GPU had three different V_t with two channel lengths, regular and +2nm, to give six choices in total – 56% were high V_t, 42% were regular V_t, and 2% were low V_t to limit leakage [13]. Transistor stacking can also be used to reduce sub-threshold leakage.

Intel is a year ahead on process technology but has opened up as a foundry to other companies. Intel's 22nm FinFET process has supply voltages from 0.75 to 1.2V and three channel lengths: 30, 34, and 40nm [21]. A longer channel length reduces leakage but such transistors have increased gate capacitance and are slower. The 30nm channel length transistors have 3× higher saturation drain current and 3000× higher leakage versus 40nm.

FinFETs are triple-gated to substantially reduce sub-threshold leakage. This enables lower supply voltage to reduce power, as it allows lower V_t to maintain device performance. The 22nm low power FinFETs have 50% better performance than 32nm low power single-gated transistors [21].

3.8.2 Supply Voltage Scaling
EDA tool support has improved significantly for changing supply voltage and use of multiple voltage regions. Current source models for libraries enable analysis of circuits subject to IR voltage drop and at different supply voltages. Libraries do not have to be characterized at each different supply voltage unless they differ significantly. The Unified Power Format 2.0 provides better standardized support for specifying voltage regions and power modes of operation. There are also EDA tools to verify correct behavior between power modes. At low supply voltage there is greater susceptibility to noise and process variation, and the library must be re-characterized.

Fine-grained use of multiple supply voltages provides little power reduction and has significant design cost to implement [9]. Different supply voltages are best used on blocks requiring significantly different performance. Additional threshold voltages help optimize performance versus leakage power for each supply voltage. Regions with different voltage on/off behavior may need islands with always on repeaters and state retention flops. Using a power-gate ring around a processor simplifies the implementation of power gating [23].

Dynamic voltage and frequency scaling (DVFS) is used to reduce power when idle. The 40nm Bobcat core and GPU were optimized to operate between 0.8V and 1.2V [13]. DVFS is also available in synthesizable ARM cores and is the mechanism used to switch between the paired cores in the ARM big.LITTLE [22].

An active research area is near threshold voltage operation. For near threshold operation, transistor stacks are at most three, low drive strengths are disallowed, and other library changes must be made to improve yield [36]. The critical path distribution changes significantly as devices are slower at lower supply voltage.

In 32nm technology, a 0.5V supply can double energy efficiency versus 0.85V [36]. Typical ASICs may have 2× power versus custom and excellent ASIC designs with lower supply voltages.

Similar to DVFS, some modern CPUs briefly boost performance in response to demand with a turbo mode, before throttling when temperature is approaching the allowed limit [35]. On-chip temperature sensors and numerical analysis are used to manage temperature [32]. Typical ASICs do not do this and so are perhaps 1.1× slower on average. Excellent ASICs should be able to do so.

3.8.3 Process Variation

Typical ASICs are optimized to function at worst case process and operating corners to ensure good yield. Additional operating corners must now be considered due to slower devices at lower temperatures. This is due to inverse temperature dependence which is more pronounced in 22nm technology. Intel's Ivy Bridge compensates for this by estimating the coldest point from temperatures from the thermal sensors placed at hot spots [20]. Multiple operating modes for DVFS multiply the number of corners to be considered. Designing conservatively for these scenarios can result in over-design and higher power in particular. Statistical static timing analysis may help but has not proven particularly popular in industry.

Custom designs use a variety of techniques to address process variation in silicon. Delay lock loops and adjustable delay buffers help reduce clock skew. Supply voltage can be increased with DVFS to increase circuit speed of a slow design. Speed binning of chips also helps, which most ASICs cannot do due to significant cost for testing at different speeds and low price per part.

These and other related issues and the corresponding power and performance gaps for process technology and process variation have not changed significantly from previous analysis.

4. OTHER ADVANCES NEEDED IN EDA

The length of time to run through the SPR flow to analysis remains a major concern for designers. EDA tools provide swifter prototyping capabilities at the cost of lower quality of results, but million gate designs can take a week or more to run through the flow. Large designs should be broken up into smaller sub-modules to reduce runtime, but top-level runs are useful for prototyping to flush through problems in the CAD flow and to help determine initial timing constraints and other specifications for sub-modules. Top level analysis, e.g. for signal noise integrity and cross-coupling parasitic extraction, can require hundreds of gigabytes of memory and is often capacity limited.

Traditionally, custom design in-house tools have an underlying database that supports sparse loading. This can be critical for avoiding memory capacity limitations on large designs by only paging into memory from disk the portion of the netlist that needs to be traversed. Moreover, only loading a small portion of the design can reduce runtime by orders of magnitude. It enables fast internally-developed design checks and other swift incremental

netlist optimizations, such as for gater cloning [6], albeit at the cost of significant CAD development time. Adding sparse loading capability to industry standard netlist representations has been discussed, but it goes against the standard EDA vendor model of loading the full design and doing a significant amount of work on it within a single compound flow step.

An internal script to *opportunistically* space non-timing-critical cells further apart to improve yield ran in minutes using the internal database. An equivalent TCL script ran for about an hour in a vendor tool, and took similar time to develop. Support for yield as a secondary optimization constraint is limited in today's EDA tools. Vendor tools do natively support *strict* enforcement of LEF rules for abutment and spacing, and these restrictions are observed during placement.

5. EDA FLOW IMPACT OF DESIGN METHODOLOGY CHOICES

Some design choices that appear straightforward from a circuit perspective can have significant impact on the EDA flow. Major flow changes are disruptive as they can cause schedule delays and introduce additional bugs in the CAD flow. Methodologies that complicate the flow should be avoided, unless they provide substantial return on investment. Isolated, modular, incremental flow changes are best for schedule predictability – allowing a new flow step to be tested with a fallback option, before phasing out the previous approach in a manner that mitigates risk.

For example, AMD's 32nm Bulldozer core used soft-edge flip-flops (SEFs) with paired early and late clock signals, providing a transparent window of about 10% of the clock period between the two clock arrivals to provide some immunity to clock skew and clock jitter, and reduce setup time at the cost of increased hold time [31]. Each paired early and late gater had to drive the same flops, with minimal skew from the offset between these early and late clock signal arrivals, so the gaters had to be placed next to each other with very similar routes to the flops. It cost several worker years of CAD development time to support construction, analysis, and verification of this, with correct time budgeting for the transparent window with time borrowing between stages.

In subsequent designs, the late gater was integrated within the high speed SSEFs to reduce routing congestion and clock power. The single-clock SSEFs are supported readily by vendor tools, though they are slightly slower than the SEFs [6]. This simplified the CAD flow and improved design productivity.

6. CONCLUSION

Synthesis, automatic place and route offer an order of magnitude higher productivity than full custom design. To take advantage of this, custom designs are now mostly SPR except for memory and critical datapaths where absolutely necessary. The extra time to do more design space exploration also improves SPR quality.

Combining factors in Table 1 with the microarchitecture model in [9] suggests that compared to custom, an excellent ASIC may be 2× slower and 4× higher power at a tight performance constraint. In reality, today's designs are too large and time-to-market too important to afford to take full advantage of custom techniques. Synthesized ASICs can achieve similar speed to custom with high performance design techniques and ongoing improvements in SPR tools. Limited use of semi-custom design in an SPR flow bridges the power gap with techniques such as datapath and flop placement restrictions, customized library, and NDR wire sizing.

The future promises further advances in EDA tools. State-of-the-art research has significantly improved quality and tool capacity for sizing, placement, and routing. Clock mesh support should improve soon for high performance ASICs. This will help close the gaps left between ASIC and custom design method.

7. ACKNOWLEDGMENTS

Many individuals in academia and industry have contributed advice and discussion on this topic over the past decade. I'm grateful to my research advisor at UC Berkeley, Professor Kurt Keutzer, who helped greatly with earlier work in this area and guided me on this path. I thank Trevor Meyerowitz, Joseph Shinnerl, and my wife Eleyda Negrón for editing feedback. Thanks also to Kiyoji Ueno for additional details on the SPU [42].

8. REFERENCES

[1] Borkar, S., 2011. The Exascale Challenge. *PACT*. https://parasol.tamu.edu/pact11/ShekarBorkar-PACT2011-keynote.pdf

[2] Burgess, B., et al., 2011. Bobcat: AMD's Low-Power x86 Processor. *IEEE Micro*. 31, 2 (Mar. 2011), 16-25.

[3] Butler, M., et al. 2011. Bulldozer: An Approach to Multithreaded Compute Performance. *IEEE Micro*. 31, 2 (Mar. 2011), 6-15.

[4] Chang, A., et al. 2002. High-Speed Logic, Circuits, Libraries, and Layout. Ch. 4 in [8], 101-144.

[5] Chang, A., and Dally, W. 2002. Exploiting Structure and Managing Wires to Increase Density and Performance. Ch. 11 in [8], 269-287.

[6] Chinnery, D., et al. 2012. Gater Expansion with a Fixed Number of Levels to Minimize Skew. *SNUG Silicon Valley*.

[7] Chinnery, D., et al. 2012. Scan Stitching Separate Groups of Mux-D or LSSD Flops. *SNUG Silicon Valley*.

[8] Chinnery, D., and Keutzer, K. 2002. *Closing the Gap Between ASIC & Custom: Tools and Techniques for High-Performance ASIC Design*. Springer, New York, NY.

[9] Chinnery, D., and Keutzer, K. 2007. *Closing the Power Gap between ASIC & Custom: Tools and Techniques for Low Power Design*. Springer, New York, NY.

[10] Dai, W., and Staepelaere, D. 2002. Useful-Skew Clock Synthesis Boosts ASIC Performance. Ch. 8 in [8], 209-223.

[11] Eleyan, N., et al. 2009. Semi-Custom Design Flow: Leveraging Place and Route Tools in Custom Circuit Design. *ICICDT*, 143-147.

[12] Fog, A. 2012. *The microarchitecture of Intel, AMD and VIA CPUs*. http://www.agner.org/optimize/microarchitecture.pdf

[13] Foley, D., et al. 2012. A Low-Power Integrated x86-64 and Graphics Processor for Mobile Computing Devices. *JSSC*, 47, 1 (Jan. 2012), 220-231.

[14] Gerosa, G., et al. 2009. A Sub-2W Low-Power IA Processor for Mobile Internet Devices in 45nm High-K Metal Gate CMOS. *JSSC*, 44, 1 (Jan. 2009) 73-82.

[15] Harris, D., et al. 1997. The Fanout-of-4 Inverter Delay Metric. http://odin.ac.hmc.edu/~harris/research/FO4.pdf

[16] Hofstee, H. 2005. Power Efficient Processor Architecture and the Cell Processor. *HPCA*, 258-262.

[17] Hrishikesh, M., et al. 2002. The Optimal Logic Depth Per Pipeline Stage is 6 to 8 FO4 Inverter Delays. *ISCA*, 14-24.

[18] Hu, C. 2009. *Modern Semiconductor Devices for Integrated Circuits*. Prentice Hall, Upper Saddle River, NJ.

[19] ITRS, International Technology Roadmap for Semiconductors 2012 Update. 2012. http://www.itrs.net/Links/2012ITRS/Home2012.htm

[20] Jahagirdar, S., et al. 2012. Power Management of the Third Generation Intel Core Micro Architecture Formerly Codenamed Ivy Bridge. *Hot Chips*.

[21] Jan, C., et al. 2012. A 22nm SoC Platform Technology Featuring 3-D Tri-Gate and High-k/Metal Gate, Optimized for Ultra Low Power, High Performance and High Density SoC Applications. *IEDM*, 44-47.

[22] Jeff, B. 2012. Advances in big.LITTLE Technology for Power and Energy Savings. *ARM Techcon*, Santa Clara, CA.

[23] Jotwani, R. 2011. An x86-64 Core in 32nm SOI CMOS. *JSSC*, 46, 1 (Jan. 2011), 162-172.

[24] Krishnamoorthy, S. 2012. The Upcoming "Golden Age" of Placement Research. *ICCAD*.

[25] Kuon, I., and Rose, J. 2007. Measuring the Gap Between FPGAs and ASICs. *TCAD*, 26, 2 (Feb. 2007), 203-215.

[26] Kurd, N., et al. 2009. Next Generation Intel Core Micro-Architecture (Nehalem) Clocking. *JSSC*, 44, 4 (Apr. 2009), 1121-1129.

[27] Lanier, T. 2010. Exploring the Design of the Cortex-A15 Processor. *ARM Techcon*, Santa Clara, CA.

[28] Li, S., et al. 2011. Clock Generation for a 32nm Server Processor with Scalable Cores. *ISSCC*, 82-83.

[29] Liao, Y., et al. 2009. Efficient Floor-Planning Methodology for the Jasper Forest SoC on a 45 Nanometer Process. *ASICON*, 407-410.

[30] Lukefahr, A., et al. 2012. Composite Cores: Pushing Heterogeneity into a Core. *Int. Symp. on Microarchitecture*.

[31] McIntyre, H., et al. 2012. Design of the Two-Core x86-64 AMD "Bulldozer" Module in 32nm SOI CMOS. *JSSC*, 47, 1 (Jan. 2012), 164-176.

[32] Nussbaum, S. 2012. AMD "Trinity" APU. *Hot Chips*.

[33] Ozdal, M., Burns, S., and Hu, J. Algorithms for Gate Sizing and Device Parameter Selection for High-Performance Designs. *TCAD*, 31, 10 (Oct. 2012), 1558-1571.

[34] Riches, S., et al. 2007. Top-down Design with Magma's Talus Automated Chip Creation. *ARM Information Quarterly*, 6, 2 (2007), 62-65.

[35] Rotem, E., et al. 2012. Power Management of the Intel Microarchitecture Code-Named Sandy Bridge. *IEEE Micro*, 32, 2 (Mar. 2012), 20-27.

[36] Ruhl, G., et al. 2012. An IA-32 Processor with a Wide Voltage Operating Range in 32nm CMOS. *Hot Chips*.

[37] Rupley, J. 2012. "Jaguar" AMD's Next Generation Low Power x86 Core. *Hot Chips*.

[38] Rusu, S. 2010. A 45nm 8-core Enterprise Xeon Processor. *JSSC*, 45, 1 (Jan. 2010), 7-14.

[39] Smith, M. 1997. *Application-Specific Integrated Circuits*. Addison-Wesley, Boston, MA.

[40] Tsai, C. 2013. FF-Bond: Multi-bit Flip-Flop Bonding at Placement. 2013. *ISPD*.

[41] TSMC. 2013. 28nm Technology. http://www.tsmc.com/english/dedicatedFoundry/technology/28nm.htm

[42] Ueno, K., et al. 2007. A Design Methodology Realizing an Over GHz Synthesizable Streaming Processing Unit. *VLSIC*, 48-49.

[43] Ward. S. 2012. Keep it Straight: Teaching Placement how to Better Handle Designs with Datapaths. *ISPD*, 79-86.

[44] Warnock, J., et al. 2006. Circuit Design Techniques for a First-Generation Cell Broadband Engine Processor. *JSSC*, 41, 8 (Aug. 2006), 1692-1706.

[45] Xiang, H., et al. Network Flow Based Datapath Bit Slicing. 2013. *ISPD*.

[46] Zlatanovici, R., Kao, S., and Nikolić, B. 2009. Energy–Delay Optimization of 64-Bit Carry-Lookahead Adders with a 240ps 90nm CMOS Design Example. *JSSC*, 44, 2 (Feb. 2009), 569-583.

Electromigration and Its Impact on Physical Design in Future Technologies

Jens Lienig
Dresden University of Technology
Institute of Electromechanical and Electronic Design (IFTE)
01062 Dresden, Germany
www.ifte.de
Email: jens@ieee.org

ABSTRACT

Electromigration (EM) is one of the key concerns going forward for interconnect reliability in integrated circuit (IC) design. Although analog designers have been aware of the EM problem for some time, digital circuits are also being affected now. This talk addresses basic design issues and their effects on electromigration during interconnect physical design. The intention is to increase current density limits in the interconnect by adopting electromigration-inhibiting measures, such as short-length and reservoir effects. Exploitation of these effects at the layout stage can provide partial relief of EM concerns in IC design flows in future.

Categories and Subject Descriptors

B7.2[**Integrated Circuits**]: Design Aids

Keywords

Electromigration; current density; physical design; layout; interconnect reliability; short-line rules; short-length effects; Blech length; reservoir effect

1. INTRODUCTION

Excessive current density within interconnects – which if not effectively mitigated causes electromigration (EM) and electrical overstress – is a major concern for integrated circuit (IC) designers. This is a growing reliability issue in modern ICs in the wake of smaller feature sizes. Analog designers have been aware of this issue for some time, and digital designs are now being affected as well. The latest edition of the ITRS roadmap [18] predicts that all minimum-sized interconnects will be EM-affected by 2018, potentially restricting any further downscaling of wire sizes (Figure 1).

Current density verification and thus the detection of electromigration issues are already an integral part of the sign-off verification of circuit layouts. Current density violations detected during sign-off are corrected by layout modifications – by the widening of wires, for example.

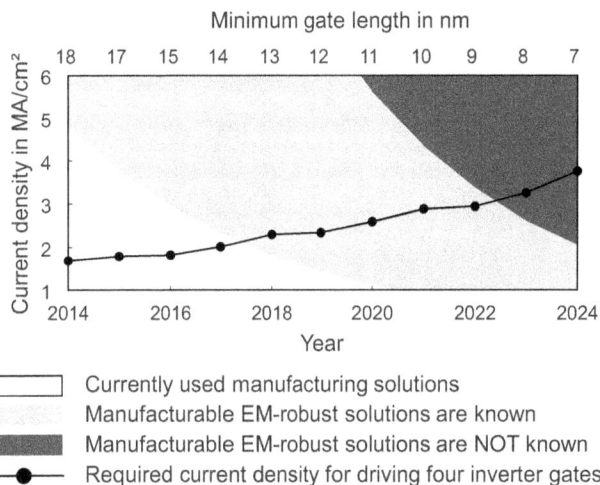

Figure 1. Expected evolution of required and maximum IC wires (ITRS [18]). While the required current density scales with frequency and reducing cross-section, the maximum tolerable current density is shrinking due to smaller structure sizes (see Section 4).

The number of late corrections needs to be capped going forward due to the restrictions imposed by complex circuitry. This can be achieved by adopting electromigration inhibiting measures, such as short-length and reservoir effects, during physical IC design to maximize current density limits. However, we need to understand the physical processes involved, and the constraints and benefits of the EM-inhibiting effects in order to successfully implement them.

In this talk we present the options available to combat the negative effects of electromigration during layout design. In Section 2 we introduce the physical process of electromigration, followed by a survey of how current density is considered in today's analog and digital design flows (Section 3). Section 4 investigates future technology nodes and their effects on current density and electromigration issues in IC layout. Section 5 presents various EM-inhibiting effects that can be exploited in current and future technologies in order to reduce the negative impact of electromigration on the circuit's reliability.

2. ELECTROMIGRATION (EM)

Current flow through a conductor produces two forces to which the individual metal ions in the conductor are exposed. The first is an electrostatic force F_{field} caused by the electric field strength in the metallic interconnect. Since the positive metal ions are shielded to some extent by the negative electrons in the conductor, this force can be ignored in most cases. The second force F_{wind} is generated by the momentum transfer between conduction electrons and metal ions in the crystal lattice. This force acts in the direction of the current flow and is the main cause of electromigration (Figure 2).

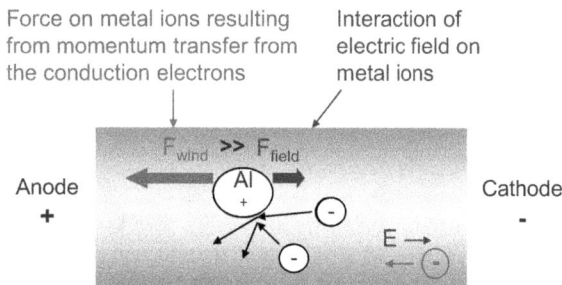

Figure 2. Two forces act on metal ions which make up the lattice of the interconnect material. Electromigration is the result of the dominant force, that is, the momentum transfer from the electrons which move in the applied electric field.

In a homogeneous crystalline structure, because of the uniform lattice structure of the metal ions, there is hardly any momentum transfer between the conduction electrons and the metal ions. However, as this symmetry does not exist at grain boundaries and material interfaces, momentum is transferred much more vigorously in these zones from the conductor electrons to the metal ions. And since the metal ions in these zones are bonded much more weakly than in a regular crystal lattice, once the electron wind has reached a certain strength, atoms become separated from the grain boundaries and are transported in the direction of the current. This direction is also influenced by the grain boundary itself, as atoms tend to move along grain boundaries.

If the direction of an excessive current is kept constant for a longer period, voids and hillocks appear in the wire. Analog circuits, or power supply lines in digital circuits, are thus particularly susceptible to electromigration. When the current direction varies – for example, in digital circuits with their alternating capacitive charging and discharging in conductors – there is a certain amount of compensation due to material backflow (self-healing effect). Nonetheless, interconnect failures are still possible, with thermal migration playing a major role.

Furthermore, the susceptibility of wires to electromigration is a function of grain size and thus of the distribution of grain sizes. Smaller grains encourage material transport, because there are more transport channels than in coarse-grained material. This causes voids to appear at the points of transition from coarse to fine grains, since at these points atoms flow out faster than they flow in. Conversely, where the structure turns from fine grains to coarse, hillocks tend to form, since the inflowing atoms cannot disperse fast enough through the coarse structure.

These types of variations in grain size appear in interconnects at every contact hole or via. Because the current here commonly encounters a narrowing of the conductive pathway, contacts and vias are particularly susceptible to electromigration.

Diffusion processes caused by electromigration can be divided into grain boundary diffusion, bulk diffusion and surface diffusion (Figure 3). In general, grain boundary diffusion is the major

migration process in aluminum wires [33][34], whereas surface diffusion predominates in copper interconnects [9][10][26][40].

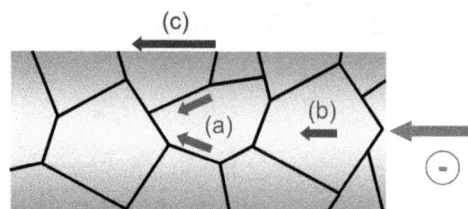

Figure 3. Illustration of various diffusion processes within the lattice of an interconnect: (a) grain boundary diffusion, (b) bulk diffusion, and (c) surface diffusion.

Detailed investigations of the various electromigration failure mechanisms can be found in [4][9][34][42].

Many electronic interconnects, for example in integrated circuits, have an intended mean time to failure (MTTF) of at least 10 years. The failure of a single interconnect caused by electromigration can result in the failure of the entire circuit. At the end of the 1960s the physicist J. R. Black developed an empirical model to estimate the MTTF of a wire segment, taking electromigration into consideration [3]:

$$MTTF = \frac{A}{J^n} \cdot \exp\left(\frac{E_a}{k \cdot T}\right) \qquad (1)$$

where A is a constant based on the cross-sectional area of the interconnect, J is the current density, E_a is the activation energy (for example, 0.7 eV for grain boundary diffusion in Al [3][42], 0.9 eV for surface diffusion in Cu [17]), k is the Boltzmann constant, T is the temperature and n a scaling factor. It has been established through studies on Al and Cu interconnects that void-growth-limited failure is characterized by $n = 1$, while void-nucleation-limited failure is best represented by $n = 2$ [9][24].

Equation (1) shows that current density J and (to a lesser extent) the temperature T are deciding factors in the physical design process that affect electromigration.

3. CONSIDERATION OF EM IN TODAY'S PHYSICAL DESIGN FLOWS

3.1 Analog Design

Current strengths in analog circuits vary widely depending on the application: sensor applications require only a few nano-amps, whereas wires in power circuitries need to carry up to several amps continuously. In addition to this broad spectrum of currents, analog circuits may have to operate reliably at very high temperatures. The number of nets in circuits prone to electromigration is growing due to technology downscaling and increasingly complex designs. Issues arising from the joule heating of conductors need to be addressed for the same reasons. Yet another characteristic of analog circuits is that currents in power and signal nets are on the same order of magnitude. Hence, analog designers have been aware of electromigration issues for the past couple of years.

A proprietary design flow that incorporates the comprehensive consideration of current density in analog IC design is presented in [28] (Figure 4). It includes current characterization, current propagation in design hierarchies, current-aware layout Pcells, current-aware route planning [27] as well as current density verification [19][20].

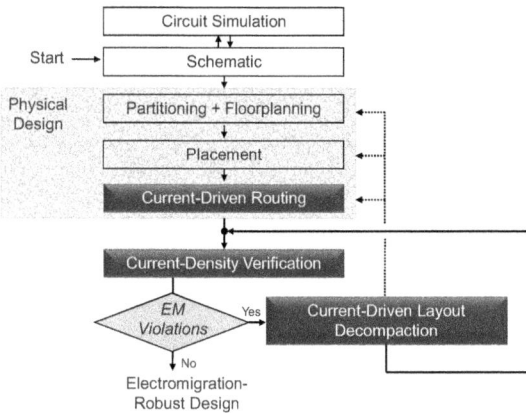

Figure 4. An electromigration-aware analog design flow [28].

Extensive manual intervention in commercial synthesis tools has been needed up to now in order to include the impact of current densities in physical design. A first current-density-aware routing tool has been available from Pulsic Ltd. [12]. It widens wires based on terminal currents. It is however restricted to direct currents (DC).

Verification tools for current densities, such as Cadence Virtuoso Power Systems [13], Synopsys CustomSim [15] and Apache Totem MMX [14], have been available for some time. They extract a netlist from the layout (including parasitics). This netlist is then used to simulate the currents in all wires. If any of the resulting current densities exceed an EM-relevant boundary, a violation is detected and highlighted.

3.2 Digital Design

Digital circuits are characterized by net classes that exhibit different susceptibilities to electromigration [30]. We can assume that power nets carry moderately constant currents and that their current directions are consistent. Clock and signal nets, on the other hand, conduct alternating currents (AC). Due to self-healing mechanisms, these alternating currents affect wire lifetimes less than direct currents [29][32]. Thus, the different net classes of digital circuits require different current density limits. Beyond that, digital circuits stand out because of their large number of nets. The complexity of analog circuits does not exceed a few thousand nets, whereas digital circuits typically have millions of them.

Methods employed in analog design to deal with electromigration cannot be applied in digital circuits – other solutions, which can handle the complexity of digital designs, are needed.

In principle, the digital design flow consists of a series of synthesis steps, which methodically concretize circuit geometry (Figure 5a) [23]. Simultaneously, verification steps ensure the circuit acquires the required electrical characteristics and functions, and meets the reliability and manufacturability criteria (Figure 5b).

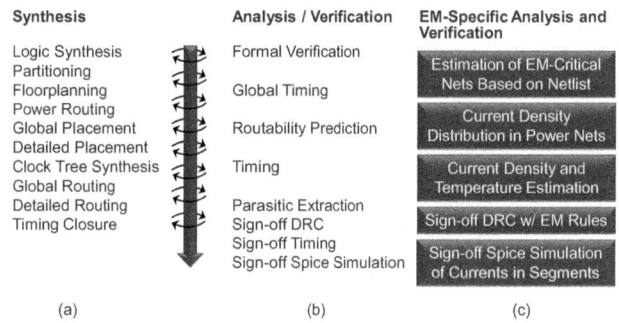

Figure 5. Synthesis-analysis loops in the design flow for digital circuits. The critical steps – physical synthesis (a) and analysis (b) – are shown, supplemented by options to address current density and other electromigration issues (c).

Figure 5c shows flow options to analyze the impact of electromigration on circuit reliability. These have only been partly supported to date by layout tools. However, "Sign-off DRC w/ EM-rules" and "Sign-off Spice Simulation" with subsequent current density verification are now standard functions in state-of-the-art digital layout tools (for example, Synopsys IC Compiler, Cadence Encounter). These functions are also available as stand-alone verification tools (for example, Apache Totem MMX, Mentor Calibre PERC, Tanner HiPerVerify).

An early stage "Estimation of EM-Critical Nets" is presented in [21]. It determines the worst-case bounds on segment currents, so nets can be separated into critical and non-critical sets. Only the set of critical nets, which is typically considerably smaller, requires subsequent special consideration during layout generation.

The "inner workings" of the aforementioned analysis steps are almost identical in the tools. They are based on three global current density limits to identify EM violations: maximum allowable peak, average, and RMS current densities. The actual current density value in each wire segment is determined by transient or steady state Spice simulations at transistor level. EM violations are detected if these actual local current densities exceed a specific limit. Power and signal nets are verified separately.

4. EM IN FUTURE TECHNOLOGIES

The global trend in size reduction leads to improved circuit performance, efficiency at higher circuit frequencies and smaller footprints. Line widths will continue to decrease over time, as will wire cross-sectional areas. As shown in Table 1, the cross-sectional area shrinks from about 1,000 nm² in 2014 to less than 500 nm² in 2018. Currents are also reduced (see Table 1 and Figure 6, left) due to lower supply voltages and shrinking gate capacitances. However, as current reduction is constrained by increasing frequencies, the more marked decrease in cross-sectional areas (compared to current reduction) will give rise to increased current densities J in ICs going forward (Figure 6, right).

To make matters worse, the maximum tolerable current densities are shrinking at the same time due to smaller structure sizes[1] [17]. As a result, the ITRS [18] indicates that all minimum-sized interconnects

---

[1] Small voids and other material defects, which could have been tolerated in earlier technology nodes, cause increasingly dramatic damage or resistance change to the wires with shrinking metal structures. Thus, maximum tolerable current densities will have to decrease in order to maintain reasonable interconnect reliability.

will be EM-affected by 2018, potentially limiting any further downscaling of wire sizes (Figure 6, yellow barrier).

Furthermore, the total length of interconnect per IC will continue to increase. As a consequence, reliability requirements per length unit of the wires need to *increase* in order to *maintain* overall IC reliability. This accepted wisdom is contradicted by the future *decrease* in interconnect reliability due to electromigration – as noted above. The ITRS thus states that no known solutions are available for the EM-related reliability requirements that we will face approximately 10 years from now (Figure 6, red barrier).

Table 1. Predicted technology parameters based on the ITRS, 2011 edition [18]; maximum currents and current densities for copper at 105°C.

Year	2014	2016	2018	2020	2022	2024	2026
Gate length (nm)	18.41	15.34	12.78	10.65	8.88	7.40	6.16
On-chip local clock frequency (GHz)	4.211	4.555	4.927	5.329	5.764	6.234	6.743
DC equivalent maximum current (µA)*	18.14	12.96	10.33	7.36	5.53	4.45	3.52
Metal 1 properties							
Width – half-pitch (nm)	23.84	18.92	15.02	11.92	9.46	7.51	5.96
Aspect ratio	1.9	2.0	2.0	2.0	2.1	2.1	2.2
Layer thickness (nm)*	45.29	37.84	30.03	23.84	19.87	15.77	13.11
Cross-sectional area (nm²)*	1,079.7	716.0	451.0	284.1	187.9	118.4	78.13
DC equivalent current densities (MA/cm²)							
Maximum tolerable current density (w/o EM degradation)**	4.8	3.0	1.8	1.1	0.7	0.4	0.3
Maximum current density (solutions unknown)**	25.4	15.4	9.3	5.6	3.4	2.1	1.2
Required current density for driving four inverter gates	1.68	1.81	2.29	2.59	2.94	3.76	4.50

*) Calculated values, based on given width W, aspect ratio A/R, and current density J in [18], calculated as follows: layer thickness $T = A/R \times W$, cross-sectional area $A = W \times T$ and current $I = J \times A$.
**) Approximated values from the ITRS figure INTC9 [18].
All remaining values are from the ITRS 2011 edition [18].

Increased interconnect resistivity caused by scattering effects in small wires will throw up further challenges [36]. Coupled with a rise in current densities, this will lead to large local temperature gradients inside the wire caused by Joule heating. It accelerates temperature-dependent electromigration and introduces additional thermomigration [22][43].

The tendency to replace SiO_2 with low-k dielectrics [18] with less stiffness reduces the stress-induced atomic backflow [39], which counteracts electromigration in short lines (see Section 5.2). Another consequence of this replacement is the increased likelihood of EM-induced compressive failures (extrusions) [41].

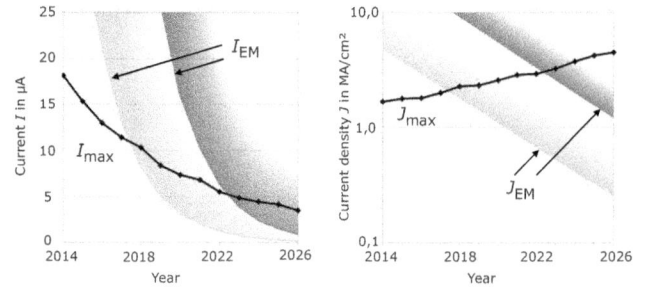

Figure 6. Expected development of currents (I_{max}, left) and current densities (J_{max}, right) needed for driving four inverter gates, according to ITRS 2011 [18] (see also Table 1). EM degradation needs to be considered when crossing the yellow barrier of currents (I_{EM}) and current densities (J_{EM}). As of now, manufacturable solutions are not known in the red area.

As already mentioned, the surge in current density is also driven by an increase in clock frequencies (Table 1), in response to the demand for enhanced performance and made possible by smaller transistors. Although higher frequencies will neither worsen nor improve electromigration issues in signal or clock nets ([38], see also Section 5.7), they will increase currents (and thus current densities) in (DC) supply nets which are already sensitive to electromigration in today's technologies.

5. ADDRESSING EM IN PHYSICAL DESIGN: WHAT ARE THE OPTIONS?

There are a number of well-known options for influencing electromigration and current density, respectively, during the physical design of an electronic circuit.

- Wire material: It is known that pure copper used for Cu-metallization is more electromigration-robust than aluminum at low operating temperatures.

- Wire temperature: In Equation (1), the temperature of the conductor appears in the exponent, that is, it strongly impacts the MTTF of the interconnect. For an interconnect to remain reliable at high temperatures, the maximum tolerable current density of the conductor must necessarily decrease. On the other hand, lowering the temperature supports higher current densities while keeping the reliability of the wire constant.

- Wire width: Given that current density is the ratio of current I and cross-sectional area A, and given that most process technologies assume a constant thickness of the printed interconnects, it is the wire width that exerts a direct influence on current density: the wider the wire, the smaller the current density and the greater the resistance to electromigration.

These effects have been covered in detail in [28]. They will be of limited use in new generation technologies because they have been largely exploited and/or their application would be counter-intuitive to the new technology node itself, that is, its reduced structure size. Hence, tolerable current density limits need to be maximized by exploiting further electromigration inhibiting factors which are described next.

5.1 Bamboo Structure

If the wire width is reduced to below the average grain size of the wire material, the resistance to electromigration increases, despite an increase in current density. This apparent contradiction is caused by the grain boundary locations. In narrow wires, the grain boundary locations are perpendicular to the whole wire (as shown in Figure 7) forming a bamboo like structure. As already mentioned, material transport occurs as much in the direction of the current flow as along grain boundaries (grain boundary diffusion, see Section 2). Because grain boundaries in this type of bamboo structure are often perpendicular to the current flow, the boundary diffusion factor is demoted, and material transport is reduced accordingly.

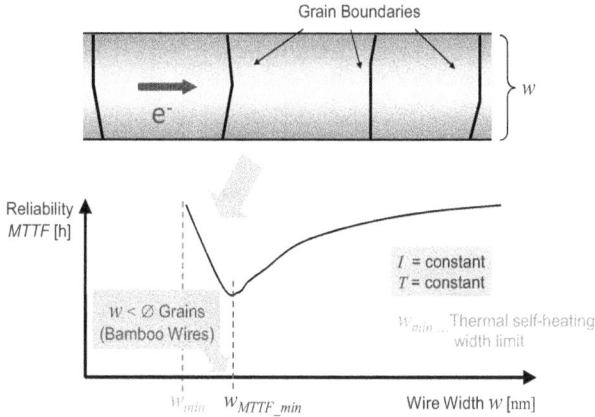

Figure 7. Reduced wire width – less than the average grain size – improves wire reliability with regard to electromigration. So-called bamboo wires are characterized by grain boundaries which are perpendicular to the direction of the electron wind and thus permit only limited grain boundary diffusion.

The bamboo structure improves reliability if the self-heating limit of the layer is taken into account. In order to exploit this, wire widths are deliberately kept narrow enough to maintain a bamboo structure. Also the wire material can be annealed selectively during IC processing in order to support bamboo formation.

However, the maximum possible wire width for a bamboo structure, which is about 850 nm for damascene copper [2], is often too narrow for signal lines carrying large-magnitude currents in analog circuits or for power supply lines. In these circumstances, slotted wires – in which rectangular holes are carved – can be used. Here, the widths of the individual metal structures between the slots support bamboo formation, and the total width of all the metal structures meets power requirements.

Based on the same principle, a fine-grain power mesh is often superimposed on the circuit. This mesh consists of a large number of wires, and the individual wire widths form a bamboo-like structure.

It must be pointed out that process anomalies often produce variations in wire widths – due to etch loss, lithography issues, and the like – and variations in wire heights – due to metal deposition fluctuations, for example – as well as impacting the via fill rate. It is vital to take these possible process fluctuations into account when determining viable interconnect dimensions.

5.2 Short-Length Effects

There are also upper limits set on the length of an interconnect to block electromigration. One of them is known as the "Blech length," and any wire that has a length below this threshold length (typically on the order of 5-50 μm) will not fail by electromigration. In this scenario, mechanical stress buildup causes a reverse migration process which reduces, or even compensates for, the effective material flow towards the anode (Figure 8).

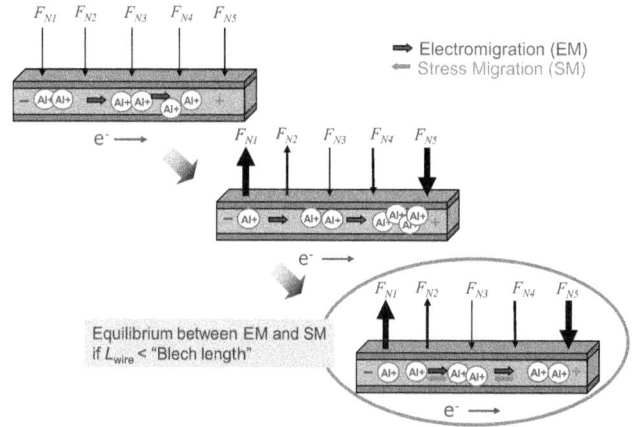

Figure 8. An illustration of stress migration caused by the mechanical stress buildup in a short wire. This reversed migration process essentially neutralizes the material flow due to electromigration.

Specifically, a wiring segment is not susceptible to electromigration if the product of the segment's current density J and the segment's length L is less than a process-technology-dependent threshold value $(JL)_{Blech}$ [4]. The critical product $(JL)_{Blech}$, often called the "Blech immortality condition," is obtained from

$$(JL) < (JL)_{Blech} = \Omega \frac{\Delta\sigma}{z^* q\rho} \qquad (2)$$

where Ω is the atomic volume of the wire material (cm³/number), $\Delta\sigma$ is the difference in the hydrostatic stress at both ends of the segment, z^* is the effective charge of the wire's material (a measure of the momentum transfer from electrons to the ions of the wire's material), q is the fundamental electronic charge ($z^*q < 0$) and ρ is the density of the wiring material [39].

The Blech length can be increased by depositing (stiff) cap layers on top of the copper metal and by using dielectrics with a higher Young's module (i.e., higher stiffness) because both support reverse stress migration. However, the cap layer also needs to suppress surface diffusion (see Section 5.6).

Besides the *Blech effect* or *Blech immortality condition* discussed above, there is another, related short-length effect. It is based on *void growth saturation* due to mechanical stress buildup. Void-related failures in wiring segments require the void to be grown to a critical size V_{fail}. Reverse stress migration in short segments, described above, can block void growth so that it cannot reach V_{fail} (Figure 9). Specifically, the void growth with its increase in resistance ΔR does not reach the failure state marked by ΔR_{fail}.

There is therefore a critical product $(JL)_{sat}$ below which immortality is obtained due to void growth saturation, given by

$$(JL)_{sat} = \frac{\rho/A}{\rho_1/A_1} \cdot \frac{\Delta R_{fail}}{R} \cdot \frac{2\Omega \cdot B}{z^*q\rho} \qquad (3)$$

where ρ/A and ρ_1/A_1 are the ratio of resistivity to cross-sectional area of the wire material (Cu) and the liner, respectively, and R is the initial wire resistance [1][37][39]. The value B is the elastic bulk modulus for the material that surrounds the wiring segment [24].

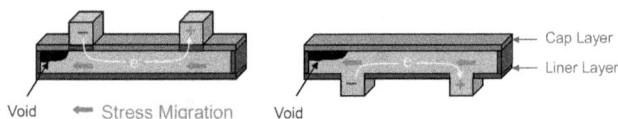

Figure 9. A stress gradient can also cause void growth saturation in short line via-above (left) and via-below configurations (right). Note the lesser impact of voids in a via-below configuration (right, see also Section 5.5).

It is important to note that, only if $(JL)_{sat} > (JL)_{Blech}$, immortality is obtained due to void grow saturation and the wire will not fail until the void size reaches V_{fail}. Furthermore, $(JL)_{sat}$ is lower for via-above than via-below copper configurations (see Section 5.5 and Figure 9). It is also lower if a low-k (low-stiffness) dielectric is used instead of SiO_2 [39]. Some of the values of the critical product (JL) for immortality obtained experimentally are, for example, 1,500 A/cm for Cu/SiO_2 via-above segments [8], 3,700 A/cm for Cu/SiO_2 via-below segments [25] and 375 A/cm for Cu/low-k via-above segments [7]. Assuming a current density J of 5×10^5 A/cm^2, the latter (low) value of 375 A/cm corresponds to a segment length of 7.5 μm, while Cu/SiO_2 via-below segments can have lengths of up to 74 μm.

Exploiting the critical product (JL) for immortality, however, requires knowledge of the effects of *linked* segments. Specifically, the entire interconnect tree, which is the set of segments that are linked without diffusion barriers – that is, within one layer of metallization – needs to be included in the calculation [6][39]. Inactive segments require special attention because they can serve as reservoir sinks (see Section 5.3) [6].

Starting with the 45 nm technology node, short-length effects can be exploited to deliberately generate EM-resistant wire segments during the routing step. This is particularly relevant for minimum-sized wires because they "automatically" become increasingly EM-resistant with shrinking feature sizes. Also, wires that are too long for short-length effects can be split into shorter, more EM-resistant segments by inserting vias. The downside is that vias are susceptible to electromigration, this must be considered at the same time (see Section 5.4).

5.3 Reservoir Effect

The so-called "reservoir effect" [31] can significantly improve the lifetime of multiple-level interconnects. Increased metal-via layer overlaps enlarge the amount of interconnect material at one of the prime points of EM failure: below, above and within the via. Reservoirs can act as sources that have to be drained before voids generated by electromigration become critical to the circuit. This

does not actually decrease the effects of electromigration itself, but prolongs the time to failure due to void growth.

Reservoirs that serve as a material source typically prolong lifetimes, and reservoirs that act as dominant sinks for migrated interconnect material can degrade lifetimes. Whether a reservoir acts as sink or source depends on the direction of current flow (Figure 10).

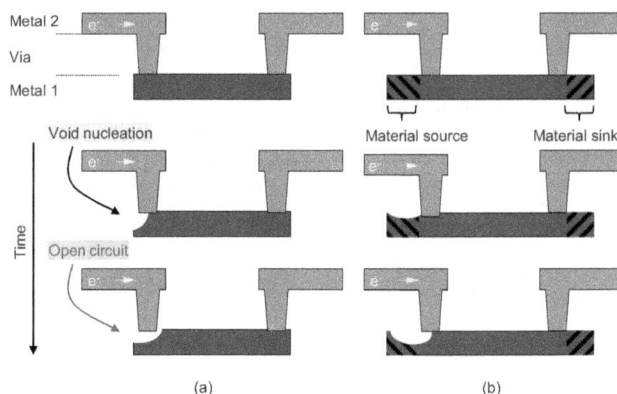

Figure 10. The reservoir effect in (b) extends the lifetime of the via configuration compared to a regular via without layer overlaps (a).

The reservoir effect needs to be considered in conjunction with short-length effects because any sink effectively increases the wire length and thus reduces any possible stress buildup. Consequently, the reservoir effect is most effectively used for nets with known, directed currents because, in this case, reservoirs acting as sources prolong time to failure and sinks can be deliberately avoided.

The reservoir effect is also leveraged by using multiple vias (see Section 5.4) since via arrays have a larger reservoir area than a single via. Experimental results in [31] indicate that the prolonged lifetime achieved is more a result of the increased reservoir area than of current sharing between vias.

5.4 Double/Multiple Vias

Particular attention needs to be paid to vias and contact holes, because generally the ampacity of a (tungsten) via is less than that of a metal wire of the same width. Moreover, migration velocities in the via material, the diffusion barrier and the metal wire differ. Hence, vias are one of the prime points of void nucleation and thus EM failure. To compensate for this increased vulnerability and to address typical manufacturing and yield issues, double or multiple vias are often deployed. The via array geometry is crucial here: multiple vias need to be arranged so that the resulting current flow and thus EM degradation is distributed as evenly as possible throughout the parallel vias [28].

5.5 Via-Above and Via-Below Configuration

The generation of voids in the vicinity of vias is strongly influenced by the geometry of the via configuration. Depending on whether a specific wire segment is connected from above or below, the configuration is called *via-above* or *via-below*, respectively. As described above, via stacks are prone to void nucleation because of the predominance of interface/surface diffusion in copper interconnects and different electromigration drift velocities. Although this drift velocity is low in the diffusion barrier, it is higher in the adjacent copper wire and tungsten via. Consequently, electrons passing through the diffusion barrier can cause voids in the material behind the barrier. The impact of excessive current density is a function of the geometry, and can cause a variety of damage [5][39]. Figure 11 shows that even a low-volume void causes a failure if placed directly underneath the via, whereas a void in a via-below configuration has to grow larger before the interconnect is destroyed.

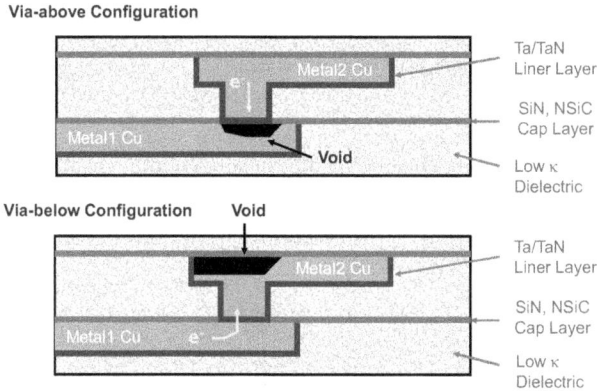

Figure 11. Via-above and via-below configurations with their different damage locations partly due to the interface/surface diffusion prevalent in copper wires.

5.6 Surface Coating / Metal Capping

As mentioned in Section 2, the Cu/cap layer interface is the dominant diffusion path for electromigration in submicron copper interconnects. Electromigration studies in copper line structures show that mass transport is dominated by diffusion at the Cu/cap layer interface, probably due to the presence of defects (voids) induced by the chemical mechanical polish (CMP) process prior to cap deposition [11][35]. The effect of a dielectric capping layer on interfacial mass transport for copper interconnects can therefore be exploited. For that purpose cap layers which increase activation energies at the interface and which provide a high degree of stiffness (see Section 5.2) will improve EM resistance. For example, Hu et al. [16] demonstrated a significant improvement in EM lifetime by coating the Cu surface with a thin 10-nm electroless layer of CoWP.

5.7 Frequency Dependency

The lifetime of a wire carrying alternating currents (AC) is significantly longer than that of a wire with direct currents (DC), due to damage-healing effects (Figure 12). Material migrating in one direction can partially migrate back to its original location under beneficial conditions. Consequently, the interconnect suffers less damage from electromigration under AC conditions. Experiments in [38] reveal that the effect of self-healing becomes noticeable above 10 Hz and increases with the switching frequency up to about 10 kHz where it reaches a saturation value. There is no improvement in circuit durability at frequencies above this point.

Tao et al. [38] report that AC lifetime at high frequencies is typically over 1,000 times longer for Al-2% Si and Al-4% Cu and over 500 times longer for Cu interconnects.

As a consequence, two different sets of maximum allowable current densities are commonly used in practical current-density verification, one for signal nets with frequencies above 10 kHz and one for the remaining (DC) nets, such as power supply lines.

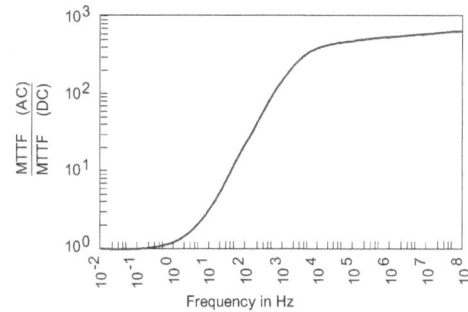

Figure 12. Mean time to failure (MTTF) if interconnect is stressed by an alternating current (AC) compared to MTTF if a directed current (DC) is applied [38].

6. OUTLOOK

Electromigration (EM) is becoming an increasingly difficult design challenge due to higher interconnect current densities. The on-going trend in IC downscaling is producing physical designs with ever-smaller feature sizes, which can easily give rise to current densities that exceed their maximum allowable value.

In order to address this problem during layout synthesis, this talk has focused on basic design issues that affect electromigration during interconnect physical design. The aim is to increase current density limits in the interconnect by utilizing electromigration-inhibiting measures, such as short-length and reservoir effects. Exploitation of these effects at the layout stage provides partial relief of EM concerns in today's design flows.

In order to use these effects in next generation design flows, we propose establishing a dependence between the applied current density limits on the specific circuit geometry and the application's mission profile. Design tools can significantly improve the EM robustness of the generated layout by utilizing EM-optimized layout configurations as constraints during synthesis steps, such as routing. We believe that this inclusion of EM-specific requirements in the physical design can provide relief from severe reliability constraints in future technologies.

ACKNOWLEDGEMENTS

The author wishes to thank Matthias Thiele and Tilo Meister for their contributions to this paper. This work was done in close collaboration with Göran Jerke at Robert Bosch GmbH and Shanthi Siemes and Uwe Hahn of GlobalFoundries Inc. Their valuable practical input is greatly appreciated.

REFERENCES

[1] V. K. Andleigh, V. T. Srikar, Y.-J. Park, C. V. Thompson, "Mechanism maps for electromigration-induced failure of metal and alloy interconnects," *J. Appl. Phys.* 86 (1999), 6737. DOI= http://dx.doi.org/10.1063/1.371750

[2] L. Arnaud, G. Tartavel, T. Berger, D. Mariolle, Y. Gobil, I. Touet, "Microstructure and electromigration in copper damascene lines," *Proc. 37th Annual Reliability Physics Symposium* (1999), 263 -269. DOI= http://dx.doi.org/10.1109/RELPHY.1999.761624

[3] J. R. Black, "Electromigration – A brief survey and some recent results," *IEEE Trans. on Electronic Devices* (April 1969), 338-347. DOI= http://dx.doi.org/10.1109/T-ED.1969.16754

[4]   I. A. Blech, "Electromigration in thin aluminum films on titanium nitride," *J. Appl. Phys.*, vol. 47 (1976), 1203–1208. http://dx.doi.org/10.1063/1.322842

[5]   Z.-S. Choi, C. L. Gan, F. Wei, C. V. Thompson, J. H. Lee, K. L. Pey, W. K. Choi, "Fatal void size comparisons in via-below and via-above Cu dual-damascene interconnects," *MRS Proc. Materials, Technology and Reliability of Advanced Interconnects*, vol. 812 (2004) F7.6. DOI= http://dx.doi.org/10.1557/PROC-812-F7.6

[6]   C. L. Gan, C. V. Thompson, K. L. Pey, W. K. Choi, "Experimental characterization and modeling of the reliability of 3-terminal dual-damascene Cu interconnect trees," *J. Appl. Phys.* 94 (2003), 1222. DOI= http://dx.doi.org/10.1063/1.1585119

[7]   C. S. Hau-Riege, A. P. Marathe, V. Pham, "The effect of low-k ILD on the electromigration reliability of Cu interconnects with different line lengths," *Proc. of the 41st Int. Reliability Physics Symp.* (2003), 173-177. DOI= http://dx.doi.org/10.1109/RELPHY.2003.1197740

[8]   C. S. Hau-Riege, A. P. Marathe, V. Pham, "The effect of line length on the electromigration reliability of Cu interconnects," *Proc. of Advanced Metallization Conf.* (2002), 169.

[9]   C. S. Hau-Riege, "An introduction to Cu electromigration," *Microel. Reliab.*, vol. 44 (2004), 195–205. DOI= http://dx.doi.org/10.1016/j.microrel.2003.10.020

[10] M. Hayashi, S. Nakano, T. Wada, "Dependence of copper interconnect electromigration phenomenon on barrier metal materials," *Microel. Reliab.*, vol. 43 (2003), 1545–1550. DOI= http://dx.doi.org/10.1016/S0026-2714(03)00273-7

[11] J. Hohage, M. U. Lehr, V. Kahlert, "A copper-dielectric cap interface with high resistance to electromigration for high performance semiconductor devices," *Microel. Engineering* (2009), 86, 408-413. DOI= http://dx.doi.org/10.1016/j.mee.2008.12.012

[12] http://pulsic.com/products/pulsic-implementation-solution/unity-analog-router/

[13] http://www.cadence.com/products/

[14] http://www.apache-da.com/products/totem/totem-mmx

[15] http://www.synopsys.com/Tools/Verification/AMSVerification/Reliability/Pages/customsim-interconnect-reliability-analysis.aspx

[16] C.-K. Hu, L. Gignac, R. Rosenberg, E. Liniger, J. Rubino, C. Sambucetti, A. Stamper, A. Domenicucci, X. Chen, "Reduced Cu interface diffusion by CoWP surface coating," *Microelectronic Engineering* 70 (2003), 406–411. DOI= http://dx.doi.org/10.1016/S0167-9317(03)00286-7

[17] C.-K. Hu, L. Gignac, R. Rosenberg, "Electromigration of Cu/low dielectric constant interconnects," *Microelectronics and Reliability* (2006), 46, 213 – 231. DOI= http://dx.doi.org/10.1016/j.microrel.2005.05.015

[18] International Technology Roadmap for Semiconductors (ITRS), 2011 Edition, Online: www.itrs.net/reports.html, 2012

[19] G. Jerke, J. Lienig, "Hierarchical current-density verification in arbitrarily shaped metallization patterns of analog circuits," *IEEE Trans. on CAD of Integr. Circuits Sys.*, vol. 23(1) (Jan. 2004), 80–90. DOI= http://dx.doi.org/10.1109/TCAD.2003.819899

[20] G. Jerke, J. Lienig, J. Scheible, "Reliability-driven layout decompaction for electromigration failure avoidance in complex mixed-signal IC designs," *Proc. Design Automation Conf.* (2004), 181–184. DOI= http://dx.doi.org/10.1145/996566.996618

[21] G. Jerke, J. Lienig "Early-stage determination of current-density criticality in interconnects," *Proc. 11th IEEE Int. Int. Symp. on Quality Electronic Design (ISQED)* (2010), 667-774. DOI= http://dx.doi.org/10.1109/ISQED.2010.5450505

[22] K. Jonggook, V. C. Tyree, C. R. Crowell, "Temperature gradient effects in electromigration using an extended transition probability model and temperature gradient free tests. I. Transition probability model," *IEEE Int. Integrated Reliability Workshop Final Report* (1999), 24 -40. DOI= http://dx.doi.org/10.1109/IRWS.1999.830555

[23] A. B. Kahng, J. Lienig, I. L. Markov, J. Hu, *VLSI Physical Design: From Graph Partitioning to Timing Closure*, Springer-Verlag, Berlin, Heidelberg, New York, ISBN 978-90-481-9590-9, 2011. DOI= http://dx.doi.org/10.1007/978-90-481-9591-6

[24] M. A. Korhonen, P. Børgesen, K. N. Tu, C.-Y. Li, "Stress evolution due to electromigration in confined metal lines," *J. Appl. Phys.* 73 (1993), 3790. DOI= http://dx.doi.org/10.1063/1.354073

[25] K.-D. Lee, E. T. Ogawa, H. Matsuhashi, P. R. Justison, K.-S. Ko, P. S. Ho, V. A. Blaschke, "Electromigration critical length effect in Cu/oxide dual-damascene interconnects," *Appl. Phys. Lett.* 79 (2001), 3236, DOI= http://dx.doi.org/10.1063/1.1418034

[26] B. Li, T. D. Sullivan, T. C. Lee, D. Badami, "Reliability challenges for copper interconnects," *Microel. Reliab.*, vol. 44 (2004), 365–380. DOI= http://dx.doi.org/10.1016/j.microrel.2003.11.004

[27] J. Lienig, G. Jerke, "Current-driven wire planning for electromigration avoidance in analog circuits," *Proc. Asia and South Pacific Design Automation Conf.* (2003), 783–788. DOI= http://dx.doi.org/10.1109/ASPDAC.2003.1195125

[28] J. Lienig, "Introduction to electromigration-aware physical design," *Proc. Int. Symp. on Physical Design (ISPD'06)*, (2006), 39-46. DOI= http://dx.doi.org/10.1145/1123008.1123017

[29] B. K. Liew, N. W. Cheung, C. Hu, "Electromigration interconnect lifetime under AC and pulse DC stress," *Proc. 27th Int. Reliab. Phys. Symp. (IRPS)* (1989), 215–219. DOI= http://dx.doi.org/10.1109/RELPHY.1989.36348

[30] J. A. Maiz, "Characterization of electromigration under bidirectional (BC) and pulsed unidirectional (PDC) currents," *Proc. 27th Int. Reliab. Phys. Symp. (IRPS)* (1989), 220–228. DOI= http://dx.doi.org/10.1109/RELPHY.1989.36349

[31] H. V. Nguyen, C. Salm, R. Wenzel, A. J. Mouthaan, F. G. Kuper, "Simulation and experimental characterization of reservoir and via layout effects on electromigration lifetime," *Microel. Reliab.*, vol. 42 (2002), 1421–1425. DOI= http://dx.doi.org/10.1016/S0026-2714(02)00162-2

[32] D. G. Pierce, E. S. Snyder, S. E. Swanson, L. W. Irwin, "Wafer-level pulsed-DC electromigration response at very high frequencies," *Proc. of the Int. Reliab. Phys. Symp. (RELPHY-94)* (1994), 198-206. DOI= http://dx.doi.org/10.1109/RELPHY.1994.307836

[33] A. G. Sabnis, "VLSI reliability," *VLSI Electronics—Microstructure Science*, London: Academic Press Ltd., vol. 22, 1990.

[34] A. Scorzoni, B. Neri, C. Caprile, F. Fantini, "Electromigration in thin-film inter-connection lines: models, methods and results," *Material Science Reports*, New York: Elsevier, vol. 7 (1991), 143–219. http://dx.doi.org/10.1016/0920-2307(91)90005-8

[35] W. Shao, S. G. Mhaisalkar, T. Sritharan, A. V. Vairagar, H. J. Engelmann, O. Aubel, E. Zschech, A. M. Gusak, K. N. Tu, "Direct evidence of Cu/cap/liner edge being the dominant electromigration path in dual damascene Cu interconnects," *Appl. Phys. Lett.*, 90 (2007), 052106, DOI= http://dx.doi.org/10.1063/1.2437689.

[36] W. Steinhögl, G. Schindler, G. Steinlesberger, M. Engelhardt, "Size-dependent resistivity of metallic wires in the mesoscopic range," *Physical Review B*, 66 (2002) 075414. http://dx.doi.org/10.1103/PhysRevB.66.075414

[37] Z. Suo, "Stable state of interconnect under temperature change and electric current," *Acta Mater.* 46 (1998), 3725. DOI= http://dx.doi.org/10.1016/S1359-6454(98)00098-6

[38] J. Tao, N. W. Cheung, C. Hu, "Metal electromigration damage healing under bidirectional current stress," *IEEE Electron Device Letters* (1993), 14, 554-556. DOI= http://dx.doi.org/10.1109/55.260787

[39] C. V. Thompson, "Using line-length effects to optimize circuit-level reliability," *15th Int. Symp. on the Physical and Failure Analysis of Integrated Circuits (IPFA)* (2008), 63-66. DOI= http://dx.doi.org/10.1109/IPFA.2008.4588155

[40] A. V. Vairagar, S. G. Mhaisalkar, A. Krishnamoorthy, "Electromigration behavior of dual-damascene Cu interconnects – Structure, width, and length dependences," *Microel. Reliab.*, vol. 44 (2004), 747–754. DOI= http://dx.doi.org/10.1016/j.microrel.2003.12.011

[41] F. L. Wei, C. L. Gan, T. L. Tan, C. S. Hau-Riege, A. P. Marathe, J. J. Vlassak, C. V. Thompson, "Electromigration-induced extrusion failures in Cu/low-k interconnects," *J. Appl. Phys.* vol. 104 (2008), 023529-023529-10. DOI= http://dx.doi.org/10.1063/1.2957057

[42] D. Young, A. Christou, "Failure mechanism models for electromigration," *IEEE Trans. on Reliability*, vol. 43(2) (June 1994), 186–192. DOI= http://dx.doi.org/10.1109/24.294986

[43] X. Yu, K. Weide, "A study of the thermal-electrical- and mechanical influence on degradation in an aluminum-pad structure," *Microelectronics and Reliability* (1997), 37, 1545 – 1548. DOI= http://dx.doi.org/10.1016/S0026-2714(97)00105-4

Data Mining In Design and Test Processes - Basic Principles and Promises *

Li-C. Wang

University of California, Santa Barbara

ABSTRACT

This talk discusses several application examples to illustrate the basic principles of applying data mining in design and test. Two types of data mining are seen in most of the applications: novelty detection and feature-based rule learning. The experience of developing a practical data mining flow is summarized. Promises are demonstrated with positive experimental results based on industrial settings.

Categories and Subject Descriptors

B.7 [**Integrated Circuits**]: Miscellaneous; H.2.8 [**Database Management**]: Database Applications—*Data mining*

General Terms

Design

Keywords

Computer-Aided Design, Data Mining, Test, Verification

1. INTRODUCTION

Data mining in design and test has become a fast-growing research area in recent years. In design and test, tremendous amounts of simulation and measurement data are generated. These data present opportunities for applying data mining.

In formulating an application, it is important to ask the question: Mining for what purpose? Many problems encountered in design automation and test are NP-hard problems. Data mining does not make a NP-hard problem easier. In fact, the power of learning is so limited that even learning a simple 3-Term DNF formulae is NP-hard [1].

Then, what is the power of data mining for, if not for solving a difficult problem? Learning a Boolean function with an almost perfect accuracy is hard [1]. However, learning with a high percentage of accuracy is feasible [2]. In other words, if one demands an almost guaranteed result, the problem becomes hard. Therefore, a practical data mining application should be formulated in such a way that guaranteed result is not required for the application to be meaningful.

An application can only be formulated based on the data availability. If the data is not readily available, the cost of collecting the data will become an important consideration. One has to demonstrate that the benefit of mining the data out-weights the cost in order for the application to make practical sense. Furthermore, the size of the data has to be large enough for data mining to make sense.

Data mining is most useful when it is applied to improve the efficiency of an existing process. For example, a person needs to examine a large number of plots to identify a few interesting ones for further analysis. It is a tedious manual process. Data mining can be used to speed up the process by automatically identifying the interesting plots.

Introduction of a data mining flow should make a target task easier for its user, not harder. Hence, design of an effective usage model is crucial. This includes effective presentation of the mining results with easy visualization to facilitate user interaction and decision making.

2. EXAMPLES AND PROMISES

Data mining algorithms operate on a collection of *samples*. Given N samples, novelty detection intends to find a few samples that are "novel" with respect to others. A popular algorithm to implement novelty detection is the SVM one-class [3]. In feature-based rule learning, two classes of samples are given, a large number of known non-novel samples and a small number of known novel samples. The objective is to uncover special properties of novel samples and represent them as rules. A popular algorithm to implement rule learning is the subgroup discovery [4].

In pre-silicon design, functional verification remains a key bottleneck. In a design cycle, the design evolves over time. Consequently, functional verification is an iterative process in which extensive simulation is run on a few relatively stable versions of the design. From one iteration to another, two assets are kept. The first comprises the test templates refined and accumulated up to the previous iteration. The second comprises the important tests identified so far.

In this context, data mining can be employed in two applications, to reduce the simulation time required to find an important test and to improve a test template for generating additional important tests. Here, a sample is a test.

The work in [5] proposes a novel test detection framework that can filter out a large number of unimportant tests before simulation, effectively reducing the simulation time by up to 90%. The data mining approach is novelty detection. The work in [6] presents a feature-based analysis approach to extract special properties of novel tests. These properties are then used to improve the test template for achieving a better coverage. The data mining approach is feature-based rule learning [7]. In both works, a test is an assembly

*This work is supported in part by Semiconductor Research Corporation, projects 2010-TJ-2093 and 2012-TJ-2268

program. The experiments were run on a low-power 64-bit Power Architecture-based processor core.

Lithography simulation is another example in pre-silicon design where simulation time can be a major concern. In this context data mining can be used to reduce the need of lithography simulation. The objective is to identify potential problematic layout spots that are to be checked with lithography simulation. Majority of the layout areas do not need lithography simulation and hence simulation time is reduced. The work in [8] proposes two approaches for such a purpose, the *supervised* approach where two classes of training samples are first labeled by a lithograph simulation and then used to learn a classifier, and the *unsupervised* approach where potential problematic samples are identified as novel samples by novelty detection. Here, a sample is a small piece of layout image based on a raster scan of a layout [8].

For processor design, one important task in post-silicon is to identify speed limiting paths as guides for performance improvement. Data mining can be applied in two applications, facilitating the identification of potential speed paths [9][10] and understanding known speed paths [11]-[14]. In these applications, a sample is a path. The work in [14] summarizes the speed path analysis research in an industrial application. Design issues were uncovered by analyzing top speed paths against a large number of non-speed paths, which otherwise were difficult to find without the proposed feature-based data mining approach [14].

In production, test cost and/or quality continue to be major concerns. In die/chip-level analysis, a sample is a die/chip. In wafer-level analysis, a sample is a wafer. The work in [15] discusses how to predict potential defective parts as novel samples. Because novelty depends on the tests used in the analysis, the work [16] discusses the test selection problem. Both works were based on test data from an SoC production line for the automotive market where quality requirement is extremely high. Higher quality usually demands more sophisticated test processes and hence, higher cost. One expensive test process is burn-in. The work in [17] discusses the potential of burn-in reduction for another product line also for the automotive market, by predicting parts that do not need long hours of burn-in (or equivalently (novel) samples that need the burn-in). Data mining is applied on test data collected after a short period of burn-in to predict the outcomes after a long period of burn-in.

3. SUMMARY

In the above applications, data mining begins by defining what a sample is. For example, a sample can be an assembly program, a layout image, a path, a die/chip or a wafer. To apply a data mining algorithm, a sample needs to be encoded with a set of features. The effectiveness of the mining largely depends on this encoding. For example, the feature set used to describe the characteristics of an assembly programs or a path can take weeks to develop. Fortunately this development effort can be seen as one-time cost. In die/chip-level analysis, features are tests. Hence, test selection becomes an important problem [16].

In all applications, we are interested in finding and/or understanding a small set of novel samples in a large population of samples. While one can formulate the problem as binary classification if examples of novel samples are provided, it is usually not effective due to the extreme imbalance between the size of novel sample set and the size of non-novel sample set, i.e. the data usually contains a lot more information on the non-novel samples while our interest is on the novel samples. To overcome this issue, the feature set needs to incorporate some domain knowledge such that the analysis can be directed to a more refined space of interest. User intervention may also be needed to guide the selection of relevant features and the mining process. Data mining is rarely a one-time task. It is often seen as an iteratively knowledge discovery process where results are interpreted and actions are taken by user from one iteration to the next.

Modern learning algorithms such as SVM are designed with guaranteed asymptotic behavior when the data size approaches infinity [18]. Its effectiveness on a sample size of thousands or tens of thousands may vary, depending on the accuracy requirement for the application. With limited data, the quality of the feature set becomes more important. Hence, in many applications it is expected that a significant part of the effort will be spent on feature set development.

4. REFERENCES

[1] Michael J. Kearns and Umesh V. Vazirani. An Introduction to Computational Learning Theory, *MIT Press*, 1994.

[2] Onur Guzey, et. al. Extracting a Simplified View of Design Functionality Based on Vector Simulation. Lecture Note in Computer Science, *LNCS*, Vol 4383, 2007, pp. 34-49.

[3] Bernhard Schölkopf, and Alexander J. Smola. Learning with Kernels: Support Vector Machines, Regularization, Optimization, and Beyond. The MIT Press, 2001.

[4] N. Lavrač, B. Kavšek, P. Flach, and L. Todorovski. Rule induction for subgroup discovery with CN2-SD *Journal of Machine Learning Research*, 5:153–188, Dec. 2004.

[5] Wen Chen, et. al. Novel Test Detection to Improve Simulation Efficiency A Commercial Experiment. *ACM/IEEE ICCAD*, 2012.

[6] Wen Chen, et. al. Novel Test Analysis to Improve Structural Coverage A Commercial Experiment. *IEEE VLSI Design Automation and Test Symposium*, 2013.

[7] Li-C. Wang, Data Learning Based Diagnosis. *ASP-DAC*, 2010, pp 247-254.

[8] Dragoljub (Gagi) Drmanac, Frank Liu, Li-C. Wang. Predicting Variability in Nanoscale Lithography Processes. *ACM/IEEE DAC*, 2009, pp. 545-550.

[9] Pouria Bastani, et. al. Speedpath Prediction Based on Learning from a Small Set of Examples. *ACM/IEEE DAC*, June 2008, pp. 217-222.

[10] Nicholas Callegari, Li-C. Wang, Pouria Bastani. Feature based similarity search with application to speedpath analysis. *IEEE ITC*, 2009.

[11] Li-C. Wang, Pouria Bastani, Magdy S. Abadir. Design-silicon timing correlation — a data mining perspective. In *ACM/IEEE DAC* 2007, pp. 384-389.

[12] P. Bastani, et. al. Statistical Diagnosis of Unmodeled Timing Effect. *ACM/IEEE DAC*, 2008, pp. 355-360.

[13] Nicholas Callegari, et. al. Classification rule learning using subgroup discovery of cross-domain attributes responsible for design-silicon mismatch. *ACM/IEEE DAC*, 2010, pp. 374-379.

[14] Janine Chen, et. al. Mining AC Delay Measurements for Understanding Speed-limiting Paths. *IEEE ITC*, 2010.

[15] Nik Sumikawa, et. al. Screening Customer ReturnsWith Multivariate Test Analysis. *IEEE ITC*, 2012.

[16] Nik Sumikawa, et. al. Important Test Selection For Screening Potential Customer Returns. *IEEE VLSI Design Automation and Test Symposium*, 2011, pp. 171-174.

[17] Nik Sumikawa, Li-C.Wang, and Magdy S. Abadir. An Experiment of Burn-In Time Reduction Based On Parametric Test Analysis. *IEEE ITC*, 2012.

[18] V. Vapnik, The nature of Statistical Learning Theory. 2nd ed., *Springer*, 1999.

SRAM Dynamic Stability Verification by Reachability Analysis with Consideration of Threshold Voltage Variation

Yang Song[1], Hao Yu[2], Sai Manoj Pudukotai Dinakarrao[2], and Guoyong Shi[1]
[1]School of Microelectronics, Shanghai Jiao Tong University, Shanghai, 200240, China
[2]School of Electrical and Electronic Engineering, Nanyang Technological University, 639798, Singapore
haoyu@ntu.edu.sg *

ABSTRACT

Dynamic stability margin of SRAM is largely suppressed at nanoscale due to not only dynamic noise but also process variation. A novel dynamic stability verification is developed in this paper based on analog reachability analysis for checking SRAM failure. In the presence of mismatch such as threshold voltage variation of all transistors, zonotope-based reachability analysis is deployed to efficiently verify SRAM failure at transistor level. The threshold voltage variation is considered by the modified input range of SRAM. As such, the suppressed stability margin and further failure region can be verified by performing a time-evolved reachability analysis with formed zonotope to distinguish safe and failure regions. One can perform efficient verification of the SRAM dynamic stability without repeated yet time-consuming Monte-Carlo simulations considering variations from all transistors. As demonstrated by numerical experiment results, the developed reachability analysis can accurately verify the SRAM dynamic stability under threshold voltage variations from all transistors. Speedup of more than $400\times$ in runtime can be achieved over the Monte Carlo approach of 500 samples with the similar accuracy.

Categories and Subject Descriptors

B.8.1 Hardware [**Performance and Reliability**]: Reliability, Testing, and Fault-Tolerance

Keywords

Verification, Reachability Analysis, SRAM Reliability

1. INTRODUCTION

Static noise margin (SNM)[1, 2] is traditionally deployed for SRAM failure characterization due to simple interpretation and measurement. As it may overestimate read failures and underestimate write failures, dynamic SRAM stability margin [3] is increasingly adopted by deploying critical

---

*This work is sponsored by MOE Tier-1 fund RG 26/10 from Singapore.

word-line pulse width to produces better estimation of failure rates. The verification of SRAM stability margin becomes even harder when technology scales down into nanoscale, which has significantly suppressed the SRAM stability margin in presence of process variations. The operation of SRAM cell shows not only nonlinear but also undetermined behavior. For example, threshold voltage variation in any transistor of one SRAM cell can bring about malfunction. A thorough verification considering threshold variations in all transistors is necessary to provide designers a close scrutiny of potential hazards. However, such verification can be computationally expensive [4].

Based on DC characteristics of inverters, stability analysis has been performed in [5] by modeling failure with normal distribution even when failure occurs in tail of normal distribution. In [6], accurate estimation is achieved without the assumption of normal distribution of failure probability but is based on the most probable failure point searching. Moreover, stochastic orthogonal polynomials are used to derive failure/yield rate without the prohibitive Monte Carlo simulation in [7].

The analytical SRAM reliability verification is also developed. Euler-Newton curve tracing [8] is utilized to find the boundary between the success and failure regions without brute-force exploration in the parameter space. The work in [9] formulates dynamic noise margin with the use of stability boundary, namely the *separatrix* which separates two stability regions in the parameter space. Note that both approaches can reach to prominent efficiency when compared with brute-force Monte Carlo approach. But the searching of boundary for failure region is limited for dimensions of two parameters, and the computational cost may be high to determine the failure of 6T-SRAM with consideration for variations from all transistors.

Reachability analysis has been widely deployed in reliability verification of dynamic circuits and systems. By exploring potential trajectories of operating points in state space [10, 11], it can conveniently provide accurately predicted boundary of multiple trajectories under an uncertainty by one time computation. As such, there is no need to simulate different trajectories to distinguish safe and unsafe cases one by one. The reachability analysis has been deployed for a number of hard analog circuit verifications [12, 13, 14]. In this paper, we introduce a reachability based verification for SRAM dynamic stability, which can take into account variations from all transistors at the same time. The proposed reachability analysis is implemented into a SPICE-like simulator with consideration of nonlinear device model. It can consider not only threshold voltage but also width variations from multiple transistors. The verification accuracy of the proposed approach is the same while computational cost is much smaller when compared with the Monte Carlo method. Experiments show that the proposed method can achieve speedups up to 481× over Monte Carlo approach with 500 samples under the similar accuracy.

The rest of this paper is organized as follows. Section 2 reviews the dynamic stability mechanisms for SRAM failures with consideration of process variation. Section 3 describes the reachability analysis for the verification of SRAM dynamic stability such as the linkage between the reachability analysis and the SRAM verification flow. The proposed verification methodology is validated by experiments in Section 4 for different SRAM malfunctions including write, read and hold failures. Conclusions are drawn in Section 5.

2. SRAM FAILURE ANALYSIS

A 6T-SRAM cell contains two cross-coupled inverters and two access transistors, which functions properly within dynamic noise margin (DNM) [9]. However, due to process variation, the mismatch among transistors may lead to functional failures. For the simplification of illustration, only variation of threshold voltage is considered in this paper, but the proposed approach is general to deal with other kinds of variations such as transistor width.

In this section, physical mechanisms of SRAM failures caused by threshold voltage variation are reviewed, including write, read and hold failures. In addition, the nonlinear SRAM dynamics is also presented as the basis for reachability based verification.

2.1 Failure Mechanisms

2.1.1 Write Failure

Write failure is defined as the inability to write data properly into the SRAM cell. During write operation, both access transistors should be strong enough to pull down or pull up the voltage level at internal nodes. As shown in Fig.1, write operation can be described on the variable plane as the process of pulling the operating point from initial state (bottom-right corner) to the target state (top-left corner). The crossing line named *separatrix* divides the variable plane into two convergent regions. Given enough time, operating point in any region will converge to the nearest stable equilibrium state either at top-left or bottom-right corner. Write operation is aimed at pulling operating point into targeted convergent region such that operating point can converge to the closest equilibrium state after operation finishes, which is shown by point B in Fig.1.

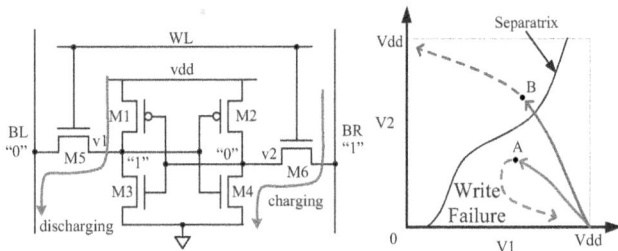

Figure 1: Illustration of SRAM write failure.

However, an increase in threshold voltage due to variation can reduce the transistor driving strength and vice verse for a decrease in threshold. The increase of V_{th} in M6 along with the decrease of V_{th} in M4 can result in difficulty to pull down v_2. On the variable plane, it becomes more difficult for operating point to move towards the target state. If operating point cannot cross the *separatrix* before access transistors are turned off, it goes back to the initial state, which means a write failure.

2.1.2 Read Failure

Read failure refers to the loss of the previously stored data. Before read operation is performed, both BR and BL are pre-charged to v_{dd}. Suppose previous internal states in SRAM are $v_1 = v_{dd}$ and $v_2 = 0$, electric charge on BR is discharged through M6 and M4 while that on BL remains the same. As such, a small voltage difference between BR and BL is generated which will be detected and amplified. In this way, data stored in the SRAM can be read. Note that access transistors need careful sizing such that their pull-up strength is not strong enough to pull the stored "0" to "1" during read operation.

On the variable plane, operating point is inevitably perturbed and pulled towards the *separatrix*. If read operation does not last too long, access transistors can be shut down before operating point crosses the *separatrix*. As such, the operating point returns to the initial state in the end, as point A in Fig.2, which means a read failure.

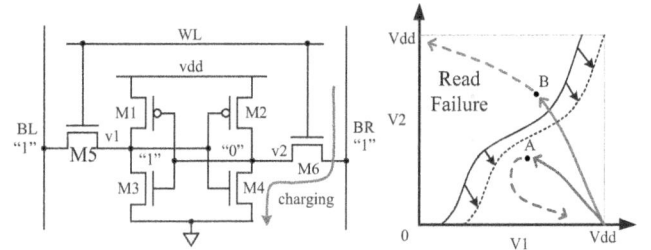

Figure 2: Illustration of SRAM read failure.

Even though all the sizing are carefully taken, threshold variations may still result in read failure. For example, variation caused by mismatch between M4 and M6 may result in unbalanced pulling strength, and v_2 can be pulled up more quickly. As a result, operating point crosses the *separatrix* before read operation ends, as point B in Fig.2.

2.1.3 Hold Failure

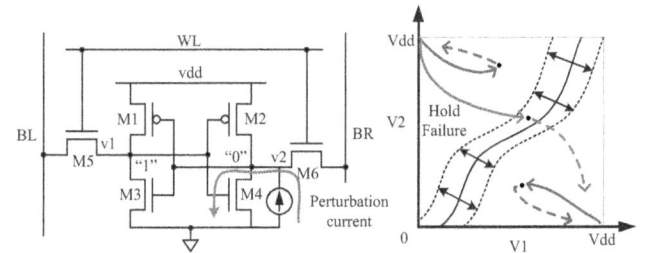

Figure 3: Illustration of SRAM hold failure.

Hold failure happens when the SRAM fails to retain the stored data. It can be caused by external noise or single event upset (SEU). The external perturbation can be modeled as noise current injected into SRAM. Similar to read operation, operating point is expected to converge back to initial state after settling down from disturbance. Otherwise, it will cross to the other convergent region.

While access transistors have no impact on the retention of SRAM data, M1-4 together can determine the likelihood of hold failure by finding the position of the *separatrix* and thus threshold variation may cause failure by perturbing the *separatrix* as shown in Fig.3. A such, one needs to verify if the SRAM is still tolerable to the injected noise in the presence of threshold voltage variation.

2.2 SRAM Nonlinear Dynamics

The nonlinear dynamics of SRAMs can be described by a differential equation (1)

$$\dot{x} = f(z(t)) \tag{1}$$

in which $z^T(t) = [x^T, u^T]$ is a vector consisting of the state variable vector $x(t)$ and the input vector $u(t)$. $f(x, u)$ describes nonlinearity of SRAMs. As for Monte Carlo based method, one needs to repeatedly launch Newton-Raphson iterations to solve (1) for sample by sample.

Moreover, the nonlinear dynamic equation can be linearized at one solved operating point as $\frac{\partial f}{\partial z}$, and $f(x, u)$ at any neighboring point can be expressed by 1st-order Taylor expansion centering at the linearized operating point (2), called as *nominal point*,

$$\begin{aligned} \dot{x} = & f(z^*) + \frac{\partial f}{\partial z}\Big|_{z=z^*} (z - z^*) + \\ & \frac{1}{2}(z - z^*)^T \frac{\partial^2 f}{\partial z^2}\Big|_{z=\xi} (z - z^*), \\ & \xi \in \{z^* + \alpha(z - z^*) | 0 \leq \alpha \leq 1\} \end{aligned} \tag{2}$$

where z^* is a nominal point and z is a neighbor point around the nominal point. For notation, the 2nd-order remainder in (2) is represented by L, which is also called as *linearization error* in this paper.

As such, in the neighborhood of the nominal point, the SRAM nonlinear dynamics can be depicted by (5)

$$\dot{x^*} + \dot{\Delta x} = f(x^*, u^*) + A\Delta x + B\Delta u + L, \tag{3}$$

in which

$$\begin{aligned} A = & \frac{\partial f}{\partial x}\Big|_{x=x^*}, B = \frac{\partial f}{\partial u}\Big|_{u=u^*} \\ \Delta x = & x - x^*, \Delta u = u - u^*. \end{aligned} \tag{4}$$

After cancelling $\dot{x^*}$ with $f(x^*, u^*)$, a linear differential equation for Δx can be obtained as

$$\dot{\Delta x} = A\Delta x + B\Delta u + L. \tag{5}$$

This equation lays the foundation for reachability analysis in the next section.

The threshold voltage variation can be implemented as additional noise current source added to the drain current of one transistor in SPICE as shown in Fig.4. As such, it becomes the 1st-order Taylor approximation of drain current by (6)

$$\begin{aligned} I_d + \Delta I_d = & \frac{1}{2}k\frac{W}{L}[V_{gs} - (V_{th} + \Delta V_{th})]^2 \\ \Delta I_d \approx & -k\frac{W}{L}(V_{gs} - V_{th})\Delta V_{th}. \end{aligned} \tag{6}$$

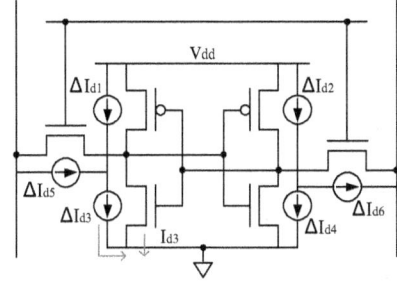

Figure 4: SRAM with threshold voltage variations modeled by additional current sources for all transistors.

The threshold voltage variation of each transistor is included in the input vector u as an independent current source. The other process variations can be conveniently considered in the similar way. Note that noise current in Fig.4 (from the internal transistor) is different from the noise current in Fig.3 (from the external environment).

3. SRAM VERIFICATION BY REACHABILITY ANALYSIS

In this section, we introduce reachability analysis for efficient SRAM dynamic stability verification considering threshold voltage variations. Reachability analysis [15] can efficiently determine a reachable region that one dynamic system can evolve with uncertain initial states or inputs. Applications of reachability analysis include formal verification of continuous, hybrid or discrete systems such as hard analog circuit verifications [12, 13, 14].

In the case of SRAM verification, variation of threshold voltage results in a number of SRAM trajectories. As such, if one models threshold voltage variations from all transistors as noise current sources added to the uncertain input, one can perform one-time reachability analysis to determine a set of reachable trajectories of SRAM. As such, one can easily determine if SRAM trajectories end up in the unsafe region with a failure. In the rest of this section, we discuss the

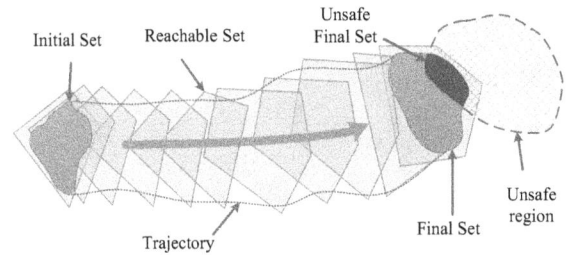

Figure 5: Reachable set and reachability analysis with a set of trajectories under uncertain input range.

reachability analysis for nonlinear circuits, and the according flow for SRAM verification.

3.1 Reachable Set

One important concept for reachability analysis is *reachable set*. A reachable set is the collection of all possible operating points or states in the variable space that a dynamic system may visit. Several approaches have been proposed to approximate reachable set by an enclosing hypercube. One simple and symmetrical type of hypercube, *zonotope* [10] is defined as

$$Z = \{x \in \mathbb{R}^{n \times 1} : x = c + \sum_{i=1}^{q} [-1,1] g^{(i)}\} \qquad (7)$$

where $c \in \mathbb{R}^{n \times 1}$ is the zonotope center and $g^{(i)} \in \mathbb{R}^{n \times 1}$ is called as a zonotope generator.

For SRAM verification, let the zonotope center equal the nominal point in order to minimize linearization error [15]. Thus variation of state vector Δx in (5) is expressed as the combination of generators

$$\Delta Z = Z - c = \{x \in \mathbb{R}^{n \times 1} : x = \sum_{i=1}^{q} [-1,1] g^{(i)}\}. \qquad (8)$$

Note that variation of input vector Δu can be expressed in the same way. Each generator of Δu represents an independent noise current source in (6), i.e., the threshold voltage variation of one transistor.

Algorithm 1: Reachability Analysis for SRAM Dynamic Stability Verification

---

Input: Initial set Z_0, system equation $\dot{x} = f(x, u)$, input sets $U_{1,...,N}$, simulation time interval h, maximum number of time steps N
Output: Final state P_N
1: $k = 0$
2: $P_N = \varnothing$
3: $queue.push(Z_0)$
4: **while** $!queue.isempty()$ **do**
5:     $Z_k = queue.pop()$
6:     $(c_k, G_k) = Z_k$
7:     compute c_{k+1} and Jacobian matrices A,B
8:     approximate linearization error L_{k+1}
9:     **if** $\mathbf{IH}(hL_{k+1}) \subseteq [-\varepsilon, \varepsilon]$ **then**
10:         $G_h = (I - hA)^{-1} G_k$
11:         $G_i = (I - hA)^{-1} hBU_{k+1}$
12:         $G_e = (I - hA)^{-1} hL_{k+1}$
13:         $G_{k+1} = G_h \oplus G_i \oplus G_e$
14:         $Z_{k+1} = (c_{k+1}, G_{k+1})$
15:         **if** $k + 1 == N$ **then**
16:             $P_N = P_N \cup projection(Z_N)$
17:             *continue*
18:         **end if**
19:         $queue.push(Z_{k+1})$
20:         $k = k + 1$
21:     **else**
22:         $[Z_k', Z_k''] = split(Z_k)$
23:         $queue.push(Z_k')$
24:         $queue.push(Z_k'')$
25:     **end if**
26: **end while**

3.2 Reachability Analysis

The complete flow for reachability analysis for SRAM verification is shown in Algorithm 1. Here, Z_0 is the initial reachable set in zonotope form, and P_N is the projection of final reachable sets onto the variable plane for observation.

In each iteration cycle (Line 5-25), the calculation of a new reachable set goes through Line 5-14 with the final set

collected in Line 15-20. Otherwise, the current reachable set is split in Line 22-24 and iteration is performed again for two new sets. Detailed discussion of the verification flow follows in the remainder of this section.

3.2.1 Initial Trajectory Queue

According to Section 2, when the operating point gets close to the *separatrix*, it may move away from it later and return to the initial state; or it may cross the *separatrix* and go ahead for the opposite state. As such, more than one trajectories are needed to account for all potential situations, which are stored in a queue. The reachability analysis is then performed for all trajectories in the queue (Line 5-25).

3.2.2 Linear Multi-Step Integration

Within each cycle, the first step is to pick up a reachable set Z_k from the trajectory queue (Line 5-6). Newton-Raphson method is deployed to calculate the new zonotope center c_{k+1} based on c_k (Line 7). At the same time, the nonlinear dynamic function in (1) is linearized around c_{k+1}.

After c_{k+1} is obtained, the next step is the calculation of the unknown zonotope generators. Based on (8), Δx in (5) can be substituted with a series of generators. As such we further solve (5) by the Implicit Euler method with discretized time-step h (9).

$$\frac{G_{k+1} - G_k}{h} = AG_{k+1} \oplus BU_{k+1} \oplus L_{k+1} \qquad (9)$$

with

$$G_k = [g_k^{(1)}, ..., g_k^{(p)}], \quad U_k = [u_k^{(1)}, ..., u_k^{(m)}] \qquad (10)$$

where G_k and U_k are sets of generators for state variables and input current sources, respectively. The operator $\oplus$ performs set addition for generator sets. Recall that A and B are the Jacobian matrices at the nominal point c_{k+1} (4).

3.2.3 New Reachable Set Formulation

The superposition principle allows to separate the solution of (9) into two parts: the homogeneous solution with respect to the initial state when there is no input (11a); and the inhomogeneous solution accounting for the system input when the initial state is the origin (11b)(11c). Note that the linearization error is treated as input. As a result, given an initial set for current time step, three sets of solutions are computed and added together by the so-called Minkowski sum (11d).

$$G_h = (I - hA)^{-1} G_k \qquad (11a)$$
$$G_i = (I - hA)^{-1} hBU_{k+1} \qquad (11b)$$
$$G_e = (I - hA)^{-1} hL_{k+1} \qquad (11c)$$
$$G_{k+1} = G_h \oplus G_i \oplus G_e. \qquad (11d)$$

This procedure is shown in Line 10-13, where a new reachable set Z_{k+1} is obtained by combining c_{k+1} with G_{k+1}.

3.2.4 Reachable Set Refinement

Approximation of linearization error L_k in Line 8 is a critical step in each iteration cycle. Linearization error (12) accounts for nonlinearity of SRAM dynamics. Here, nominal point x^* is the zonotope center for current iteration c_k and x varies within the zonotope Z_k. L_k cannot be exactly calculated but approximated for $x \in Z_k$. Detailed approximation of L_k can be found in [15].

$$L_k = \frac{1}{2}(x - x^*)^T \left. \frac{\partial^2 f}{\partial x^2} \right|_{x=\xi} (x - x^*),$$ (12)

$$\xi \in \{x^* + \alpha(x - x^*) | 0 \le \alpha \le 1\}.$$

As for SRAM, nonlinearity of the system is rather prominent in the transition area around the *separatrix* where linearization error expands rapidly. Over-expanded new reachable sets in (11d) may be too rough to be meaningful. Thus reachable set is split into smaller ones when linearization error exceeds user-defined limit so that linearization error of each new set is in an appropriate size.

A judgement condition for set splitting is shown in (13),

$$\mathbf{IH}(hL_k) \subseteq [-\varepsilon, \varepsilon]$$ (13)

in which $\mathbf{IH}()$ is the interval hull operation which converts a zonotope to a multi-dimensional interval and ε is user-defined limit vector. After the current reachable set is divided into two subsets, along with a new trajectory being created, the reachability analysis will be repeated at current time point for the two new subsets (Line 22-24).

3.2.5 Reachability Check

Finally, after reachable set at the maximum simulation step is generated, it will be projected onto the variable plane for observation. Projections of all possible final reachable sets are collected in P_N (Line 16). Dynamic stability is guaranteed if P_N ends up in the safe region.

As a summary, the developed reachability analysis by zonotope in Algorithm 1 can be efficiently deployed to check the SRAM dynamic stability in read, write and hold operations. As reachability analysis can consider mismatches from all transistors, modeled as input current sources, it is scalable to deal with large dimensioned variation problem. Moreover, the impact of all possible mismatches is characterized in zonotope with the derived failure region by one time simulation, which avoids expensive multiple runs of Monte Carlo simulations.

4. EXPERIMENTAL RESULTS

The verification of SRAM dynamic stability by reachability analysis is implemented inside one SPICE-like simulator by MATLAB. Application of zonotopes based reachability analysis is performed by MATLAB toolbox named Multi-Parametric Toolbox (MPT)[16]. Experiment data is collected on a desktop with Intel Core i5 3.2GHz processor and 8GB memory.

The setup of SRAM circuit is as follows. Supply voltage of SRAM v_{dd} is set to $1.8V$. BSIM3v3 is used for the transistor model. Three different threshold variation ranges are tested in the experiment, including 1%, 5% and 10%. Larger variation range can be considered when high-order noise model is available. Threshold voltage variation in each transistor is introduced as a noise current source in (6), whose center value is 0 and variation is $|k\frac{W}{L}(V_{gs} - V_{th})\delta V_{th}|$ where δ is the variation range. We start reachability analysis with an initial state set of $v_1 \in [1.7, 1.8]$ and $v_2 \in [0, 0.1]$. In this section, verification of different S-RAM operations are performed, followed by a comparison with the Monte Carlo based verification of multiple samplings.

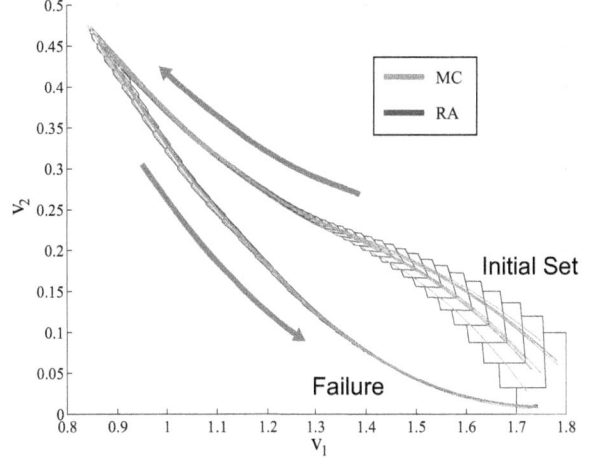

(a) Write operation fails with 50ns writing pulse.

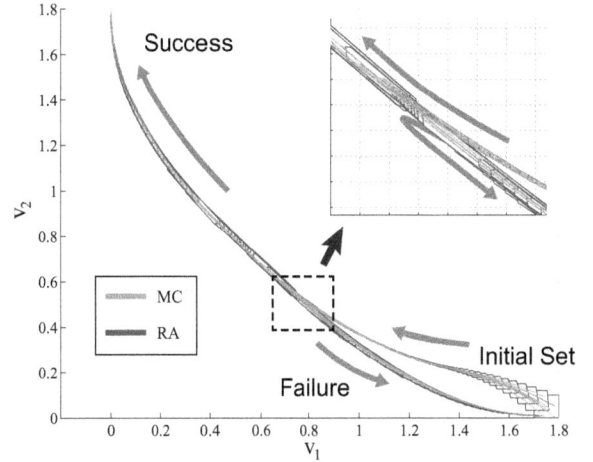

(b) Write operation fails with 60ns writing pulse.

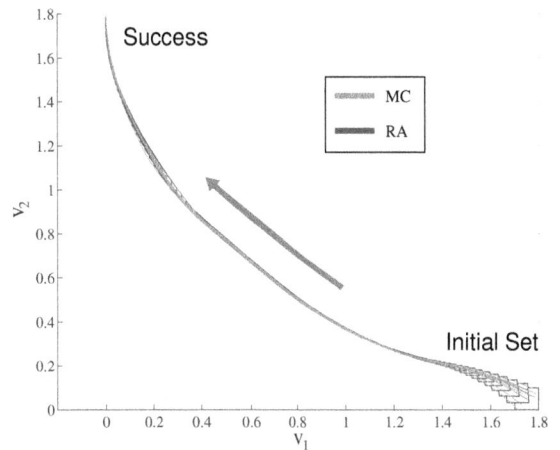

(c) Write operation succeeds with 70ns writing pulse.

Figure 6: Verification of write operation with threshold variation range of 5%.

4.1 Verification of Write Operation

First of all, the write operation is verified with consideration of threshold voltage variation. For comparison, Monte Carlo simulation is performed to verify the accuracy of reachability analysis. The duration of write signal is varied to exam SRAM behaviors under different conditions.

Verification results are shown in Fig.6 with threshold-voltage-variation range set to 5%. The trajectories simulated by Monte Carlo method are plotted in light purple and trajectories of reachability analysis are drawn in dark blue. Three different durations of write signal are tested, including 50ns, 60ns and 70ns.

In Fig.6(a) write signal lasts for 50ns. At the beginning, trajectories move towards the other corner of variable plane as data is being written into SRAM. Later, the turning point of trajectories is generated when the write signal flips to 0. Afterwards, trajectories return to initial states. As such, the data fails to be written into the SRAM, which means that write failure happens.

When the write pulse increases to 60ns in Fig.6(b), trajectories of reachable sets split around the center of the variable plane. This happens when the write signal shuts down. Some of the new trajectories move back to initial states, which means some states still fail the write operation.

Finally, when the duration increases to 70ns, all possible states finish write operation without failure. As shown in Fig.6, trajectories of Monte Carlo method remain within the reachable sets by reachability analysis with the similar accuracy. It indicates that reachability analysis can succeed in approximating the trajectory of SRAM.

4.2 Verification of Read Operation

Moreover, the read operation can be also verified by reachability analysis. Instead of observing results by variations of signal duration, we compare the verification results with different threshold voltage variations within the same duration of input signal. Duration of read signal is set to 50ns. Two different threshold variation ranges are verified, including 5% and 1%.

As shown in Fig.7, when threshold voltage variation range is 1%, all reachable sets recover back to the initial state after read operation finishes (Fig.7(a)). SRAM can function properly in this situation. But after the variation range rises to 5%, some reachable sets do not return to the initial set but deviate from others after read operation ends (Fig.7(b)). In the end, these reachable sets rest in the opposite state, which means that read failure happens. Note that the turning points in Fig.7(a) and Fig.7(b) appear when the read signal shuts down. The reason is that the read signal controls the force that pulls the trajectory towards the *separatrix*. Without the pulling strength, trajectory will move towards an equilibrium state.

The Monte Carlo trajectories are plotted in light purple and the enclosing trajectories drawn by reachability analysis are in dark blue. They match each other perfectly in other parts of the variable plane where nonlinearity is weak. There is very small error after trajectories pass through the turning point due to the linearization error in (12), which can be resolved when refinement is applied on the reachable set.

4.3 Verification of Hold Operation

Next, the hold operation is verified when an injected environmental noise to node v_2. Threshold voltage variation

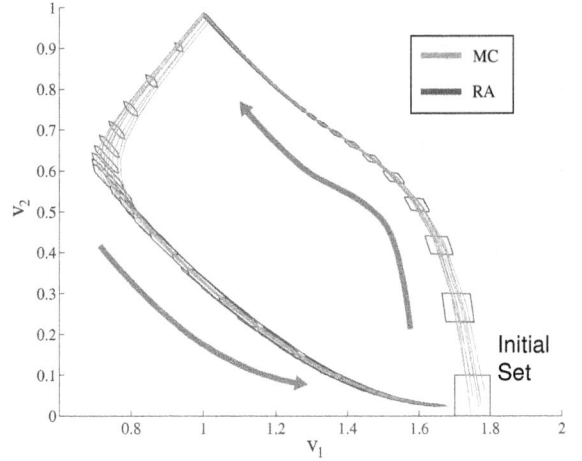

(a) Read operation succeeds with 1% threshold variation range.

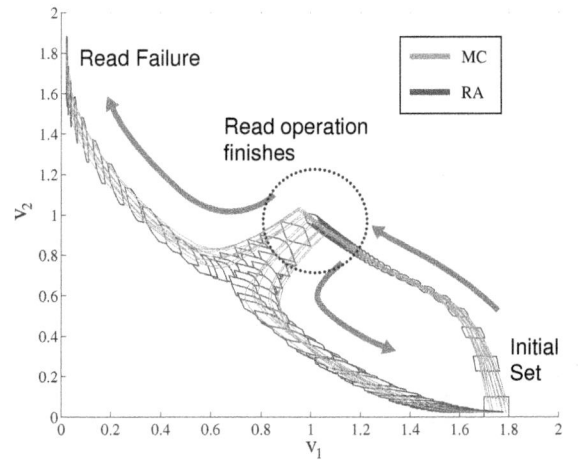

(b) Read operation fails with 5% threshold variation range.

Figure 7: **Verification of read operation with different threshold variations.**

range is set to 5% for each transistor. Duration of the injected noise is to last for 12.5ns. Two injected noises with different magnitudes are verified.

In Fig.8, the trajectories drawn by dash lines represent the case with 300uA injected noise; and other trajectories with solid lines have an injected noise of 315uA. Again, light purple trajectories are for Monte Carlo method and dark blue boxes are reachable sets by reachability analysis. When the injected noise is set to 300uA, trajectories (dash line) return to initial states after perturbation. As the noise magnitude increases to 315uA, trajectories deviate from the recovering route and converge to the opposite state. Hold failure then happens as a result. Results of Monte Carlo and reachability analysis can match with each other in the similar accuracy.

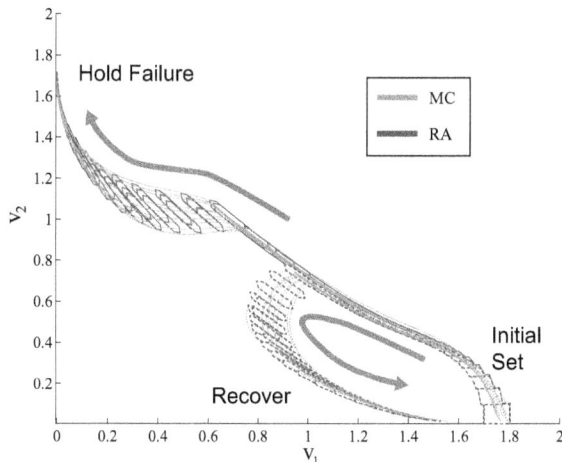

Figure 8: Verification of hold operation with different injected noise.

4.4 Comparison with Monte Carlo Method

At last, a detailed comparison between reachability analysis and Monte Carlo based verifications is made upon the write operation. For the concern of huge time consumption by Monte Carlo method, we use 500 samples which usually cost about 2 hours for a single round of verification according to our experiment. Different durations of write signal are considered as well as different threshold voltage variations in all transistors. Detailed experimental results are listed in Table 1 in which 'Pulse' refers to the duration of input signal and 'Acceleration' is the ratio of time consumption of Monte Carlo to that of reachability verification. As shown, compared with Monte Carlo, reachability analysis can achieve speedup up to more than 400 times. When write signal duration is set to 50ns (Fig.6(a)) or 70ns (Fig.6(c)), only one trajectory is generated by reachability analysis. Linearization is performed around one nominal trajectory which takes up most of the simulation time. As signal duration is set to 60ns, reachable sets are split into different parts (Fig.6(b)) and up to five trajectories are generated. The acceleration rate is roughly equivalent to the ratio of the number of Monte Carlo samples to the number of trajectories by reachability analysis. For all experiment cases listed in the table, the reachability analysis can achieve the similar accuracy as Monte Carlo method to report the failure region.

Table 1: Time consumption of SRAM verification.

Pulse (ns)	Threshold Variation	Reachability Analysis(s)	Monte Carlo(s)	Acceleration
50	1%	12.71	5635.57	443.38×
	5%	13.13	5817.71	443.01×
	10%	12.63	6078.68	481.24×
60	1%	52.70	6224.09	118.09×
	5%	52.58	6535.35	124.29×
	10%	52.68	6387.28	121.25×
70	1%	13.72	5931.76	432.32×
	5%	14.43	6245.45	432.73×
	10%	13.21	6348.54	480.45×

5. CONCLUSIONS

In this paper, we are the first to develop the reachability analysis for the verification of SRAM dynamic stability

in the presence of process variations from all transistors. By modeling variations as uncertain input currents added to the input range, the zonotope based reachability analysis is deployed to provide the system performance boundary for the estimation of SRAM dynamic stability region. As demonstrated by experimental results, when taking into account on threshold voltage variations from all transistors of 6T-SRAM memory cell, reachability analysis based verification can provide accurate trajectories with failure boundary as Monte Carlo method, but with speedup up to 481× achieved. With further implementation, stochastic reachability analysis can be used on the verification of statistical metrics including failure probability and yield rate.

6. REFERENCES

[1] E. Seevinck, F. J. List, and J. Lohstroh. Static-noise margin analysis of MOS SRAM cells. *IEEE Journal of Solid-State Circuits (JSSC)*, Jan. 1987.

[2] E. Grossar, M. Stucchi, K. Maex, and W. Dehaene. Read stability and write-ability analysis of SRAM cells for nanometer technologies. *IEEE Journal of Solid-State Circuits (JSSC)*, Nov. 2006.

[3] S. O. Toh, Z. Guo, and B. Nikolić. Dynamic SRAM stability characterization in 45nm CMOS. In *IEEE Symposium on VLSI Circuits (VLSIC)*, Jun. 2010.

[4] H. Yu and S. X.-D. Tan. Recent advance in computational prototyping for analysis of high-performance analog/rf ics. In *IEEE International Conf. on ASIC (ASICON)*, Oct. 2009.

[5] K. Agarwal and S. Nassif. Statistical analysis of SRAM cell stability. In *ACM/IEEE Design Automation Conference (DAC)*, 2006.

[6] D.E. Khalil, M. Khellah, N.S. Kim, Y. Ismail, T. Karnik, and V.K. De. Accurate estimation of SRAM dynamic stability. *IEEE Transactions on Very Large Scale Integration (VLSI) Systems*, Dec. 2008.

[7] F. Gong, X. Liu, H. Yu, S. X.-D. Tan, J. Ren, and L. He. A fast non-monte-carlo yield analysis and optimization by stochastic orthogonal polynomials. *ACM Transactions on Design Automation of Electronic Systems (TODAES)*, Jan. 2012.

[8] S. Srivastava and J. Roychowdhury. Rapid estimation of the probability of SRAM failure due to mos threshold variations. In *IEEE Custom Integrated Circuits Conference (CICC)*, Sep. 2007.

[9] W. Dong, P. Li, and G. M. Huang. SRAM dynamic stability: theory, variability and analysis. In *IEEE/ACM International Conference on Computer-Aided Design (ICCAD)*, 2008.

[10] A. Girard. Reachability of uncertain linear systems using zonotopes. In *Int. conf. on Hybrid Systems: computation and control (HSCC)*. Springer, 2005.

[11] M. Althoff. Reachability analysis and its application to the safety assessment of autonomous cars. In *PhD Dissertation, TUM*, 2010.

[12] M. Althoff and et.al. Formal verification of phase-locked loops using reachability analysis and continuization. In *IEEE Int. Conf. of Computer-aided-design (ICCAD)*, 2011.

[13] G. Frehse, B.H. Krogh, and R.A. Rutenbar. Verifying analog oscillator circuits using forward/backward abstraction refinement. In *IEEE Design, Automation and Test in Europe (DATE)*, Mar. 2006.

[14] Y. Song, H. Fu, H. Yu, and G. Shi. Stable backward reachability correction for pll verification with consideration of environmental noise induced jitter. In *IEEE/ACM Asia and South Pacific Design Automation Conference (ASPDAC)*, Jan. 2013.

[15] M. Althoff, O. Stursberg, and M. Buss. Reachability analysis of nonlinear systems with uncertain parameters using conservative linearization. In *IEEE Conf. on Decision and Ctrl.*, 2008.

[16] M. Kvasnica, P. Grieder, and M. Baotić. Multi-Parametric Toolbox (MPT). MPT 2.6.3 is available at http://control.ee.ethz.ch/~mpt/.

PushPull: Short Path Padding for Timing Error Resilient Circuits

Yu-Ming Yang
Dept. of Electronics Eng. &
Inst. of Electronics
National Chiao Tung University
yuming.yyang@gmail.com

Iris Hui-Ru Jiang
Dept. of Electronics Eng. &
Inst. of Electronics
National Chiao Tung University
huiru.jiang@gmail.com

Sung-Ting Ho
Dept. of Electronics Eng. &
Inst. of Electronics
National Chiao Tung University
gsy2i7y14@hotmail.com

ABSTRACT

Modern IC designs are exposed to a wide range of dynamic variations. Traditionally, a conservative timing guardband is required to guarantee correct operations under the worst-case variation, thus leading to performance degradation. To remove the guardband, resilient circuits are proposed. However, the short path padding (hold time fixing) problem in resilient circuits is severer than conventional IC design. Therefore, in this paper, we focus on the short path padding problem to enable the timing error detection and correction mechanism of resilient circuits. Unlike recent prior work adopts greedy heuristics with a local view, we determine the padding values and locations with a global view. Moreover, we propose coarse-grained and fine-grained padding allocation methods to further achieve the derived padding values at physical implementation. Experimental results show that our method is promising to validate timing error resilient circuits.

Categories and Subject Descriptors

B.7.2 [**Integrated Circuits**]: Design Aids–*Placement and routing*

General Terms

Algorithms, Performance, Design, Reliability.

Keywords

Resilient circuits; dynamic variations; delay padding; hold time fixing; timing analysis; engineering change order

1. INTRODUCTION

Due to a wide range of dynamic variations, e.g., supply voltage droops, process variations, temperature fluctuations, soft errors, and transistor aging degradation, the timing characterization is extremely difficult in modern IC designs. Therefore, in conventional design, designers conservatively (pessimistically) reserve a timing guardband to ensure correct functionality even under the (rare) worst-case circumstance. However, this reserved guardband may severely degrade circuit performance, i.e., limit the clock frequency.

Recently, several resilient circuits have been proposed to eliminate the guardband by error detection and correction

This work was partially supported by Synopsys and NSC of Taiwan under Grant No's. NSC 101-2220-E-009-044 and NSC 101-2628-E-009-012-MY2.

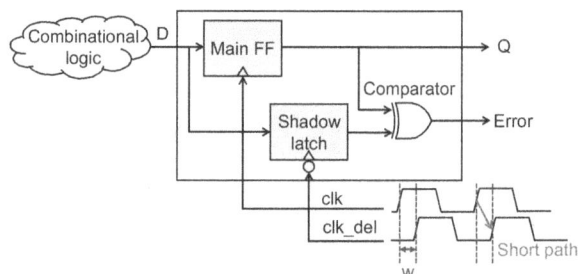

Figure 1. The error-detection part of a Razor flip-flop proposed in [1].

[1][2][3][4][5]. For example, Figure 1 illustrates one error-detection circuit, the Razor flip-flop proposed in [1]. One extra storage element, the shadow latch, is augmented to sample the output of a combinational logic by a delayed clock. The main flip-flop and shadow latch outputs are compared to generate a timing-error signal. If the output of the combinational logic transitions late, a timing error (discrepancy) is detected. Error correction is then performed through instruction replay.

However, these resilient circuits require a significant hold time margin for short paths. The resilient circuit may detect a false timing error if the result of the next computation is propagated through a short path and sampled by the delayed clock. To avoid false error detection, short paths should exceed the error detection window, i.e., the phase difference between the delayed clock and the normal clock (w in Figure 1). The error detection window induces an extra hold time margin requirement. This issue also exists in new forms of resilient circuits [2][3][4][5]. In fact, short path padding (hold time fixing) is an inevitable and essential task in conventional IC design. A circuit with hold violations cannot operate correctly. This short path issue becomes even more challenging in resilient circuits due to this extra hold time margin. In order to validate the error detection and correction mechanism of resilient circuits, we focus on short path padding in this paper.

To resolve this padding problem, prior works typically insert buffers to lengthen short paths, e.g., [6][7][8][9][10][11][12][13]. Among them, conventional delay padding is combined with clock skew scheduling to minimize the clock period at the logic resynthesis stage, e.g., [7][8][9]. Their goal is to determine the padding delay for each path rather than to decide where to insert delay. In contrast, several short path padding methods determine the positions to insert delay [6][10][11][12][13]. Shenoy *et al.* solve this problem by linear programming [6]. Lin and Zhou transform this problem into a network flow problem (but resulting in larger padding delay) in [10]. Liu *et al.* reveal that linear programming is time-consuming and not applicable to large-scale circuits by empirical data in [12]. Hence, recently, two greedy heuristics are proposed in [11][12][13]. One greedy rule is to pad the gate with the largest setup slack, trying not to hurt the longest path delay. The other is to pad at the gate passed by most hold violating paths, to reduce the total padding delay.

However, we found that these greedy heuristics based on local views may not pad short paths well. Figure 2(a) gives an input design, where gates g_1, g_2, g_3 incur hold violations. After iteratively padding delay on the gate either with the largest setup slack (see Figure 2(b)) or with most hold violating paths (see Figure 2(c)), we still have an unresolved hold violation at gate g_2. In fact, all hold violations can be cleaned as shown in Figure 2(d). It can be seen that padding with local views may consume all setup slacks thus leaving some hold violations unfixed (see Figure 2(b)(c)). Moreover, even if we find an optimal padding solution (like [6]), we may still fail at physical implementation because the available buffer delays are discrete. For example, if one buffer offers either 0.15-unit or 0.25-unit delay, we cannot fulfill the padding task on gate g_2 in Figure 2(d).

Based on the above observations, we develop a two-stage short path padding algorithm to overcome these difficulties: Stage 1 tries to minimize the total padding delay with a global view and determines padding locations, while Stage 2 allocates load/buffers to attain the padding at post-layout to handle the discrete cell library. Our features include:

- Finding the padding values with a global view: The greedy heuristics proposed by prior works may fail to fix all hold violations due to local views. Instead, in Stage 1, we compute the padding flexibility of the fanout cone of each gate. With this global view, we determine the padding value for each gate accordingly.

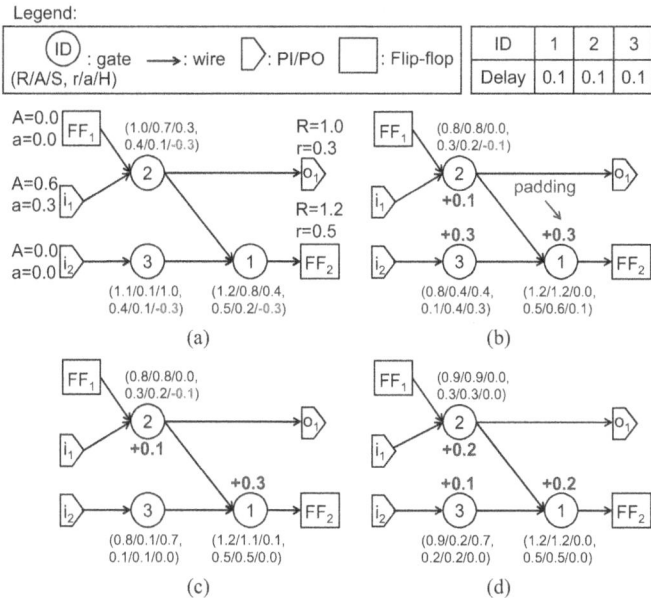

Figure 2. Short path padding. (a) Input design. (R/A/S, r/a/H) indicates the setup required time (R), setup arrival time (A), setup slack (S), hold required time (r), hold arrival time (a), and hold slack (H) of a gate. (b) Padding from gates with largest setup slacks: g_3 (+0.3) → g_1 (+0.3) → g_2 (+0.1), total padding delay = +0.7, unfixed. (c) Padding from gates with most hold time violating paths: g_1 (+0.3) → g_2 (+0.1), total padding delay = +0.4, unfixed. (d) The optimal short path padding: g_1 (+0.2), g_2 (+0.2), and g_3 (+0.1), total padding delay = +0.5.

- Coarse-grained and fine-grained delay padding at post-layout: Because the available resource of padding is uncertain at early stages, unlike prior work determines the padding values at logic resynthesis, we further realize delay padding at the post-layout stage. Since the amount of delay offered by a cell library is discrete, to achieve the delay padding determined in Stage 1, we perform coarse-grained padding followed by fine-grained padding. Coarse-grained delay padding is done using spare cells.[1] Fine-grained delay padding is done by dummy metal insertion since dummy metal offers an abundant resource of capacitance [14] and can be tuned.

Experiments are conducted on the IWLS 2005 benchmark circuits [19] through the resilient circuit design flow. Our results show that we can clean all hold violations with the shortest runtime, while prior work may either fail to clean all violations or incur long runtime. In addition, our coarse-grained and fine-grained delay padding methods can successfully achieve the derived padding values at post-layout.

The remainder of this paper is organized as follows. Section 2 briefly introduces the resilient circuit design flow, describes the timing model and gives the problem formulation. Section 3 presents the overview of our short path padding framework. Section 4 derives setup/hold slack properties and details padding value determination. Section 5 presents load/buffer allocation. Section 6 shows experimental results. Finally, Section 7 concludes this paper.

2. PRELIMINARIES AND PROBLEM FORMULATION

In this Section, we briefly introduce the design flow for resilient circuits, describe the timing model, and give the problem formulation.

2.1 The Resilient Circuit Design Flow

Figure 3 shows a sample design flow to integrate timing error resilient circuits into a design. After logic synthesis and timing analysis based on a conservative clock period (considering a timing guardband), the target clock period and the error detection window w are determined. S_{th} (respectively, H_{th}) means the ratio of the target clock period (respectively, the error detection window) over the conservative clock period. The timing suspicious flip-flops, whose longest path delays exceed the target clock period, will be replaced by resilient circuits. Before the replacement, the design is resynthesized where the suspicious flip-flops are assigned with an extra hold time margin to cover the error detection window w. After the replacement, placement and routing are applied. Because of the significant hold time margin, hold violations may still exist in a placed and routed resilient design. Finally, short path padding is performed.

As mentioned in Section 1, the short path issue is not only inevitable for conventional IC design but also more challenging in resilient circuits due to this extra hold time margin. Therefore, short path padding is a must to validate timing error resilient circuits.

2.2 Timing Model

The cell timing model used in this paper is based on Synopsys' Liberty library [15]. The calibrated delay values of each library cell are stored in lookup tables and indexed by its input slew and

---

[1] A design is usually sprinkled with redundant (spare) cells at placement. Incremental design changes can be done by rewiring spare cells.

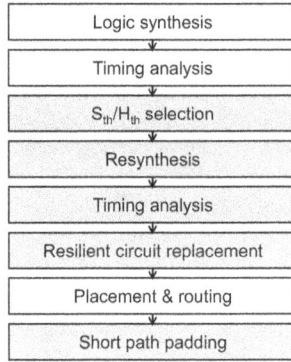

Figure 3. The design flow of resilient circuits.

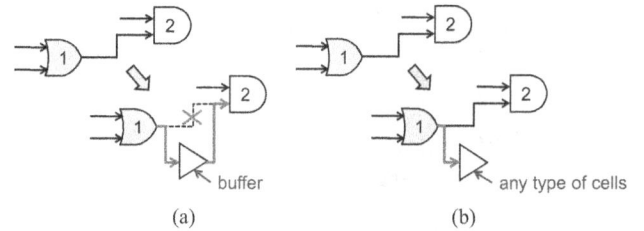

Figure 4. Delay padding. (a) Buffer insertion (padding the wire between gates g_1 and g_2). (b) Extra load hook-up (padding gate g_1).

output capacitance. The wire delay of each net is lumped into the delay of its driving gate. The output capacitance of a gate includes wire loading, the input capacitance of its fanout gates, and its output pin capacitance. In addition, the output capacitance of each cell is bounded by the maximum load capacitance defined in the cell library. In [16], Chen *et al.* observe loading dominance phenomenon: The change on the gate delay is dominated by output capacitance. Later, experimental results also show that the impact of input slew on the gate delay is quite small.

2.3 Problem Formulation

In order to validate the error detection and correction mechanism of resilient circuits, we focus on the short path padding problem which is formulated as follows.

The Short Path Padding Problem: Given a placed and routed resilient design, the cell library, spare cells, dummy metal information, the target clock period and the error detection window, our goal is to pad short paths such that the padding overhead is minimized and setup/hold timing constraints are satisfied.

Since the reported timing is somewhat inaccurate and the available resource for padding is uncertain at early stages, we perform short path padding (hold time fixing) at the post-layout stage. To lengthen short paths, we may insert buffers (see Figure 4(a)) or introduce extra load capacitance (see Figure 4(b)). The inserted delay can be provided by cells and metal. A design is usually sprinkled with spare cells (redundant cells) at placement. In addition, dummy metal offers an abundant resource of capacitance [14] and can be tuned. Hence, padding at the post-layout stage can then be done by rewiring spare cells and dummy metal. Because of loading dominance, the amount of delay increment and the corresponding amount of load/buffers inserted can be directly converted to each other by table lookup. Later, experimental results show that the impact of input slew on the padding delay capacitance conversion is quite small.

3. PUSHPULL: THE DELAY PADDING FRAMEWORK

In this Section, we present the overview of our short path padding framework, PushPull, as shown in Figure 5. Our framework consists of two stages: padding value determination and load/buffer allocation. Finally, timing analysis is applied to verify our framework.

In the padding value determination stage, we first collect the available padding resources including spare cells and dummy metal. Second, we calculate the padding flexibility of the whole fanout cone of each gate and check the feasibility of padding.

Third, with this global view, the padding delay of each gate is decided accordingly. The fanout padding flexibility calculation and padding value decision steps are repeated until all hold violations are resolved or no more violations can be eliminated. Then, we further reduce the total padding delay on gates and resolve unfixed hold violations by padding wires.

In the load/buffer allocation stage, the padding values on gates (respectively, wires) are realized by introducing extra load (respectively, inserting buffers). To achieve the assigned padding values, we propose coarse-grained delay padding using spare cells followed by fine-grained delay padding using dummy metal. At the coarse-grained padding, we adequately select spare cells for each padding gate/wire. If the selected spare cells cannot match the required padding delay of some gate/wire, the remaining padding delay is fixed by dummy metal insertion during fine-grained padding.

Moreover, by adjusting the target clock period and the error detection window, our framework can be applied to ECO hold time fixing for general designs.

4. PADDING VALUE DETERMINATION

We propose a padding value determination algorithm to determine the padding values and locations with a global view in this Section.

Basically, the more total padding delay, the more total padding overhead. Hence, we first target to minimize the total padding delay and then convert the padding delay of each gate/wire to padding load/buffers. However, challenges are twofold: One is to find good locations to pad delay; the other is not to hurt the setup time.

Conceptually, padding on gates close to primary inputs can easily satisfy the timing constraints, but may increase the total padding values. Padding on gates shared by many short paths can lower total padding values, but may violate the timing constraints. As shown in Figure 2, if we individually pad gates g_2 and g_3 with 0.3-

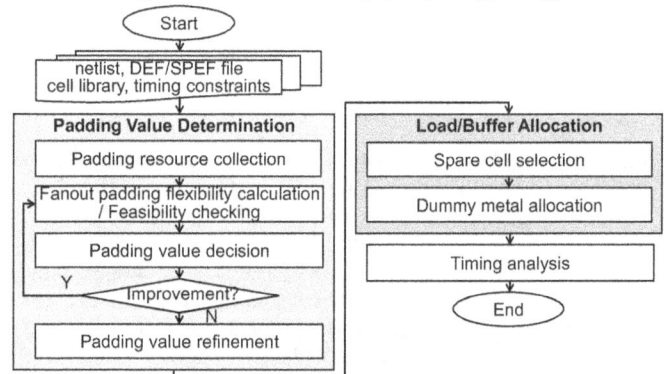

Figure 5. The overview of our short delay padding framework.

unit delay, the timing constraints are satisfied, but the total padding value is somewhat large (+0.6). If we pad 0.3-unit delay on gate g_1 first, the short path through g_2 to primary output o_1 is unresolved (see Figure 2(b)(c)). Thus, to tackle these challenges, we shall determine the padding values and locations with a global view.

4.1 Padding Resource Collection

Since our short path padding method is applied at the post-layout stage, first of all, the available padding resource is collected. The available resource to pad each gate (respectively, wire) includes the spare cells and dummy metal located within the bounding box of its fanout net (respectively, the investigated wire). We have the following definition to constrain the maximum padding capacitance.

Definition 1: The *maximum padding capacitance* $C_{max}(i)$ of gate g_i is the minimum of the maximum output capacitance defined in the cell library and its available padding resource. $C_{max}(i)$ is 0 for a primary output or a flip-flop input.

The *maximum padding capacitance* $C_{max}(i,j)$ of the wire between gates g_i and g_j is defined similarly. $C_{max}(i)$ and $C_{max}(i,j)$ give upper bounds but still preserve flexibilities to set padding values. In some cases, the bounding boxes of padding gates/wires heavily overlap or the padding value cannot be fulfilled at Stage 2, $C_{max}(i)$ and $C_{max}(i,j)$ can be adjusted.

4.2 Fanout Padding Flexibility Calculation and Feasibility Checking

To determine the padding values and locations with a global view, we first calculate the padding flexibility of the whole fanout cone of each hold violating gate.

A design is represented by a directed graph $K = (G, E)$, where each node $g_i \in G$ represents a gate associated with gate delay $D(i)$, and each edge $e(i, j) \in E$ represents the wire between gates $g_i, g_j \in G$. In the following, we derive the slack properties used in this paper. The *setup* constraints indicate the timing requirement on *long* paths, while the *hold* constraints indicate that for *short* paths.

Definition 2: In [17], the *setup arrival time* $A(i)$ of the output signal of node $g_i \in G$ is computed as

$$A(i) = \max_j\{A(j)|e(j,i) \in E\} + D(i), \qquad (1)$$

while the *setup required time* $R(i)$ of g_i is computed as

$$R(i) = \min_k\{R(i,k)|R(i,k) = R(k) - D(k), e(i,k) \in E\}. \qquad (2)$$

Definition 3: In [17], the *hold arrival time* $a(i)$ of the output signal of node $g_i \in G$ is computed as

$$a(i) = \min_j\{a(j)|e(j,i) \in E\} + D(i), \qquad (3)$$

while the *hold required time* $r(i)$ of node g_i is computed as

$$r(i) = \max_k\{r(i,k)|r(i,k) = r(k) - D(k), e(i,k) \in E\}. \qquad (4)$$

Definition 4: In [17], the *setup edge slack* $S(i, j)$ is the slack of edge $e(i, j) \in E$ contributed from node g_j back to node g_i,

$$S(i, j) = R(i, j) - A(i). \qquad (5)$$

The *setup node slack* $S(i)$ of node $g_i \in G$ is the slack of node g_i,

$$S(i) = \min_j\{S(i,j)|e(i,j) \in E\} = R(i) - A(i). \qquad (6)$$

Definition 5: In [17], The *hold edge slack* $H(i, j)$ is the slack of edge $e(i, j) \in E$ contributed from node g_j back to node g_i,

$$H(i, j) = a(i) - r(i, j). \qquad (7)$$

The *hold node slack* $H(i)$ of node $g_i \in G$ is the slack of node g_i,

$$H(i) = \min_j\{H(i,j)|e(i,j) \in E\} = a(i) - r(i) \qquad (8)$$

For example, as shown in Figure 2(a), $H(2, 1) = 0.1 - 0.4 = -0.3$, $H(2, o_1) = 0.1 - 0.3 = -0.2$, and $H(2) = -0.3$.

Definition 6: The *maximum padding delay* $P_{max}(i)$ of gate g_i is the padding delay converted from $C_{max}(i)$. $P_{max}(i)$ is 0 for a primary output or a flip-flop input. The *safe padding value* $P_{saf}(i)$ of gate g_i is computed as

$$P_{saf}(i) = \min\{S(i), |\min\{0, H(i)\}|, P_{max}(i)\}. \qquad (9)$$

Lemma 1: The setup constraint is satisfied when the delay of a node g_i on a short path is increased by t, $t \leq P_{saf}(i)$.

We define the fanout padding flexibility $P_F(i)$ for each gate to reflect the maximum padding value allowed on its whole fanout cone. For a hold satisfying gate or a primary output (a flip-flop input is considered as a pseudo primary output), the flexibility is zero. For a hold violating gate g_i, $P_F(i)$ is the difference between g_i's current hold slack and the minimum updated hold edge slack over all fanout edges if each fanout is padded with the maximum allowable value.

Definition 7: The fanout padding flexibility $P_F(i)$ of node $g_i \in G$ is computed as

$$P_F(i) = \begin{cases} 0, & g_i \in PO \text{ or } H(i) \geq 0; \\ \min\{0, \min_{e(i,j) \in E}\{H'(i,j)\} - H(i)\}, & otherwise, \end{cases} \qquad (10)$$

where $H'(i, j) = H(i, j) + P_F(j) + P_{saf}(j)$.

$H'(i)$ and $S'(i)$ represent the updated slacks if g_i's fanout cone is padded with the maximum allowable delay. $P_{saf}(i)$ is dynamically updated accordingly.

$$H'(i) = \min_j\{H'(i,j)|e(i,j) \in E\}. \qquad (11)$$
$$S'(i) = \min_j\{S(i,j) - (S'(j) - S(j)) - P_{saf}(j)|e(i,j) \in E\}. \qquad (12)$$

By definition, the fanout padding flexibility is thus calculated from primary outputs (and flip-flop inputs) toward primary inputs (and flip-flop outputs). Consider the case shown in Figure 2(a). According to Equations (10)(11)(12), we have

$P_F(o_1) = 0.0$, $P_F(FF_2) = 0.0$;
$P_F(1) = \min\{0, (-0.3 + 0.0 + 0.0)\} - (-0.3) = 0.0$,
$H'(1) = -0.3$, $S'(1) = 0.4$;
$P_F(2) = \min\{0, (-0.2+0.0+0.0), (-0.3+0.0+0.3)\} - (-0.3) = 0.1$,
$H'(2) = \min\{(-0.3+0.0+0.3), (-0.2+0.0+0.0)\} = -0.2$,
$S'(2) = \min\{(0.4 - 0.0 - 0.3), (0.3 - 0.0 - 0.0)\} = 0.1$;
$P_F(3) = \min\{(-0.3+0.0+0.3), 0.0\} - (-0.3) = 0.3$,
$H'(3) = \min\{(-0.3+0.0+0.3)\} = 0.0$;
$S'(3) = \min\{(1.0 - 0.0 - 0.3)\} = 0.7$.

Moreover, we have the following lemma to check the padding feasibility.

Lemma 2: If $|\min\{0, H(i)\}| > P_F(i)$, $g_i \in PI$, the hold violations cannot be resolved by padding on gates. If $S(i, j) < |\min\{0, H(i,j)\}|$, $e(i, j) \in E$, the hold violations cannot be resolved by padding on wires.

4.3 Padding Value Decision

After the fanout padding flexibility is calculated with a global view in Section 4.2, the padding value is decided accordingly. The padding value of each hold violating gate is derived in the topological order [18]. For each hold violating gate, the fanout padding flexibility can be considered as the maximum allowable delay padded on its fanout cone. Then, the hold violating gate only needs to be padded to fix the remaining negative hold slack, i.e., the difference between the safe padding value and the fanout padding flexibility.

$$P(i) = \max\{P_{saf}(i) - P_F(i), 0\},\quad (13)$$

where $P_{saf}(i)$ represents the safe padding value after gate g_i's fanin gates are padded. When the padding value of a gate is decided, the increased delay affects the arrival time of its fanout gates. The fanout edge slack of the padding gate should be updated accordingly.

$$S(i,j) = R(i,j) - A(i) - P(i).\quad (14)$$
$$H(i,j) = P(i) + a(i) - r(i,j).\quad (15)$$

After updating the fanout edge slacks of each padding gate, the setup and hold node slacks of its fanout gates are also updated by Equations (6) and (8).

Figure 6(a) gives an example of padding value decision. Assume $P_{\max}(2) = 0.5$, $P_{\max}(3) = 0.4$, and $P_{\max}(1) = 0.4$, respectively. Based on the fanout padding flexibilities, we have

$P_{saf}(2) = \min\{0.3, |-0.3|, 0.5\} = 0.3, P(2) = 0.3 - 0.1 = 0.2,$
$S(2,1) = 1.1 - 0.7 - 0.2 = 0.2, H(2,1) = 0.2 + 0.1 - 0.4 = -0.1;$
$P_{saf}(3) = \min\{0.3, |-0.3|, 0.4\} = 0.3, P(3) = 0.3 - 0.3 = 0.0,$
$S(3,1) = 1.1 - 0.1 - 0.0 = 1.0, H(3,1) = 0.0 + 0.1 - 0.4 = -0.3;$
$S(1) = \min\{1.0, 0.2\} = 0.2, H(1) = \min\{-0.3, -0.1\} = -0.3,$
$P_{saf}(1) = \min\{0.2, |-0.3|, 0.4\} = 0.2, P(1) = 0.2 - 0.0 = 0.2.$

After the above padding value decision, the short path from g_3 to g_1 still has a negative hold slack, -0.1, because of the over-estimated fanout padding flexibility. This short path can be resolved by applying another iteration of fanout padding value calculation plus padding value decision. The fanout padding flexibility calculation step and the padding value decision step are repeated until all hold violations are resolved or no more violations can be eliminated. With the iterative manner, this procedure can determine the padding values and locations with a global view. As shown in Figure 6(b), all short paths are resolved, and the result is same as the optimal solution (see Figure 2(d)).

4.4 Padding Value Refinement

Now, we further reduce the total padding delay on gates and resolve unfixed hold violations by padding wires.

In the padding value decision step, the padding locations are decided as close to primary outputs as possible. For a circuit with forked short paths, the total padding value is increased if the padding location is not determined on the gate where two or more short paths fork. Figure 7(a) gives an example, gate g_4 has forked paths. After determining the padding values by our padding value decision method (see Section 4.3), the padding values and locations are indicated beside each gate, and the total padding

delay is 0.5. In fact, the total padding delay can be further reduced to 0.4 by changing the padding values and locations as shown in Figure 7(b).

At refinement, we further reduce the padding values by pushing the padding values toward the gates where two or more short paths fork. To accomplish this task, we define the reverse padding value, the added safe padding value, and the refined padding value as follows.

Definition 8: The *reverse padding value* $P_{rev}(i)$ of gate g_i is computed as

$$P_{rev}(i) = \begin{cases} P(i), if\ g_i\ has\ only\ one\ hold\ violating\ fanin; \\ 0, otherwise. \end{cases}\quad (16)$$

The reverse padding value of each gate g_i is to record how much padding can be propagated backward to its fanin gate. To avoid propagating padding values to joined paths, we consider the case that g_i has only one fanin with a hold violation. A fanin of g_i has a hold violation if the hold edge slack is smaller than the padding value of g_i. Furthermore, the padding value can be fully propagated in this case. The refined padding value of gate g_i is constrained by its setup slack and its maximum padding delay $P_{\max}(i)$.

Definition 9: The *added safe padding value* $P_{add}(i)$ of gate g_i is computed as

$$P_{add}(i) = \min\{S(i), P_{max}(i) - P(i)\}.\quad (17)$$

Definition 10: The *refined padding value* $P_{ref}(i)$ of gate g_i is computed as

$$P_{ref}(i) = P(i) + \min\{P_{add}(i), \min_j\{P_{rev}(j)|H(i,j) < P(i)\}\}.\quad (18)$$

Based on the above definitions, the refined padding values are calculated in the reverse topological order, and thus the total padding delay can be reduced.

Sometimes, hold violations cannot be fully cleaned by padding on gates (extra load hook-up) due to insufficient setup slacks or maximum output capacitance constraints. In this case, we may further apply padding on wires (buffer insertion) after the above refinement.

Definition 11: The *wire padding value* $P(i,j)$ of $e(i,j)$ is

$$P(i,j) = \min\{S(i,j), |\min\{0, H(i,j)\}|\}.\quad (19)$$

The wire padding value is determined in the topological order. According to the timing library, the final padding delay of each gate/wire is then converted to an amount of padding load/buffers.

5. LOAD/BUFFER ALLOCATION

In this Section, to achieve the padding delay of each padding gate/wire determined by the first stage, we propose coarse-grained

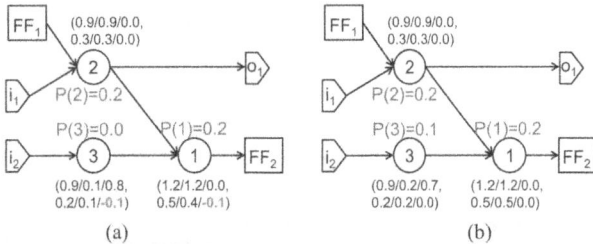

Figure 6. The padding values of gates in Figure 2. (a) The padding value $P(i)$ of node $g_i \in G$ after the first iteration of padding value decision: g_3 (+0.0) → g_2 (+0.2) → g_1 (+0.2). (b) The padding value $P(i)$ of node $g_i \in G$ after the second iteration of padding value decision: g_3 (+0.1).

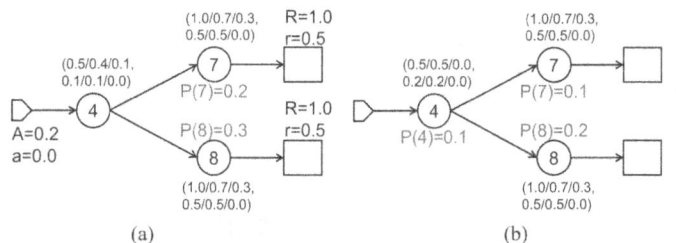

Figure 7. Padding refinement. (a) The padding delay after padding value decision: g_7 (+0.2), g_8 (+0.3). Total padding delay is 0.5. (b) The ideal padding value: g_4 (+0.1), g_7 (+0.1), g_8 (+0.2). Total padding delay is 0.4.

and fine-grained padding allocation methods. Since the available cell capacitances/delays are discrete for a given cell library, spare cells may not match the required padding load/buffer for a padding gate/wire. Therefore, the coarse-grained padding is done by spare cells. Because dummy metal accommodates an abundant resource of capacitance [14] and can be tuned, the remaining padding capacitance is resorted to fine-grained padding which is done by dummy metal insertion. For example, if spare cells offer either 0.15-unit or 0.25-unit delay, a 0.2-unit padding delay can be done by a spare cell of 0.15-unit delay plus a dummy metal of 0.05-unit delay.

5.1 Spare Cell Selection

At the coarse-grained padding step, we first generate spare cell candidates for each padding gate/wire. For each padding gate (respectively, wire), we extract the available spare cells located within the bounding box of its fanout net (respectively, the investigated wire). As shown in Figure 8(a), the available spare cells of padding gate g_2 are s_1, s_2 and s_3, while that of the wire between gates g_2 and g_3 is s_2. We then decide how to pad the gate/wire by these spare cells. The amount of delay offered by the spare cells and related rewiring should be as close to the determined padding delay as possible, but it should not exceed the determined padding value to avoid setup violations.

We reduce the task of finding suitable spare cell candidates for a single padding gate/wire to the subset sum problem, which can optimally be solved by dynamic programming [18]. The step size of dynamic programming is chosen to keep a balance between precision and efficiency. To facilitate dynamic programming, the increased delay contributed by each spare cell plus the corresponding rewiring is rounded up according to the step size. For example, as shown in Figure 8(a), the assigned padding delay of a padding gate g_2 is 0.25, and its available spare cells are s_1, s_2, and s_3. Figure 8(b) lists the dynamic programming table for the subset sum problem, where the step size used here is 0.05. Spare cells s_2 and s_3 are recorded as g_2's spare cell candidates. Similarly, the spare cell candidates for a padding wire can be extracted.

However, the subset sum solutions for different padding gates/wires may compete for the same spare cell, as shown in Figure 9(a). To deal with the resource competition problem, we record multiple subset sum solutions within a user-defined tolerance as spare cell candidates. For example, $\{s_1\}$ and $\{s_2, s_3\}$ in Figure 8(b) are both recorded when the tolerance is 0.05.

As mentioned above, there is a resource competition problem among spare cells. Several sets of spare cell candidates are recorded for each padding gate/wire. Here, we determine a feasible assignment so that each spare cell is assigned to at most one padding gate/wire. The spare cell selection problem is NP-hard, which can be reduced from the set packing problem [18]. To do it efficiently, first, each padding gate/wire is assigned to its best subset sum solution. If there are conflicts, the conflicted padding gates/wires and their multiple subset sum solutions are extracted. These conflicted padding gates/wires are sorted in ascending order of the number of their recorded subset sum solutions. Each gate is then assigned to their best and available set of spare cell candidates in the sorted order. (see Figure 9(b))

5.2 Dummy Metal Allocation

As mentioned in Section 5.1, at the coarse-grained padding step, we select spare cells for each padding gate/wire. If the selected spare cells cannot match the required padding delay, the remaining padding is converted to an amount of capacitance, fixed by dummy metal insertion during the fine-grained padding step.

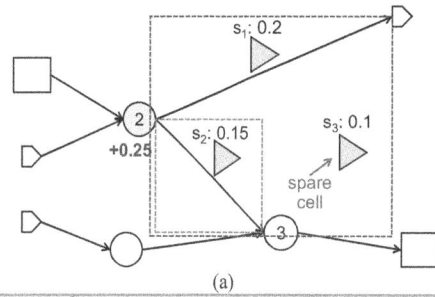

(a)

		Padding value					
		0.00	0.05	0.10	0.15	0.20	0.25
Spare cells	$\varnothing$	0.00 $\varnothing$	0.00 $\varnothing$	0.00 $\varnothing$	0.00 $\varnothing$	0.00 $\varnothing$	0.00 $\varnothing$
	$\{s_1\}$	0.00 $\varnothing$	0.00 $\varnothing$	0.00 $\varnothing$	0.00 $\varnothing$	0.20 $\{s_1\}$	0.20 $\{s_1\}$
	$\{s_1, s_2\}$	0.00 $\varnothing$	0.0 $\varnothing$	0.00 $\varnothing$	0.15 $\{s_2\}$	0.20 $\{s_1\}$	0.20 $\{s_1\}$
	$\{s_1, s_2, s_3\}$	0.00 $\varnothing$	0.0 $\varnothing$	0.10 $\{s_3\}$	0.15 $\{s_2\}$	0.20 $\{s_1\}$	0.25 $\{s_2, s_3\}$

(b)

Figure 8. Space cell candidates. (a) For each padding gate (wire), spare cells inside the bounding box of its fanout net (the investigated wire) are extracted. (b) The spare cell candidates are identified by subset sum using dynamic programming.

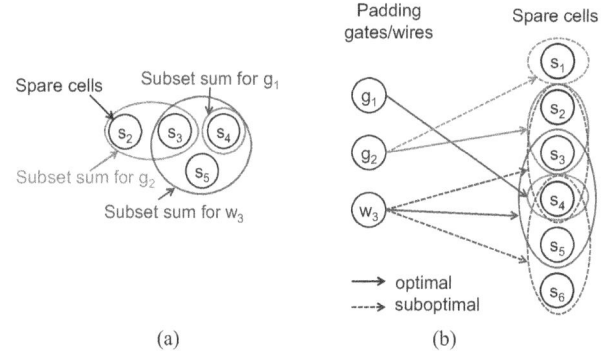

(a)   (b)

Figure 9. Spare cell selection. (a) The competition among spare cells. (b) Spare cell selection. g_1: $\{s_4\}$, g_2: $\{s_2, s_3\}$, w_3: $\{s_5, s_6\}$.

The amount of available dummy metal of each padding gate (respectively, wire) is the unoccupied routing resource upon the bounding box of its fanout net (respectively, the investigated wire). Different padding gates/wires may compete for the same metal resource if their corresponding bounding boxes overlap. We first assign the dummy metal in the independent bounding boxes. If there still are padding gates/wires with unfixed padding capacitance, we resort to maximum network flow [18]. The flow network contains a source node s, a sink node t, a node g_i for each padding gate/wire, and a node d_i for dummy metal of each overlapping region. An edge connects s to each g_i, and its capacity is the remaining padding capacitance of g_i. An edge connects each overlapping region d_i to t, and its capacity is the amount of offered dummy metal of d_i. An edge connects g_i and d_i if g_i's bounding box covers d_i, and its capacity is infinite. For example, as shown in Figure 10(a), after assigning independent dummy metal, we have gates g_1, g_2 and wire w_3 with remaining unfixed padding capacitances 0.15, 0.2, and 0.1, respectively. The dummy metal of overlapping regions can offer capacitances 0.25 and 0.2. The corresponding flow network and the maximum flow are shown in Figure 10(b). Based on the obtained flow, we can assign dummy metal to fix the remaining padding capacitance accordingly. If there are still unfixed padding capacitances after dummy metal insertion, we go back to the first stage to adjust $C_{max}(i)/C_{max}(i,j)$

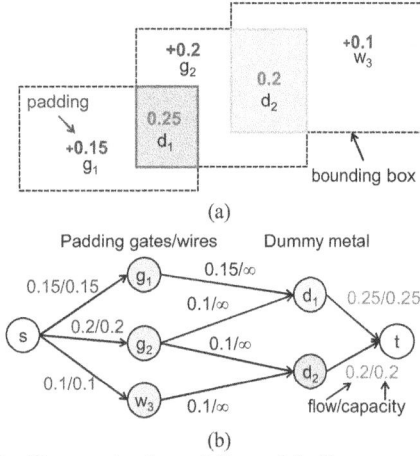

Figure 10. Fine-grained padding. (a) Dummy metal for conflicted padding gates/wires. (b) Dummy metal allocation.

and reassign padding values until no more padding can be done (via the feasibility check described in Lemma 2).

6. EXPERIMENTAL RESULTS

We implemented our algorithm in the C++ programming language and executed the program on a platform with an Intel Xeon 3.8 GHz CPU and with 16 GB memory under CentOS 5.5. Experiments are conducted on the IWLS 2005 benchmark circuits [19] through the resilient circuit design flow (see Section 2.1). Table 1 lists benchmark statistics, where 'Circuit' indicates the circuit name, '#Gate' lists the combinational logic gate count, '#FF' means the number of flip-flops, '#SFF' is the number of timing suspicious flip-flops, 'Conservative clock period' means the clock period considering a timing guardband, and 'THS' represents the total negative hold slack contributed from suspicious flip-flops. Each circuit is synthesized, placed and routed based on 55-nm technology using state-of-the-art commercial tools [20][21]. We also use these tools to verify the circuit timing. 'S_{th}=75%' and 'H_{th}=25%', the values used in our experiments are typical settings in modern designs.

Table 2 compares our padding value determination method with optimal [6] and state-of-the-art work [11][12][13]. 'Padding delay' indicates the total assigned padding delay (including gates and wires). 'LP' uses linear programming (we implemented [6] using CPLEX [22]). 'Greedy 1' greedily pads from the gate with the largest setup slack [12], while 'Greedy 2' greedily pads from the gate passed by most hold violating paths [11][12][13]. All

Table 1. Benchmark statistics.

Circuit	#Gate	#FF	#SFF	Conservative clock period (ns)	THS (ps)
s1196	301	19	1	1.0	152.5
s1423	486	74	45	1.0	4,916.9
s5378	739	162	37	1.0	3,852.8
s9234	555	132	24	1.0	1,929.0
s13207	748	213	14	1.0	371.2
s15850	428	128	29	1.0	2,114.6
s38584	7,890	1,159	194	3.4	108,759.0
des_perf	51,349	8,808	1,190	2.9	583,579.0
b19	72,872	5,541	2,737	3.8	1,481,800.0

methods pad short paths without hurting setup time. Because of the global view, our method cleans all hold violations for each case. LP is time consuming for large-scale circuits, but interestingly LP is efficient for small cases. In contrast, Greedy 1 and Greedy 2 may either fail to clean all hold violations or suffer from long runtime. Because of the local view, Greedy 1 and Greedy 2 may induce inefficient padding and thus require many iterations. Moreover, we conducted one experiment to test the impact of input slew on padding delay capacitance conversion. Consider the error rate of the converted padding capacitance with and without input slew consideration. For each padding gate/wire, error rate = |1 - (padding capacitance without slew)/(padding capacitance with slew)|×100%. The average error rate over all cases is quite small, only 0.56%.

Table 3 compares our load/buffer allocation method with [6]. 'LP+Mapping' is the heuristic proposed in [6] to map the linear programming results to available cells. As mentioned in Section 1, even [6] finds the optimal padding, the directly mapped results may still incur hold violations. In contrast, based on our coarse-grained and fine-grained load/buffer allocation methods, we can successfully achieve the padding values assigned by Stage 1 for all cases. The combination of spare cells and dummy metal provides the flexibility to allocate load/buffers, and thus dummy metal can indeed solve the discrete buffer delay problem well. It can be seen that our algorithm allocate spare cells and dummy metal well and the runtime is acceptable in large cases.

Based on Tables 1–3, it can be seen that our short path padding framework, padding value determination and load/buffer allocation, is promising to solve the short path padding (hold time fixing) problem for resilient circuits.

7. CONCLUSION

Resilient circuits are recently proposed to mitigate dynamic variations. In this paper, to enable the timing error detection and

Table 2. Stage 1: Padding value determination. S_{th}=75%, H_{th}=25%.

Circuit	LP [6]				Greedy 1: Largest setup first [12]					Greedy 2: Most path sharing first [11][12][13]					Ours: PushPull				
	TNS_1 (ps)	THS_1 (ps)	Padding delay (ps)	Runtime (s)	TNS_1 (ps)	THS_1 (ps)	Padding delay (ps)	#Ite.	Runtime (s)	TNS_1 (ps)	THS_1 (ps)	Padding delay (ps)	#Ite.	Runtime (s)	TNS_1 (ps)	THS_1 (ps)	Padding delay (ps)	#Ite.	Runtime (s)
s1196	0.0	0.0	152.5	0.03	0.0	0.0	338.4	5	0.02	0.0	0.0	173.8	3	0.01	0.0	0.0	152.5	1	0.02
s1423	0.0	0.0	4,126.8	0.15	0.0	43.5	4,968.7	56	0.45	0.0	43.5	4,802.8	35	0.24	0.0	0.0	4,746.9	2	0.05
s5378	0.0	0.0	3,661.3	0.06	0.0	79.3	4,678.1	60	0.79	0.0	79.3	4,555.4	43	0.58	0.0	0.0	3,722.7	2	0.08
s9234	0.0	0.0	1,459.6	0.05	0.0	12.0	1,878.3	33	0.30	0.0	12.0	2,158.5	27	0.26	0.0	0.0	1,647.5	1	0.05
s13207	0.0	0.0	371.2	0.03	0.0	34.5	762.3	22	0.21	0.0	34.5	621.5	18	0.21	0.0	0.0	371.2	2	0.07
s15850	0.0	0.0	1,305.1	0.03	0.0	0.0	1,662.8	43	0.25	0.0	0.0	2,161.5	41	0.27	0.0	0.0	1,510.8	2	0.04
s38584	0.0	0.0	108,476.0	6.96	0.0	0.0	225,026.0	780	115.19	0.0	85.8	130,703.0	512	80.37	0.0	0.0	143,764.1	2	1.04
des_perf	0.0	0.0	583,579.0	60.59	0.0	19,270.5	795,022.0	3,414	6,401.86	0.0	3,864.1	595,088.0	2,257	4,202.46	0.0	0.0	583,579.0	2	9.46
b19	NA	NA	NA	timeout	0.0	108,078.0	6,128,710.0	21,902	37,473.30	0.0	65,746.7	1,729,230.0	6,220	10,978.40	0.0	0.0	1,675,688.0	4	21.68

TNS_1: total negative setup slack after padding value determination.
THS_1: total negative hold slack after padding value determination.
#Ite.: number of iterations.
timeout: runtime exceeds 12 hours.

Table 3. Stage 2: Load/Buffer allocation.

Circuit	LP+Mapping [6]			Ours: PushPull		
	TNS$_2$ (ps)	THS$_2$ (ps)	Runtime (s)	TNS$_2$ (ps)	THS$_2$ (ps)	Runtime (s)
s1196	0.0	0.3	0.03	0.0	0.0	0.33
s1423	0.0	826.4	0.09	0.0	0.0	0.43
s5378	0.0	302.3	0.05	0.0	0.0	0.54
s9234	0.0	631.3	0.04	0.0	0.0	0.41
s13207	0.0	161.2	0.03	0.0	0.0	0.61
s15850	0.0	352.0	0.02	0.0	0.0	0.38
s38584	0.0	29,970.7	0.58	0.0	0.0	1.96
des_perf	0.0	51,146.0	41.99	0.0	0.0	12.36
b19	NA	NA	NA	0.0	0.0	43.33

TNS$_2$: total negative setup slack after load/buffer allocation.
THS$_2$: total negative hold slack after load/buffer allocation.

correction mechanism of resilient circuits, we focused on the severe short path padding problem in resilient circuits. Unlike greedy heuristics adopted by recent prior work, we determined the padding values and locations with a global view. Moreover, to further realize the determined padding values at physical implementation, we proposed coarse-grained and fine-grained load/buffer allocation by using spare cells and dummy metal. Experimental results showed the efficiency and effectiveness of our method. Our coarse-grained and fine-grained allocation methods can successfully achieve the derived padding values which may be infeasible when only discrete delays/capacitances are used. In addition, our framework can be generalized to ECO hold time fixing. Future work includes adopting Composite Current Source (CCS) timing model and achieving multi-corner timing closure.

8. REFERENCES

[1] D. Ernst *et al.*, "Razor: a low-power pipeline based on circuit-level timing speculation," *MICRO*, pp. 7–18, 2003.

[2] D. Blaauw *et al.*, "RazorII: In situ error detection and correction for PVT and SER tolerance," *ISSCC*, pp. 400–401, 2008.

[3] K. A. Bowman *et al.*, "Energy-efficient and metastability-immune timing-error detection and instruction-replay-based recovery circuits for dynamic-variation tolerance," *ISSCC*, pp. 402–403, 2008.

[4] D. Bull *et al.*, "A power-efficient 32 bit ARM processor using timing-error detection and correction for transient-error tolerance and adaptation to PVT variation," *IEEE JSSC*, vol. 46, no. 1, pp. 18–31, Jan. 2011.

[5] K. Bowman *et al.*, "A 45nm resilient microprocessor core for dynamic variation tolerance," *IEEE JSSC*, vol. 46, no. 1, pp. 194–208, Jan. 2011.

[6] N. V. Shenoy *et al.*, "Minimum padding to satisfy short path constraints," *ICCAD*, pp. 156–161, 1993.

[7] J.P. Fishburn, "Clock skew optimization," *IEEE TC*, vol. 39, no. 7, pp. 945–951, 1990.

[8] R.B. Deokar and S.S. Sapatnekar, "A graph-theoretic approach to clock skew optimization," *ISCAS*, vol. 1, pp. 407–410, 1994.

[9] S.-H. Huang *et al.*, "Clock period minimization with minimum delay insertion," *DAC*, pp. 970–975, 2007.

[10] C. Lin and H. Zhou, "Clock skew scheduling with delay padding for prescribed skew domains," *ASP-DAC*, pp. 541–546, 2007.

[11] S. Yoshikawa, "Hold time error correction method and correction program for integrated circuits," *US Patent* 6,990,646, 2006.

[12] Y. Sun *et al.*, "Method and apparatus for fixing hold time violations in a circuit design," *US Patent* 7,278,126, 2007.

[13] Y. Liu *et al.*, "Re-synthesis for cost-efficient circuit-level timing speculation," *DAC*, pp. 158–163, 2011.

[14] Y. Kim *et al.*, "Simple and accurate models for capacitance increment due to metal fill insertion," *ASP-DAC*, pp. 456–461, 2007.

[15] Liberty library modeling: The semiconductor industry's most widely used library modeling standard. http://www.opensourceliberty.org/.

[16] Y.-P. Lin *et al.*, "ECO timing optimization using spare cells," *ICCAD*, pp. 530–535, 2007.

[17] G. D. Hachtel and F. Somenzi, *Logic Synthesis and Verification Algorithms*, Springer-Verlag, 2006.

[18] J. Kleinberg and E. Tardos, *Algorithm Design*, Addison Wesley, 2006.

[19] IWLS 2005 benchmarks. http://iwls.org/iwls2005/benchmarks.html.

[20] Synopsys Design Compiler. http://www.synopsys.com.

[21] Cadence SoC Encounter. http://www.cadence.com.

[22] IBM ILOG CPLEX Optimizer. http://www.ilog.com/products/cplex/.

Dawn of Computer-aided Design
– from Graph-theory to Place and Route –

Atsushi Takahashi

Tokyo Institute of Technology

2-12-1-S3-58 Ookayama, Meguro-ku, Tokyo 152-8550, Japan

atsushi@lab.ss.titech.ac.jp

ABSTRACT

The research area of computer-aided-design emerged soon after integrated circuit had emerged. Memorial works in the dawn of computer-aided design are introduced briefly.

Categories and Subject Descriptors

B.7.2 [**Hardware, Integrated Circuits**]: Design Aids—*Layout, Placement and Routing*

General Terms

Algorithms, Theory

Keywords

principal partition, left edge algorithm, minimum width channel routing, via minimization

1. REMINISCENCES

One of epoch making works in the history of computer-aided-design (CAD) and electronic-design-automation (EDA) development is the left edge algorithm invented by Hashimoto and Stevens which was presented at DAC in 1971 [1]. In those days, not a few researchers in circuit-theory and graph-theory moved to CAD and EDA field. These researchers felt that the research in this field was required and requested since more than hundreds attendees were in the session room in design-automation conference (workshop), while there was even only several attendees in the session room in graph-theory related conferences.

For example, Professor Kajitani started his academic career on graph-theory where electrical circuits were major applications. One of his main contributions was a discovery of the "principal partition of a graph" which solves the problem of finding the minimum set of voltages and currents that describes all variables in a circuit [5]. Some graph theoretical unsolved problems were related to principal partition [2], and these works were followed by many researches, and one session of international symposium on circuits and systems (ISCAS) in 1982 was held with the title of "Theory and Applications of Principal Partition." From late seventies, Professor Kajitani shifted his interest to placement and routing. The minimum width and height channel routing algorithm which gives a very fundamental theorem was presented at

design automation conference (DAC) in 1979 [4]. Also, Professor Kajitani provided a polynomial time algorithm that minimizes the number of vias at international conference on circuits and computers (ICCC) in 1980 [3], and this result stimulated a number of related researches. The minimization of the number of vias in two-layer Manhattan routing had been believed an NP-hard problem, and many approximation algorithms have been proposed at those days. However Professor Kajitani knew that the minimum vias problem can be transferred to maximum cut problem, and knew that maximum cut problem which is NP-hard in general was proved to be solvable in polynomial time if the graph is planar.

The research area of computer-aided-design emerged soon after integrated circuit had emerged. From the dawn of computer-aided design, the development of algorithms needs strong theoretical analysis which is still highly required.

2. ACKNOWLEDGMENTS

The author would like to express my deepest gratitude to Professor Yoji Kajitani for his frontier spirit and pioneer works that accelerate our research, and Kajitani research family who are continuously improving themselves through friendly rivalry. Also, the author would like to thank Professor Shuji Tsukiyama, Professor Masao Yanagisawa, and Professor Shuichi Ueno for their prompt adequate advice, and Professor Shigetoshi Nakatake and Professor Yasuhiro Takashima for their kind arrangements and advice.

3. REFERENCES

[1] A. Hashimoto and J. Stevens. Wire routing by optimizing channel assignment within large apertures. In *Proceedings of the 8th Design Automation Workshop (DAC)*, pages 155–169, 1971.

[2] Y. Kajitani. The semibasis in network analysis and graph theoretical degrees of freedom. *IEEE Transactions on Circuits and Systems*, 26(10):846–855, 10 1979.

[3] Y. Kajitani. On via minimization of routings on a 2-layer board. In *Proceedings of the first IEEE International Conference on Circuits and Computers (ICCC)*, pages 295–298, 1980.

[4] T. Kawamoto and Y. Kajitani. The minimum width routing of a 2-row 2-layer polycell-layout. In *Proceedings of the 16th Conference on Design Automation (DAC)*, pages 290–296, 1979.

[5] G. Kishi and Y. Kajitani. Maximally distant trees and principal partition of a linear graph. *IEEE Transactions on Circuit Theory*, 16(3):323–330, 8 1969.

Practicality on Placement Given by Optimality of Packing

Shigetoshi Nakatake
The University of Kitakyushu
1-1 Hibkino, Wakamatsu, Kitakyushu, Fukuoka 808-0135, Japan
nakatake@kitakyu-u.ac.jp

Categories and Subject Descriptors

B.7.2 [**INTEGRATED CIRCUITS**]: Design Aids

General Terms

Algorithms,Theory

Keywords

Packing, Placement, BSG, Sequence-pair

1. FROM SEQUENCE-PAIR TO ISPD

Kajitani and his group are one of pioneers of 2D rectangle packing algorithm. I introduce the history of the birth of the algorithm partially excerpted from Kajitani's home page (written in Japanese).

In 1991, Japan Advanced Institute of Science and Technology (JAIST) was established in Ishikawa. Kajitani spent four years (1992–1996) had been working there in parallel with the Tokyo Institute of Technology. He kicked-off the laboratory with Fujiyoshi, while Murata and I were the first students. We were researching about data structures representing a general rectangle placement with enthusiasm, and struggled to propose Bounded Sliceline Grid (BSG) and Sequence-pair (SP). In his home page, the situation at that time was described as follows;

> We were not talking about "we luckily hit these proposals". We started with Nakatake's question; "Does a repetition of the minimal structure cover a general structure?" We seriously discussed about this question and Murata proposed a structure described in an oblique line grid. Kajitani noticed that this grid is a permutation matrix and any substitution matrix is represented by a permutation-pair from his knowledge of group theory. Hence, Sequence-pair (SP) was born. It is interesting to note that this provides an observation that "diversity of placement" is equivalent to "diversity of group".

Since SP and BSG looked applicable to somewhat industrial products, we decided to take a patent, and applied for patents in Japan and the United States. Writing the applicant document by hand, we paid their funds from individuals

pockets. As an international conference, at first, we made a presentation in ICCAD, in 1995. ICCAD chose SP as one of best in the layout area of "The Best of ICCAD: 20 Years of Excellence in Computer-Aided Design" which has 41 papers (SP is only one from Japan). SP is really fundamental, so any placement follows our idea as long as it does not have a limited structure like a slicing one. However, there could be various topological representations depending on how to separate the topology information and physical information. As a result, a lot of variant papers have appeared changing the representation or the purpose. In ISPD, one or more sessions occupied by only such variants had been organized for a while.

2. FROM PACKING TO PRACTICAL DESIGN

In 1995, a famous evening newspaper (Asahi Shimbun) published our packing algorithm in a front page. Murata started up a company of EDA tools based on SP. Thus, SP has affected lots of people so far. This reason may be in the following theorem which is the most significant contribution of SP.

> A solution space induced by a set of all sequence-pairs consists of $(n!)^2$ sequence-pairs, each of which can be mapped to a packing in $O(n^2)$ time, and at least one of which corresponds to an optimal solution of rectangle packing problem.

Fujiyoshi largely contributed to the proof of this theorem. It was proved such that any placement corresponds to a sequence-pair, as well as any sequence-pair does to a placement. Following this theorem, a stochastic searching method such as simulated annealing enables us to seek a compact 2D rectangle packing in a reasonable time. >From the viewpoint of algorithm, employing a stochastic approach may not be elegant. But,I believe that the formulation of a solution space including an optimal solution inspires researchers who were tired of thinking of heuristics, and drives their imagination.

Such a way of solving a 2D rectangle packing problem encouraged to develop various applications; building block design, device-level placement, analog layout, and floorplan. Most of research interests have been to seek for another representation forming a smaller solution space including an optimal solution. They have contributed to extend the scalability, so that hundreds of blocks can be currently placed in a practical time. As a successful application, Sequence-pair has remarkably accelerated to develop analog placement. As seen the above theorem, any placement corresponds to a sequence-pair. This implies that any layout constraint for analog designs such as symmetry and proximity constraints can be represented in terms of the order of rectangles in a sequence-pair. As as result, we have develop practical tools by collaborating with an EDA company. I believe that we have contributed to the synergy of theoretical field and practical one to develop the placement technology.

As well, since it can deal with rectangles well, an extension with the cubes and polygons had been studied. Hence, not just EDA of ICs, the extension such as the packing of home delivery and the cut of the cloth, had been applied to other areas.

In this talk, introducing several applications of 2D rectangle packing, I would like to celebrate Kajitani's contribution as a pioneer. (I list a major papers of Kajitani related to 2D rectangle packing in references.)

3. REFERENCES

[1] H. Murata, K. Fujiyoshi, S. Nakatake, Y. Kajitani, "Rectangle-packing-based module placement", Proc. of ICCAD'95, pp. 472-479, 1995.

[2] S. Nakatake, K. Fujiyoshi, H. Murata, Y. Kajitani, "Module placement on BSG-structure and IC layout applications", Proc. of ICCAD'96, pp. 484-491, 1996.

[3] H. Murata, K. Fujiyoshi, S. Nakatake, Y. Kajitani, "VLSI module placement based on rectangle-packing by the sequence-pair", IEEE Transactions on CAD, Vol. 15, No. 12, pp. 1518-1524, 1996.

[4] H. Murata, K. Fujiyoshi, T. Watanabe, Y. Kajitani, "A mapping from sequence-pair to rectangular dissection", Proc. of ASP-DAC'97, pp. 625-633, 1997.

[5] S. Nakatake, M. Furuya, Y. Kajitani, "Module Placement on BSG-Structure with Pre-Placed Modules and Rectilinear Modules", Proc. of ASP-DAC'98, pp. 571-576, 1998.

[6] T. Izumi, A. Takahashi, Y. Kajitani, "Air-pressure model and fast algorithm for zero-wasted-area layout of general floorplan", IEICE Transactions Vol. 81-A, No. 5, pp. 857-865, 1998.

[7] S. Nakatake, K. Sakanushi, Y. Kajitani, M. Kawakita, "The channeled-BSG: a universal floorplan for simultaneous place/route with IC applications", Proc. of ASP-DAC'98, pp. 418-425, 1998.

[8] K. Sakanushi, S. Nakatake, Y. Kajitani, "The multi-BSG: stochastic approach to an optimum packing of convex-rectilinear blocks", Proc. of ICCAD'98, pp. 267-274, 1998.

[9] T. Izumi, A. Takahashi, Y. Kajitani, "Air-Pressure-Model-Based Fast Algorithms for General Floorplan", Proc. of ASP-DAC'98, pp. 563-570, 1998.

[10] S. Nakatake, K. Fujiyoshi, H. Murata, Y. Kajitani, "Module packing based on BSG-structure and IC layout applications", IEEE Transactions on CAD, Vol. 17, No. 6, pp. 519-530, 1998.

[11] H. Yamazaki, K. Sakanushi, S. Nakatake, Y. Kajitani, "The 3D-packing by meta data structure and packing heuristics", IEICE Transactions Vol. 83-A, No. 4, pp. 639-645, 2000.

[12] Z. Wu, K. Sakanushi, Y. Kajitani, "Reuse of VLSI Layout Topology by Parametric BSG", Proc. of APCCAS'00, pp. 817-820, 2000.

[13] K. Sakanushi, Y. Kajitani, "The Quarter-State Sequence (Q-Sequence) to Represent the Floorplan and Applications to Layout Optimization", Proc. of APCCAS'00, pp. 829-832, 2000.

[14] H. Yamazaki, K. Sakanushi, Y. Kajitani, "Optimum Packing of Convex-Polygons by A New Data Structure Sequence-Table", Proc. of APCCAS'00, pp. 821-824, 2000.

[15] S. Nakatake, Y. Kubo, Y. Kajitani, "Consistent floorplanning with super hierarchical constraints", Proc. of ISPD'01, pp. 144-149, 2001.

[16] Y. Kubo, S. Nakatake, Y. Kajitani, M. Kawakita, "Chip size estimation based on wiring area" Proc. of APCCAS'02, pp. 113-118, 2002.

[17] H. Miyashita, Y. Kajitani, "On the equivalence of the sequence pair for rectangle packing to the dimension of partial order", Proc. of APCCAS'02, pp. 367–370, 2002.

[18] Y. Kubo, S. Nakatake, Y. Kajitani, M. Kawakita, "Explicit Expression and Simultaneous Optimization of Placement and Routing for Analog IC Layouts", Proc. of VLSI Design'02, pp. 467-472, 2002.

[19] C. Zhuang, Y. Kajitani, K. Sakanushi, L. Jin, "An Enhanced Q-Sequence Augmented with Empty-Room-Insertion and Parenthesis Trees", Proc. of DATE'02, pp. 61-68, 2002.

[20] K. Sakanushi, Z. Wu, Y. Kajitani, "Recognition of floorplan by Parametric BSG for reuse of layout design", IEICE Transactions Vol. 85-A, No. 4, pp. 872-879, 2002.

[21] Y. Kajitani, K. Sakanushi, "THE PRIME-GRAPH AND Q-SEQUENCE FOR CODING THE FLOORPLAN", Proc. of ISCAS'02, pp. III-771-III-774, 2002.

[22] S. Nakatake, Y. Kubo, Y. Kajitani, "Consistent floorplanning with hierarchical superconstraints", IEEE Transactions on CAD, Vol. 21, No. 1 , pp. 42-49, 2002.

[23] K. Sakanushi, Y. Kajitani, D. P. Mehta, "The quarter-state-sequence floorplan representation", IEEE Transaction, Vol. 50-A, No. 3, pp. 376-386, 2003.

[24] N. Fu, S. Nakatake, Y. Takashima, Y. Kajitani, "Abstraction and optimization of consistent floorplanning with pillar block constraints", Proc. of SP-DAC'04, pp. 19-24, 2004.

[25] T. Nojima, X. Zhu, Y. Takashima, S. Nakatake, Y. Kajitani, "Multi-level placement with circuit schema based clustering in analog IC layouts", Proc. of ASP-DAC'04, pp. 406-411, 2004.

[26] X. Zhang, Y. Kajitani, "Space-planning: placement of modules with controlled empty area by single-sequence", Proc. of ASP-DAC'04, pp. 25-30, 2004.

[27] T. Nojima, Y. Takashima, S. Nakatake, Y. Kajitani, "A device-level placement with multi-directional convex clustering", Proc. of GLSVLSI'04, pp. 196-201, 2004.

[28] H.-A. Zhao, C. Liu, Y. Kajitani, K. Sakanushi, " EQ-Sequences for Coding Floorplan", IEICE Transactions Vol. 87-A, No. 12, pp. 3233-3243, 2004.

[29] X. Zhang, Y. Kajitani, "Theory of T-junction floorplans in terms of single-sequence", Proc. of ISCAS'04, No. 5, pp. 341-344, 2004.

[30] H.-A. Zhao, C. Liu, Y. Kajitani, K. Sakanushi, "A Compact Code for Representing Floorplans", Proc. of MWCAS'04. I-433-I-436, 2004.

[31] Z. Zhou, S. Dong, X. Hong, Y. Wu, Y. Kajitani, "A new approach based on LFF for optimization of dynamic hardware reconfigurations", Proc. of ISCAS'05, Vol. 2, pp. 1210-1213, 2005.

[32] R. Liu, S. Dong, X. Hong, Y. Kajitani "Fixed-outline floorplanning with constraints through instance augmentation", Proc. of ISCAS'05, Vol. 2, pp. 1883-1886, 2005.

[33] T. Yan, Q. Dong, Y. Takashima, Y. Kajitani, "How does partitioning matter for 3D floorplanning?", Proc. of GLSVLSI'06, pp. 73-78, 2006.

[34] Y. Kajitani, "Theory of placement by numDAG related with single-sequence, SP, BSG, and O-tree", Proc. of ISCAS'06, pp. 4471-4474, 2006.

[35] T. Nojima, N. Ono, S. Nakatake, T. Fujimura, K. Okazaki, Y. Kajitani, "Adaptive Porting of Analog IPs with Reusable Conservative Properties", Proc. of ISVLSI'06, pp. 18-23, 2006.

On the Way to Practical Tools for Beyond Die Codesign and Integration

Hung-Ming Chen
Department of Electronics Engineering and Institute of Electronics
National Chiao Tung University, Hsinchu, Taiwan
hmchen@mail.nctu.edu.tw

ABSTRACT

Package and board designs were usually considered inferior parts in semiconductor supply chain, compared with major stream digital and AMS/RF chip designs. This thought has been gradually reverted due to profit margin reduction in lack of consideration for those "beyond die" parts. In recent years, Prof. Kajitani has devoted himself in this particular part of researches towards his retirement and beyond (we are honored to be invited to work together), he has generated numerous useful thoughts (legacy of object coding) in board routing automation. Although not aware of all his inventions in this field in Japan, we have been sure that Prof. Kajitani's brilliant thoughts have influenced many researchers and students, including us in Taiwan. During his several visits to Taiwan, he was invited to give talks to express his interests in non-maze routing and other more topics. Besides well-known sequence pair representation in floorplanning and placement advances, he has made other influential contributions in beyond die codesign and integration. The ultimate goal of Prof. Kajitani (and our research group) is to try to generate practical tools for non-die layout design automation, and it has been unsurprisingly uneasy task. In this talk, we try to reveal some of the development paths and challenges we have encountered, also showing the records in this cross-nation collaboration towarding this still-continuing mission to possible practicality. We also hope this talk can somewhat help pave the road to the real success of beyond die design automation, regardless of our do-the-best attempt and very little outcome so far.

Categories and Subject Descriptors

B7.2 [**Integrated Circuits**]: Design Aids

General Terms

Algorithms, Design, Theory

Keywords

Beyond die design automation, electronic design automation, EDA, computer-aided design, CAD

Coding the Objects in Place and Route CAD

Yoji Kajitani

Japan Advanced Institute of Science and Technology
Nomi, Ishikawa-ken, 923-1292, Japan
+81-761-51-1396
kaji-you@jaist.ac.jp

ABSTRACT

Computer aided design of any style starts with coding the objects. The objects are blocks and nets. The code is, in fact, a permutation of object labels. A solemn fact is that the code is a linear array while the objects are floating on the multiple layer or 3D space so that potentially very restricted data can be on the code. Given this opportunity at ISPD 2013, I would like to disclose an unpublished story which stands solely on the concept of *permutation*. It was motivated by a question about a small gap found in classic channel routing. The interest is in how unrelated place and route matters be related with each other.

Categories and Subject Descriptors

B.7.2 [Integrated Circuits]: Design Aids

General Terms

Algorithm, Design.

Keywords

Placement, Routing, Permutation, Channel, Sequence-pair

1. PERMUTATION

Once the world is discretized, the input to the digital computer is conveyed by a code which is actually a sequence of labels if the object names are fixed. The information of a code is in the deviation from the normal state. If the normal state is I=(12345), P=(12435) contains a significance that reasons why 3 and 4 are reversed. It might be more for (24513) and maximally for (54321). Here it is taken for granted that sequence I is the identity so natural not necessary to be cited particularly and it is the postulate that *there is a reason for a rebel*. EDA, so much complicated these days, still is within the framework of permutations.

The term *permutation* reminds us of a lot. In EDA, explicitly are those of pin assignment, channel routing, cyclic routing, PLA design, swapping in simulated annealing, Sequence-pair, and many others, no wonder since the *optimal solution* is a choice from all possible solutions, which are usually generated by permuting the sequence.

2. CHANNEL ROUTING

A channel here is a rectangular area on the 2-layer grid where terminals are assumed on the peripheral. The request is: Construct

Permission to make digital or hard copies of all or part of this work for personal or classroom use is granted without fee provided that copies are not made or distributed for profit or commercial advantage and that copies bear this notice and the full citation on the first page. To copy otherwise, or republish, to post on servers or to redistribute to lists, requires prior specific permission and/or a fee.
ISPD'13, March 24–27, 2013, Stateline, Nevada, USA.
Copyright © 2013 ACM 978-1-4503-1954-6/13/03...$15.00.

disjoint networks inside, each connecting all terminals of the same label. If minimization of the number of columns or rows is the target, a way for optimal routing is known in [1], an ancient publication one third century ago.

However, what we need actually is in the success ratio of connections for a given area, or the minimum area=(number of columns)x(number of rows). It is a hard problem in general but optimal solutions for some special class is fixed. The one shown in Figure 1(a) is such a case which is the one called opposite-type permutation channel. It is defined by a permutation P=(24513) such that terminals on the top is along P while those on the bottom are along the natural order of I. Figure 1(b) shows a routing which is optimum in all (row number, column number, area).

Even for this simplest case, we need extra columns! It is a weird situation for channel routing in hierarchical IC design, which has been popular since seventies. Unable to fix the exact dimension of the lowest stage, how could we build a tower? Also it is a necessary technique in board routing to arrange the nets for non-cross bus routing. We do not like to prepare extra area to cover the unknown worst case.

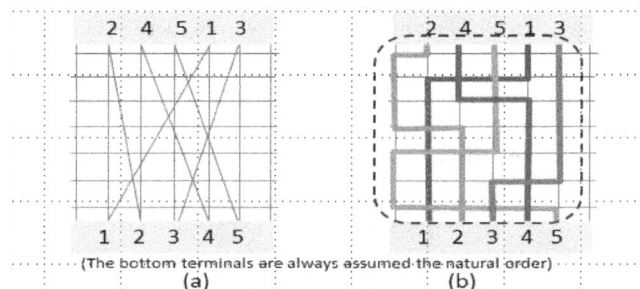

Figure 1 (a) Opposite-type channel by P=(24513). (b) An optimum square routing using n+1 columns and n+2 rows.

Take another type, side-type channel, where terminals are laid on the left side along P as shown in Figure 2.

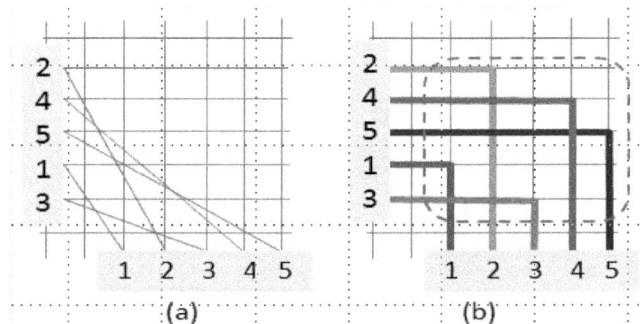

Figure 2 (a) Side-type channel by P. (b) Square routing 100% success with no extra column nor extra row, guaranteed.

These two are similar as a function but the difference in design is essential. In contrast to the opposite-type, the side-type channel routing achieves 100% by an intuitive way: Extend perpendicular wires from both sides and connect two of the same label at the intersection, the end. A question about this non-symmetric difference between opposite-type channels and side-type channels is the start of this talk.

It is a course of reasoning to come to an idea of routing the opposite-type channel by applying side-type channel routing. It is possible as seen in Figure 3(a) and (b) by slanting the grid. Routing 100% is as easy as for a side-type channel. The method is called *slant routing* while routing as shown in Figure 1 and 2 is called *square routing*.

However, people would say, it is not a solution; it is out of the presumed world (too much a trick of Christopher Columbus's Egg). So we need to invent a way to transform the solution back to the square world.

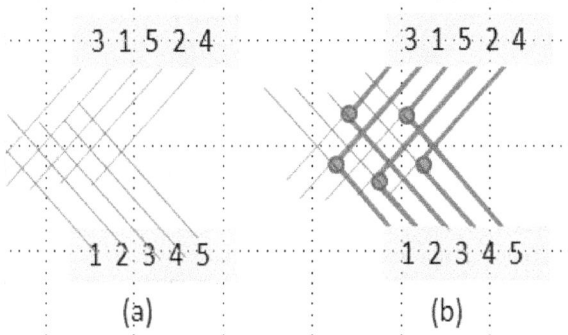

Figure 3 (a) Slanting the grid. (b) Slant routing.

In slant-routing, a good property is observed that *two paths cross each other exactly once if and only if their numbers are reversed in the permutation*. Note that in square routing, there is always a pair of nets that cross more than once unless P=I. The cross points are visualized in Figure 4(a). A cross point corresponds to an exchange. Though implicit, a cross generates a new permutation. Extract such midway permutations and represent them together with the crosses to make a table of exchanges as shown in Figure 4(b).

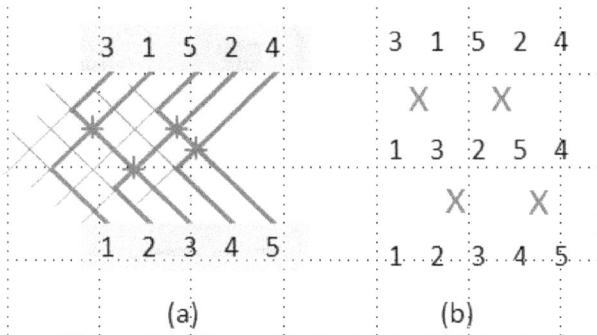

Figure 4 (a) Crosses are exchanges. (b) A table of exchanges that sort the permutation P=(31524)

This analysis reminds us of the too well-known bubble-sort. The conventional bubble-sort applies one exchange to the left most adjacent pair, one at a time. While here, disjoint pairs are concurrently exchanged, so it should be called *concurrent bubble-sort*. The table hints us to make up the channel routing by tracing

back the same labels from the bottom using 45 degree edges like the routing shown in Figure 5(a). It completes the channel with no extra column! How slim it is!

However, we remember that, in spite of many preceding works on the use of slant edges, this style has not been popular maybe by reasons for manufacturability or by electrical features.

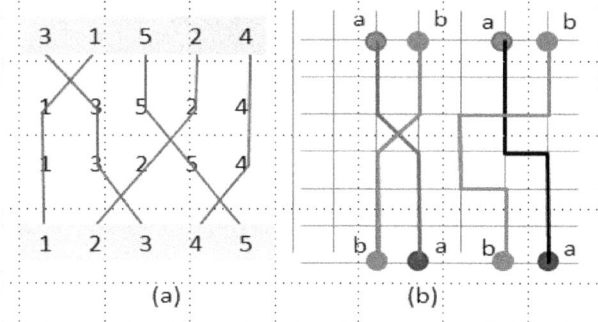

Figure 5 (a) Slim opposite-type routing. (b) The slant cross structure to the same request is better than square.

This is true but only partly since in board routing, slant edges are rather recommended. More, the author would like to remark that so far no systematic straightforward design method has been with explicit merits. This proposing method is straightforward, no need of analysis in advance. And, as for the electrical features, see Figure 5(b) which is suggesting that the slant cross routing is electrically and size point of view far better than the square cross routing.

3. PLACEMENT

Go back to Figure 3(b) where bends are named by the net label. (Note the difference from Figure 4 where we focused the cross points.) Abstract the positional relation between pairs of bends to ABLR (Above, Below, Left-of, and Right-of) relation by the rule:

Bend x is left-of bend y if the left-cone of y contains x.

There are four statements with respect to relative positions but are omitted. Then we notice that this is the relation generated by the sequence pair (12345)(31524), or simplified (31524) omitting identity I.

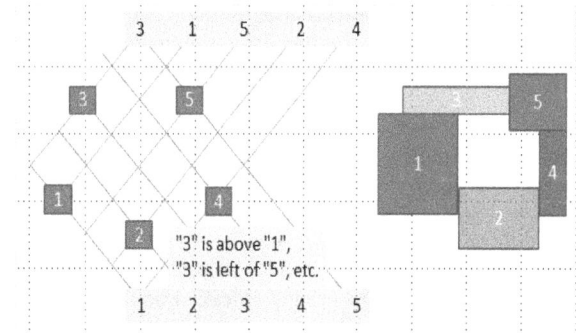

Figure 6 Relative positions of bends to lead the H and V partial orders of ABLR relations, and an example placement.

We have thus studied the fact that *one permutation generates one slant-channel routing and one placement*. Two figures in Figure 7 share the same DNA. The route shows four cross-resolutions that correspond to four vertical relations of the placement.

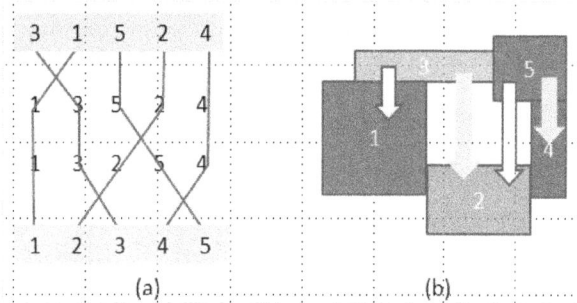

Figure 7 Sister channel routing and placement born of the same mother P=(31524)

4. OTHER PERMUTATION TOPICS

4.1. Generalized Sequence-Pair

We remark that a channel routing corresponds to an ABLR relation and vice versa. The match-maker is the permutation.

Then comes a question: A channel usually allows repetition of terminals (multi-terminal nets) such as Q=(31242) while the permutation does not. Then, what placement does this Q correspond? Can we handle L-shape blocks by the sequence pair Q (and I)? See Figure 8 which poses a question.

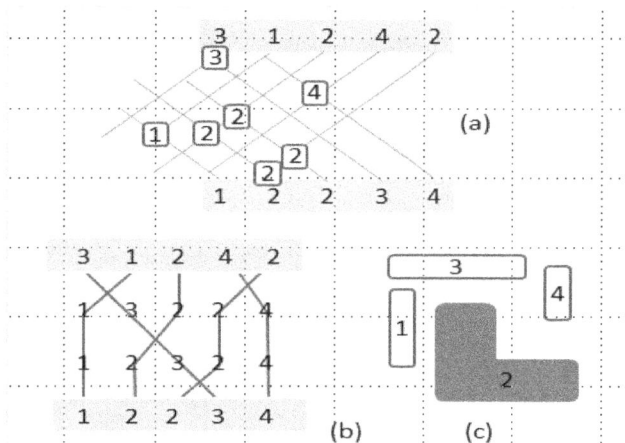

Figure 8 (a) Slant channel routing for a permutation with repetition. (b) The slim routing using slant edges. (c) Is this the corresponding placement that allow L-shape blocks?

4.2. iPS placement

An incremental design goes the way as: Find an initial feasible solution and improve the current solution incrementally keeping the feasibility. Here *feasibilty* means for a placement to satisfy the given constraint whole. However,often it is the hardest core in incremental design to find one initial feasible solution.

In placement, a smart strategy manages to find a feasible solution this way that way, successfully and often not. It is our eagerest desire if there is an *almighty placement* which is feasble for any ABLR constraint to start with.

A diagonal placement in Figure 9 (a) answers this request. As an iPS (Induced Pluripotent Stem) placement, it grows to a

new placement step by step accepting one constraint and another up to satisfy whole the given constraint. The placement in (b) is a placement to satisfy the added constraint by reversing 2 and 3.

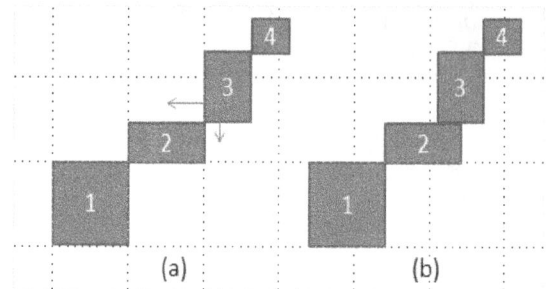

Figure 9. (a) The iPS placement that satisfies all the ABLR relations simultaneouly: Any block k is above or right-of any preceding block j(<k). (b) A new placement which satisfies the added constraint induced by (32).

4.3. Circular permutation

Planar (1-layer) routing process is, simply to say, decomposed into two phases: cross resolution (topological routing) and capacity fulfillment (river routing). Once a topological routing is found, its feasibility with respect to the capacity is rather easy to be checked.

Topological routing is often generated by permutations. The most studied class of problems is when terminals are arranged in a line or on a circle. One example is shown in Figure 10(a) which is defined by a circular permutation allowing repetition of net labels. We know the fact that *it is feasible if and only if the adjacent matching reduction results in the empty sequence* where the *adjacent matching reduction* is to delete all consecutive terminals of the same label in the circular sequence. We observe that this is feasible.

Then, an interesting problem is how the feasibility of topological routing is discussed for the case when one or more islands with terminals exist inside the bordered area. One example is shown in Figure 10(b). How do you see if this is feasible or not?

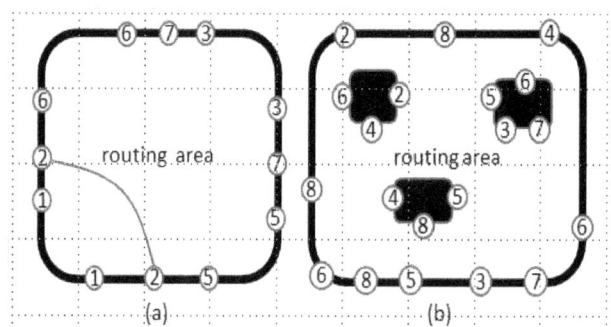

Figure 10 (a) Circular routing problem by circular permutation (126673375521) (b) Is this feasible?

5. HOW THINGS WENT ON

The visual image and formal description come to the human mind alternatively, helping each other to deepen the idea. Computer Aided Physical Design is especially attractive for us because both aspects are meaningful in computation and drawing.

BSG and Sequence–pair in early stage were typical of such event. BSG came to Nakatake's brain somehow in some morning in 1994 at JAIST with his scratch of several BSG boxes. There had been a number of non-slicing pictures so far studied but his sketch and his explanation made me believe its possibility to extend to the infinity. Though Fujiyoshi and Nakatake implemented the idea overnight to generate enormous number of placements for simulated annealing, a modern technique those days, still we have a doubt that BSG is a too radical language to make people, including ourselves, believe its consistency. Drawing the BSG image on a pair of slant lines (like Figure 4) was initiated by Murata. It was still a theory on image but gave a hint that a pair of ordered labels is enough to convey a consistent set of ABLR relations. Then, the author remembered the theory of groups where the permutation group is the central concept. It was popular in 60-80's, even included as a required subject in electrical engineering department. For popularity to adopt the term from popular mathematics, the name should be the permutation-pair. In group theory however, it is conventional to represent a pair of permutations by a single permutation regarding one as an operation. However, since labeling is essential in applications and one permutation could be used as the name tag, he named the pair as "Sequence-Pair".

Acknowledgements

The author has to apologize for telling many things without mentioning related preceding works. Any result cannot be alone. The pair of permutations is a two century long standing concept in discrete mathematics. Precious ideas are often not written out explicitly anywhere. For a half century long, the author has been belonging to several schools, longest to Tokyo Institute of Technology. Also he worked at Japan Advanced Institute of Science and Technology, U of Kitakyushu, and labs of friendly professors, inside and outside, Taiwan, USA, and China. Even a student gave a hint in case he wonders why he thinks differently from his intension. Reasons he did not cite due references are only because there are not all data at hand and he is not sure of missing but necessary contributions for him. Too many hidden contributions have been great for him.

References

[1] T. Kawamoto and Y. Kajitani, The minimum width routing of a 2-row 2-layer polycell-layout, In Proceeding of the 16th Design Automation Conference, San Diego, 1979

Circuit and PD Challenges at the 14nm Technology Node

James Warnock

IBM Systems and Technology Group, T.J. Watson Research Center
P.O. Box 218, Yorktown Heights, NY 10598
(914) 945-2620, jwarnock@us.ibm.com

ABSTRACT

As traditional CMOS scaling comes to an end, the industry is moving towards new 3D finFET multigate structures as device engineers stand the silicon transistors up on their sides. Digital circuit designers working in the 14nm technology node will face significant new challenges from additional design constraints and new sources of variability associated with this non-planar transistor structure. In addition, computational lithography and the need for double patterning at the 14nm node will drive up the complexity and difficulty of the physical design implementation, pushing designs towards more uniform and regular structures, even as wire RC and reliability issues drive increasing demand for uniquely customized solutions. New design tools and methodologies will therefore be needed to meet these circuit and PD challenges at the 14nm node.

Categories and Subject Descriptors

B.7.1 [**Integrated Circuits**]: Types and Design Styles – *advanced technologies, microprocessors and microcomputers, VLSI.* B.7.2 [**Integrated Circuits**]: Design Aids – *layout, placement and routing.* B.8.0 [**Performance and Reliability**]: General

General Terms

Performance, Design, Reliability

Keywords

14nm; CMOS; scaling; digital circuit design; finFET; trigate; double patterning; CMOS physical design; PD.

1. SCALING CHALLENGES

As technology scaling pushes towards the 14nm node and then beyond, it is generally expected that the industry will move away from the planar FET structures which have served the industry so well for the past 40 years or so[1]. The classical CMOS scaling paradigm, described by Dennard[2], and used to predict CMOS circuit power and performance over a period of several decades has already run up against some practical limits, and as various parameters reach their practical scaling limits, circuit designers must be prepared to cope with various problems associated with this lack of scaling. Figure 1 shows how supply voltage scaling has broken down in recent technology generations.

The breakdown of scaling, and the need for ever more exotic techniques to provide increased transistor performance and density will lead the industry into the multi-gate regime[3], creating new challenges for circuit and physical design engineers.

Copyright is held by the author/owner(s).
ISPD'13, March 24–27, 2013, Stateline, Nevada, USA.
ACM 978-1-4503-1954-6/13/03.

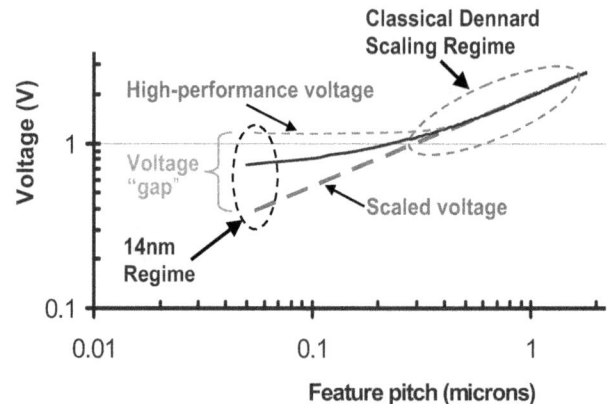

Figure 1. Typical supply voltage scaling trend.

In addition the growing "voltage gap", or the difference between the actual chip voltage and the ideally scaled voltage, will mean that tough reliability issues are likely to be encountered when implementing high-performance designs.

Finally, as if the added complexity associated with multi-gate finFET devices were not enough, 14nm lithography is expected to rely heavily on double patterning lithography (DPL), as shown in figure 2, due to the fact that feature size is generally scaling faster than the wavelength of light used for imaging.

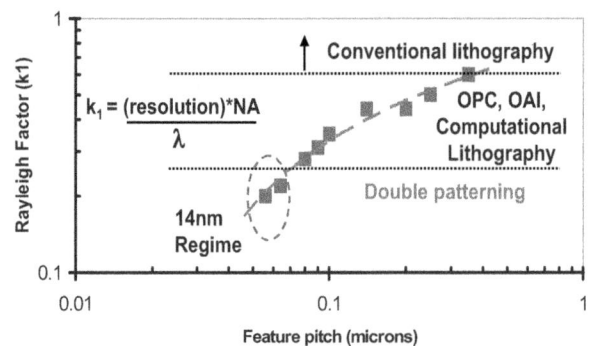

Figure 2. Increasing lithographic complexity

The cumulative impacts of finFET device structures, performance requirements and wire RC, design for reliability and manufacturability, and DPL, will pose considerable challenges for the implementation of efficient design solutions.

2. FINFET MULTIGATE DEVICES

Multi-gate finFETs, (figure 3), have already appeared at the 22nm node[4], and are expected to be in use more widely across the industry by the 14nm node. It is expected that variability due to Random Dopant Fluctuations (RDF) will be improved[5,6]. Also, sub-threshold slopes are expected to improve, helping to allow lower device threshold voltages which in turn will facilitate lower voltage operation. However variability from gate line edge roughness will still be present, along with new sources of variability from fin thickness or height variations.

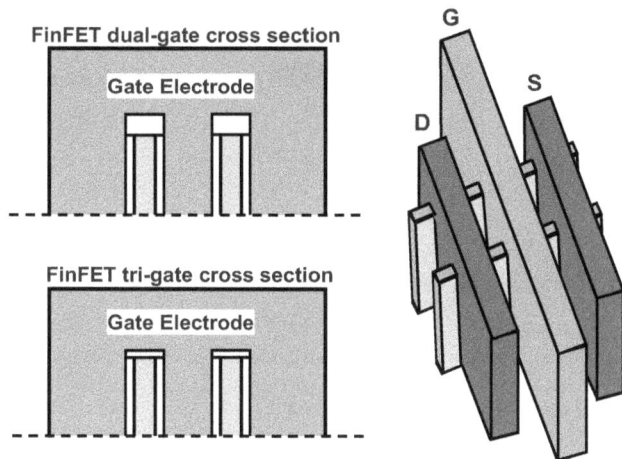

Figure 3. Multigate finFETs.

Another fundamental aspect of a finFET technology is the resulting quantization of device width, since all devices must contain an integral number of fins, as illustrated in figure 4.

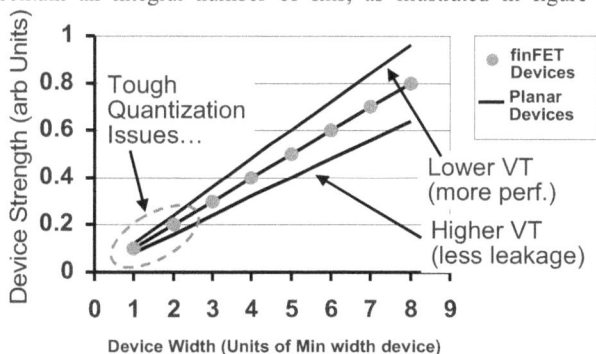

Figure 4. FinFET quantization.

This is likely to be more of a problem for small-width devices. SRAM array cells and perhaps register file cells will require careful design/technology co-optimzation to ensure the required window of functionality[7]. Also, if the technology uses undoped fins, in order to reduce the impact of random dopant fluctuations, then designers may lose another degree of flexibility, ie the ability to select higher or lower VT devices, or they may have to face a higher variability for high-VT devices with doped fins.

3. 14nm PD ISSUES

The introduction of finFETs imposes some new restrictions on the physical design. In order to minimize the impact of width quantization and provide maximum current drive per unit area, a very fine fin pitch is desirable, leading to a "sea of fins" approach. In this case library cells need to be designed so that cell heights are matched to both the fin and metal pitches, as shown in figure 5, in addition to the constraint imposed by gate and metal pitches. This means that the set of possible cell images is defined with the basic fin and metal pitches, and cannot be easily changed later.

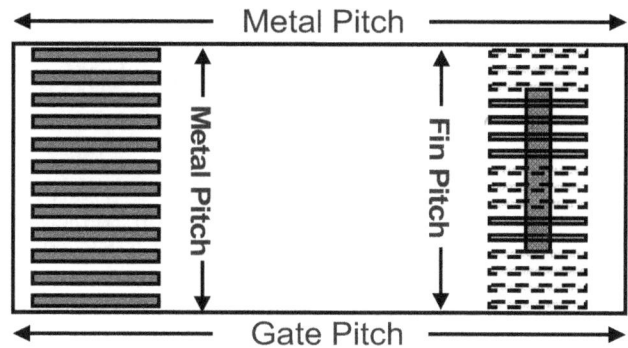

Figure 5. Sample library cell image

New PD and CAD issues[8] will also arise as DPL is used widely for patterning at the 14nm node. The design process may rely on post-layout decomposition, or CAD tools may be color-aware, aiming to produce color-correct designs by construction. The former may require a complicated rule set to guarantee error-free decomposition, whereas the latter will require new infrastructure and sophisticated algorithms to ensure efficient implementation. The need for many non-minimum-width wires, and multiple vias for performance and reliability further increases the difficulty of creating DPL-compatible designs.

Layout-based design analysis and extraction tools will also need enhancements to be able to handle DPL process variability. Correlated or anti-correlated effects due to mis-alignment between colors may result in color-dependent systematic shifts in line-to-line capacitance, for example. Color-aware methodologies will give designers the ability to predict and control these effects.

4. CONCLUSIONS

As both device technology and lithography run up against scaling limits, the 14nm technology generation will see disruptive changes in both transistor structure and in lithographic patterning techniques. The impacts of these changes, along with performance, reliability, and yield considerations will create many new challenges for circuit engineers working on the next generation of high-speed designs.

5. ACKNOWLEDGMENTS

The author would like to thank L. Liebmann for comments and material on optical lithography, and L. Sigal for a critical reading of the manuscript.

6. REFERENCES

[1] M. Bohr, IEDM Tech. Dig., p. 1 (2011)
[2] R.H. Dennard et al, IEEE J. Sol.-St. Circ. **SC-9**, p. 256 (1974).
[3] E. Nowak et al, IEEE Circ. & Devices Mag. **20**, p. 20 (2004).
[4] C. Auth et al, 2012 Symp. on VLSI Tech. Digest p. 131.
[5] C. Shin et al, IEEE Trans. On Elect. Dev. **56**, p. 1538 (2009).
[6] X. Wang et al, ESSDERC Tech. Dig., p. 113 (2012).
[7] V.S. Basker et al, 2010 Symp. on VLSI Tech. Digest, p. 19.
[8] D.Z. Pan et al, ICICDT Tech. Dig. p. 122 (2010).

Optical Lithography Extension with Double Patterning

Shigeki Nojima
Toshiba Corporation Semiconductor & Storage Products Company
2-5-1, Kasama, Sakae-ku, Yokohama 247-8585, Japan
shigeki.nojima@toshiba.co.jp

ABSTRACT

Design shrinking beyond the resolution limits of 193 nm optical lithography requires the introduction of double patterning technology (DPT) for the semiconductor industry. Two consecutive lithography and etching process (LELE; Litho-Etch-Litho-Etch) and self-aligned double patterning (SADP) process are major two options of DPT.

In LELE process, layout patterns are decomposed into two groups and each group is placed on a photo mask. The original layout is reconstructed on a wafer through two lithography with two those photo masks. The two lithography processes are independent because of the intermediate process in between, such as etching and resist freeze. Lithography conditions for each mask are optimized on source mask optimization (SMO) technique to enlarge lithography margin. Conventional SMO technique can be applied to the decomposed patterns since two lithography processes have no interaction as discussed above.

In SADP process, a layout is also decomposed into two groups as same as the methodology for LELE process. Either decomposed group is transferred on a wafer, which is named as mandrel. Sidewalls are deposited on the mandrel patterns and the following processes such as etching and mandrel removal reconstruct the original layout.

Design rules for DPT consist of two parts. One is for an original layout which is drawn by designers. The other is for each decomposed layout printed by a lithography process. Both design rules have interactions because the original layout which follows the former design rules is decomposed into two groups which follow the latter design rules.

In this paper, we will discuss the relation between the two rules above. Furthermore, we will discuss the situation on triple or quadruple patterning.

Categories and Subject Descriptors

B.6.3 [Design Aids]: Optimization; J.6 [COMPUTER-AIDED ENGINEERING]: manufacturing (CAM)

General Terms

Design, Performance, Reliability

Keywords

Lithography, Double patterning, SMO, DFM

A Structured Routing Architecture and its Design Methodology Suitable for High-throughput Electron Beam Direct Writing with Character Projection

Rimon Ikeno[1], Takashi Maruyama[2], Satoshi Komatsu[1], Tetsuya Iizuka[3], Makoto Ikeda[1], and Kunihiro Asada[1]

[1] VLSI Design and Education Center (VDEC), The University of Tokyo
[2] e-Shuttle, Inc.
[3] Dept. of Electrical Eng. and Information Systems, Graduate School of Eng., The University of Tokyo
[1,3] 2-11-6, Yayoi, Bunkyo-ku, Tokyo 113-0032, JAPAN
[2] 2-10-23 Shinyokohama, Kohoku-ku, Yokohama, Kanagawa 222-0033, Japan
ikeno@silicon.u-tokyo.ac.jp

ABSTRACT

To improve throughput of Electron Beam Direct Writing (EBDW) with Character Projection (CP) method, we propose a structured routing architecture (SRA) where VIA placement and wire-track interchange is restricted so as to reduce possible layout patterns in VIA and metal layers. CP exposure is accelerated by the increased figure numbers of VIAs and metal segments in each CP character due to the reduced character variations. We demonstrate a design flow that enables routing on the interconnect layer with the limited flexibility using a commercial routing tool and a few own programs. We also present a methodology to design character stencils with improved area efficiency by character superposition. Our experimental results proved the architecture's feasibility in achieving the target EBDW performance in 14nm technologies.

Categories and Subject Descriptors

B.7.1 [**Integrated Circuits**]: Types and Design Styles – *Advanced technologies*

General Terms

Algorithms, Design, Experimentation

Keywords

Layout Design, Interconnect, Design for Manufacturability, Electron Beam Direct Writing, Character Projection

1. INTRODUCTION

1.1 Electron Beam Direct Writing and Character Projection Method

Electron Beam Direct Writing (EBDW) is expected as a low-cost solution for high-resolution lithography in advanced semiconductor manufacturing technologies with its maskless feature. However, EBDW is not suitable for high-volume production due to its low exposure throughput. This is why application of EBDW has been limited to low-volume production

Permission to make digital or hard copies of all or part of this work for personal or classroom use is granted without fee provided that copies are not made or distributed for profit or commercial advantage and that copies bear this notice and the full citation on the first page. To copy otherwise, or republish, to post on servers or to redistribute to lists, requires prior specific permission and/or a fee.
ISPD'13, March 24– 27, 2013, Stateline, Nevada, USA.
Copyright 2013 ACM 978-1-4503-1954-6/13/03...$15.00.

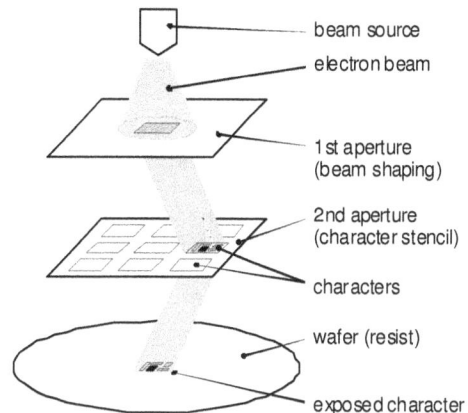

Figure 1: Concept of Electron Beam Direct Writing (EBDW) equipment with Character Projection (CP) method capability

or test chip fabrication even with the existing high-speed EBDW techniques like Variable Shaped Beam (VSB) exposure [1].

One promising enhancement of EBDW for higher exposure throughput is Character Projection (CP) method that exposes a "character" in each EB shot while VSB exposes only a rectangle at once [2][3]. A character is basically composed of a layout pattern which frequently appears on the target chip. This enables multiple layout figures to be exposed in an EB shot. As shown in Figure 1, CP method uses two apertures; one for beam formation and the other for character shaping. With the help of the first aperture, a character on the second aperture ('*character stencil*') is chosen, and the figures in the character are exposed on the target wafer. It is also possible to narrow the beam through the first aperture to expose only a part of the selected character.

A CP character example is presented in Figure 2 along with its decomposed rectangles that might be exposed separately when applying VSB method to the same layout pattern. The number of the VSB rectangles gives the comparison of EB shot counts between VSB and CP methods. In this example, VSB requires 6 EB shots to expose the layout pattern, while CP exposes it in one shot. Then, 6 times better throughput is achieved for this layout pattern by CP method compared to VSB method.

In general, we may define '*CP efficiency*', E_{CP}, as the ratio of EB shot counts by VSB and CP methods ($\equiv$ VSB shot count / CP shot count) for a layout design. A higher E_{CP} means a higher CP throughput compared to VSB. Obviously, E_{CP} is the average

69

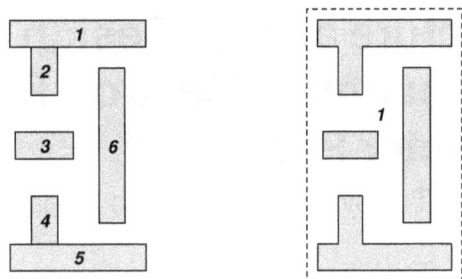

(a) Variable Shaped Beam (VSB)   (b) Character Projection (CP)

Figure 2: Shot count comparison of VSB and CP methods

number of figures exposed in one EB shot. Then, more figures need to be included in CP characters for a higher E_{CP}.

1.2 Efforts for CP Efficiency Improvement

Many researches have been carried out to improve E_{CP} of arbitrary layout patterns with limited CP characters, because the number of the characters is bounded by the character stencil area.

Most active research area of E_{CP} improvement is one related to cell layout in the standard-cell based layout design. Inanami et al. reported $E_{CP} > 10$ with major standard cells in their examples as CP characters [4]. E_{CP} can be further improved by shooting more than one cells at once as a CP character [5][6][7][8]. There, cells are often moved to fit to predefined cell cluster patterns. Finally, such cell clusters are exposed with an EB shot for each, and thus the total EB shot count is reduced.

As for interconnect design, most researches have focus on the metal layers with tile-routing approaches where the metal wires are drawn over square tiles so as either to occupy a tile from one end to the other, or to have a cut point at one end to start another wire from the next tile [9][10]. CP characters are composed as combinations of such limited segment shapes, which result in limited character variations.

However, application of CP method to VIA layers has not been studied intensely despite its high degree of freedom. Du et al. introduced an area-efficient stencil design for VIA patterns that allows three VIAs in each shot [10]. Also, we proposed a VIA CP strategy with character sets consisting of only one-dimensional VIA arrays along with an area-efficient stencil design scheme, and reported CP efficiency of 4~6 VIA/shot in average [11].

In this study, we focus on both VIA and metal layers, and discuss our new strategy for high-speed CP exposure of these layers. We propose a structured routing architecture to improve the layout regularity by its limited layout flexibility.

2. STRUCTURED ROUTING ARCHITECTURE AND STRENCIL DESIGN FOR CHARACTER PROJECTION

2.1 Structured Routing Architecture

In standard-cell based place-and-route (P&R) design, inter-cell signal nets are routed in the metal layers where either vertical or horizontal routing tracks are defined. Also, VIA layers are used to connect two metal layers below and above each VIA layer.

Then, we introduce a structured routing architecture (SRA) into the routing layers as shown in Figure 3. There, each routing track is given an index called '*color*' periodically, and each set of the tracks sharing a common color is called a '*track group*'. VIAs are

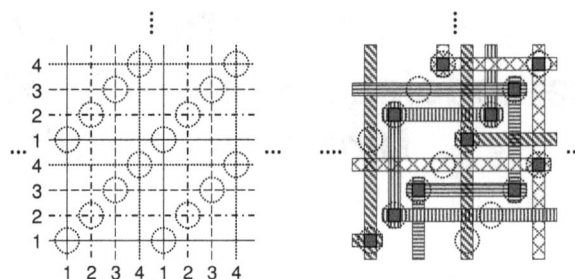

(a) Structure routing tracks   (b) Structured routing example

Figure 3: Basic concept of structure routing architecture

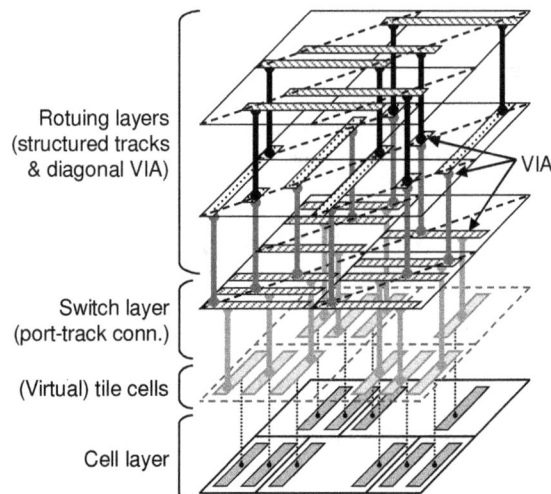

Figure 4: Metal layers for inter-cell wire routing

allowed only on the cross points of vertical and horizontal tracks sharing a common color (circles in Figure 3), and this results in the restricted VIA arrangement on the diagonal lines. Figure 3 is an example of SRA where track group number N_T is 4.

By the VIA placement constraint above, track group transitions are not allowed during routing in SRA, and thus a specific track group must be defined for each signal before routing. This should be a big difference from many commercial routing tools that run coarse routing processes before doing such track assignment.

In addition to this, the track group constraint requires that all cell ports connected to a signal must be aligned to the assigned track. However, this requirement cannot be satisfied because track alignment to cell ports varies in each cell instance. Then, in SRA, we reserve a few metal and VIA layers between the cell layer and the routing layers as '*switch layer*' as shown in Figure 4. Also, the cell layer is restructured into virtual '*tile cells*' that have a regular square shape with the size of the track group number N_T. The switch layer connects the cell ports and their target tracks within each tile cell.

Because a tile cell contains only one possible routing track for each track group, all signal nets connected to a tile cell may not have conflicting track groups. This restriction must be honored in the track group assignment.

Due to the design strategies discussed so far, SRA requires really limited VIA and metal layout patterns, which result in quite less number of CP characters compared to the ordinary layout design style. Then, higher CP exposure throughput can be achieved by employing VIA and metal character sets with more figures (VIAs or metal segments) in each character without exploding the number of CP characters.

2.2 VIA Character Set and Stencil Design

Considering the diagonal VIA arrangement in SRA, it is quite natural to expose such diagonal VIA arrays as CP characters. Figure 5 (a) presents such VIA-array character examples with 5 potential VIA sites (grids). A VIA-array character set consists of 2^M characters, when the grid number in a character is M. The characters are identified by binary codes, whose digits represent VIA existence ('1') or absence ('0') at the corresponding grids,

Stencil area suppression is important so as to increase the integrated characters on the limited stencil area. Figure 5 (a) also

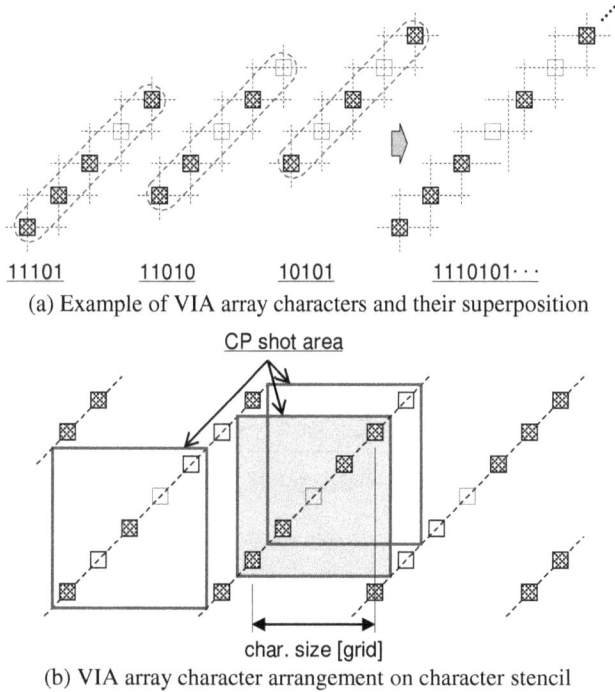

11101   11010   10101   1110101···

(a) Example of VIA array characters and their superposition

(b) VIA array character arrangement on character stencil

Figure 5: VIA array character set and VIA stencil design

Algorithm 1: Ordering VIA/metal characters for superposed arrangement

```
M:      VIA grid number / Metal track number
D:      Largest figure for a digit 1(VIA)/2(metal)
ARRAY:  Storage for the arranged character array
CLIST:  List of (D+1)^M characters
        ("00..00", "00..01",  .., "DD..DD")
ORDER:  Storage for each character's order
        (Initialize with all 0)

1:  char = "D0..00"         # Last char
2:  cord = (D+1)^M          # Last char's order
3:  ARRAY = char            # Last char in array
4:  label search top        # Search routine top
5:  ORDER{char} = cord      # Set char order
6:  ct = substr(char, 0 ,M-1) # Next char's tail
7:  for i=D downto 0 do     # Next char's head
8:    char = append(i, ct)  # Candidate char
9:    if ORDER{c0} == 0 then # If not ordered yet
10:     ARRAY = append(i, ARRAY)# Extend array
11:     cord--              # Decr. char order
12:     goto search_top     # Go top & set order
13:   end if
14: end for                 # Exit if no cand.
```

depicts a way to save the character stencil area by superposing the characters to share common parts among several characters. There, the VIA array characters are superposed by 1-grid shift between each consecutive character pair. A set of binary codes with any bit length can be ordered in such overlapping manner using a simple deterministic procedure presented in Algorithm 1.

The resultant long VIA chain is divided into shorter arrays, and finally, they are placed on the stencil as in Figure 5 (b). There are spaces between the arrays so that a CP shot does not expose unnecessary VIAs nearby and exposes only the desired VIAs.

2.3 Metal Character Set and Stencil Design

In the structured metal layers in SRA, metal wires are composed of only three segment patterns; 1) *blank*, 2) *pad*, and 3) *span* shown in Figure 6 (a). Then, the CP characters for the metal layers are made as diagonal arrangements of these patterns (Figure 6 (b)), which are identified by ternary codes with the segment pattern indices as their trits. A metal character set consists of 3^M characters, when a character occupies M tracks.

The metal characters are also ordered and superposed on others using Algorithm 1 for stencil area saving. The resultant long segment array is divided and arranged on the stencil as in Figure 6 (c) with some spaces so as to avoid the CP shot conflictions.

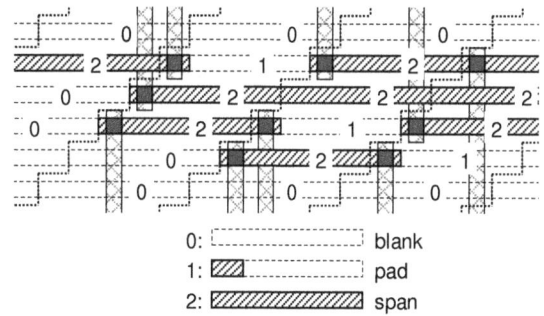

0: ┈┈┈┈┈┈┈┈┈┈ blank
1: ▨┈┈┈┈┈┈┈ pad
2: ▨▨▨▨▨▨▨ span

(a) Metal segment patterns and ternary code assignments

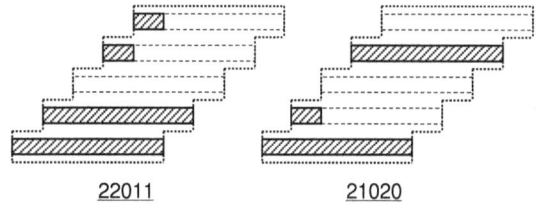

22011   21020

(b) Examples of metal characters and character codes

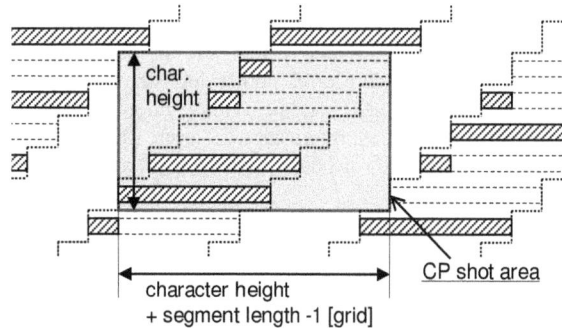

character height
+ segment length -1 [grid]

(c) Metal character arrangement on character stencil

Figure 6: Metal character set and metal stencil design

2.4 Character Set Planning with Stencil Area Constraint

In our study, we assume the stencil area of 9,000 μm^2 from a projected specification of CP equipments for 14nm technologies [8]. Then, assuming the track pitch as 0.05 μm, the stencil area constraint is translated into grids as below. VIA and metal character sets have to be planed within this constraint.

$$\text{(Max stencil area)} = 9,000 / 0.05^2 = 3.6 \times 10^6 \text{ [grid]}$$

As for the VIA characters, the total character number is 2^M, where M is the grid number in each character as discussed in section 2.2. Considering the superposed arrangement of the VIA characters and the spaces between the VIA arrays on the stencil, one character occupies M grids in average. Then, the stencil area for a VIA character set is formulated as below.

$$\text{(VIA stencil area)} = 2^M \times M \text{ [grid]}$$

Solving $2^M \times M < 3.6 \times 10^6$, we get M=17 at the maximum.

The number of metal characters and the average area occupied by a character is 3^M and $M+L-1$, respectively, where M is the track number in a character and L is the metal segment length. Then, the stencil area required for the metal character set is as below.

$$\text{(Metal stencil area)} = 3^M \times (M+L-1) \text{ [grid]}$$

Basically, the segment length L is defined by the span between the diagonal VIA sites that reflects the track group number, N_T. In this study, we adopt $N_T = L = 9$, because the library cell height is 9 grids and the virtual tile cell conversion goes smoothly if N_T equals to a multiple of the cell height.

Solving $3^M \times (M+9-1) < 3.6 \times 10^6$, we get M≤11. Then our metal characters are composed with 11 tracks with 9-grid segments.

The parameters of VIA and metal character stencils used in this study are summarized in Table 1.

3. STRUCTURED ROUTING DESIGN METHODOLOGY

3.1 Routing Flow Overview

As noted in section 2.1, SRA requires "track assignment" to be done first, despite that many commercial routing tools assign the routing tracks after some coarse routing processes like "global routing". In Figure 7, our routing design flow for SRA utilizing a commercial routing tool is presented. The routing tool is used in processes (iii) and (v), while the other operations are processed by our own programs. Details of each process are discussed in the following sections.

3.2 Tile-cell Netlist Generation

In process (i), the cell-based netlist of the target design is converted into the virtual tile-cell netlist discussed in section 2.1. The process requires cell placement information in the layout design and port location information in each library cell. The resultant tile-cell netlist is handed to processes (ii) and (iv).

Table 1: Parameters of VIA and metal character stencils

Layer	VIA	Metal
Character size [grid]	17	11
Character number	131,072	177,147
Stencil area [grid]	2,228,224	3,365,793

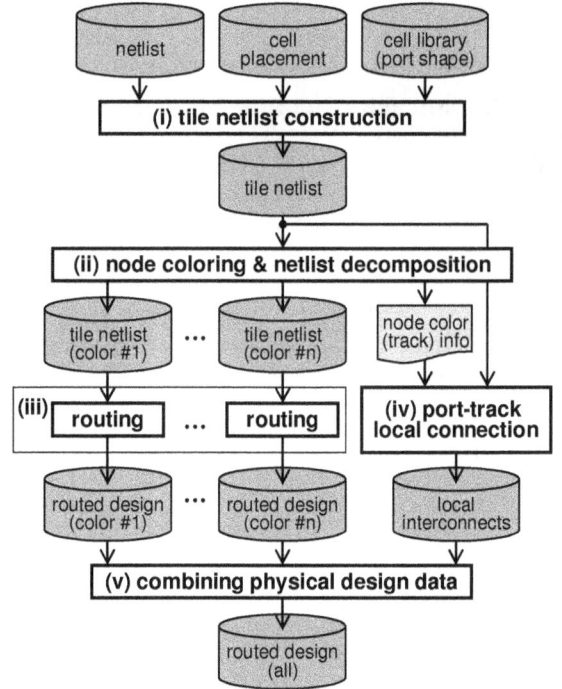

Figure 7: Routing flow for the structured routing architecture

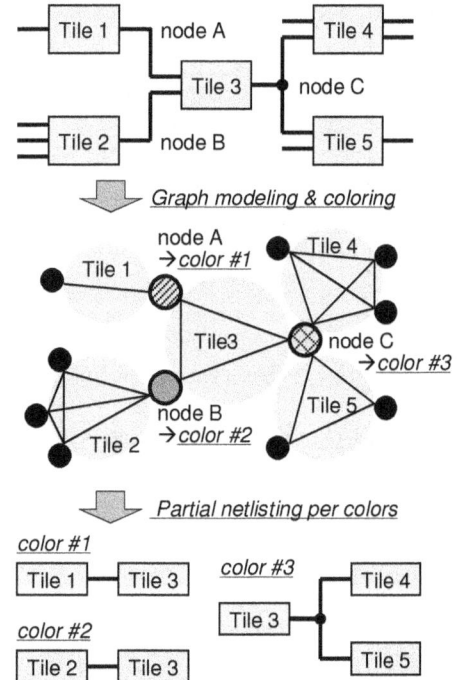

Figure 8: Track (color) assignment based on graph coloring

3.3 Track Assignment by Graph Coloring

In process (ii), the track groups are assigned to the signal nets in the tile-cell netlist. As discussed previously, this must be done so that all nets connected to a tile cell do not have group conflicts.

We solve it as a graph-coloring problem that assigns colors to all vertices in a graph so that any pairs of adjacent vertices have different colors. The tile-cell netlist is modeled as a graph structure shown in Figure 8. There, the graph vertices (nodes) represent the signal nets in the netlist, and the tile cells are modeled as partial complete graphs (cliques). With this graph-modeling, all signal nets connected to a tile cell are adjacent to each other through the clique edges, and thus they would have different colors after the color assignment. The colored netlist is then divided into the partial netlists that contain only the nets and the tile cells with a common color (bottom of Figure 8).

3.4 Interconnect Routing by Track Groups

In process (iii), "detailed routing" is carried out by a commercial routing tool for each partial netlist given in process (ii). Since there is no interference anticipated among different track groups, the detailed routing processes can be run independently in parallel. Finally, the partial routing results are superposed onto a single layout design for the whole netlist as shown in Figure 9.

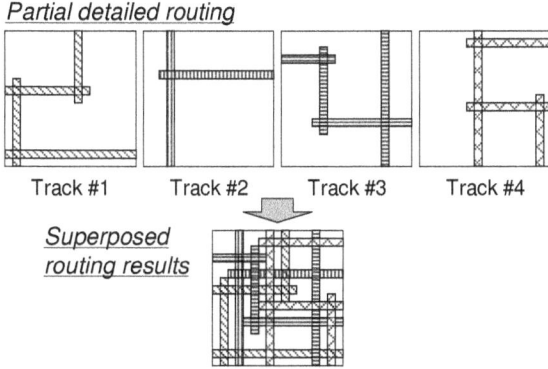

Figure 9: Partial detailed routing results and their superposition for the whole-netlist routing

3.5 Local Links from Cell Ports to Tracks

In process (iv), local connections from the tile-cell ports to their corresponding tracks are routed inside each tile using fixed-length metal segments in the switch layer as shown in Figure 4. The arbitrary connection (switching) from the cell ports to the routing tracks at the top is realized by VIA arrangement.

In Figure 10, two implementations of four-ports to four-tracks switching are presented as examples.

In '1-step switching' implementation where one VIA layer is involved, the sole VIA layer is used to realize all combinations of port to track connections. Its combinatorial number is $4! = 24$, and thus we need 24 CP characters for the VIA layer.

On the other hand, in '2-step switching' implementation with two VIA layers and one middle metal layer, the port to track interchanges are realized using the two VIA layers. For example, if

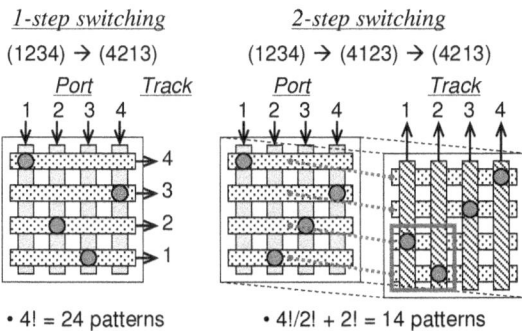

Figure 10: Local connection of cell ports to routing tracks by 1-step and 2-step switching layers

the upper VIA layer (the right figure) owes the interchange between 1st and 2nd tracks, the required variations of VIA placement in the layer are just two in number (only inside the gray box). Then, the lower VIA layer needs to have $12 / 2 = 6$ VIA arrangement variations. Consequently, only $12 + 2 = 14$ VIA characters are needed in total, which are ten-characters less than the 1-step switching implementation. A general formulation of this '2-step switching' scheme is described hereafter.

The set of all possible permutations from N ports to N tracks is identical to a symmetric group S_N for a finite set with N symbols. The cardinality of such symmetric group is given as:

$$|S_N| = N! = N \times (N-1) \times \cdots \times 2 \times 1$$

A permutation σ in S_N can be expressed as a product (composition) of cyclic permutations like below.

$$\sigma(k_N, k_{N-1}, \cdots, k_1) = c_N^{kN} \circ c_{N-1}^{kN-1} \circ \cdots \circ c_2^{k2} \circ c_1^{k1}$$

Here, c_n ($n = 1, 2, \cdots, N$) is the cyclic permutation with length n;

$$c_n : (1, 2, 3, \cdots, n-1, n, n+1, \cdots, N)$$
$$\rightarrow (2, 3, 4, \cdots, n, 1, n+1, \cdots, N)$$

and c_n^{kn} is the multiple cyclic permutation of c_n by k_n times, where k_n ($k_n = 0, 1, \cdots, n-1$) are defined uniquely for each σ.

Next, let us suppose another expression of σ in a decomposed form (product of two permutations) as below.

$$\sigma(k_N, k_{N-1}, \cdots, k_2) = \sigma_1(k_N, \cdots, k_{m+1}) \circ \sigma_2(k_m, \cdots, k_1)$$

Here. σ_1 and σ_2 are the composed permutations with the following formulae. They form two permutation groups G_1 and G_2.

$$\sigma_1(k_N, \cdots, k_{m+1}) = c_N^{kN} \circ c_{N-1}^{kN-1} \circ \cdots \circ c_{m+1}^{km+1}$$
$$\sigma_2(k_m, \cdots, k_2) = c_m^{km} \circ c_{m-1}^{km-1} \circ \cdots \circ c_1^{k1}$$

This describes how a permutation σ in the symmetric group S_N is expressed by a product of two permutations from groups G_1 and G_2 for each. Permutations in G_1 and G_2 are implemented as the first and the second switch (VIA) layers, respectively, by placing VIAs at the grid points corresponding to the permutations.

Considering the element numbers in the two permutation groups;

$$|G_1| = N \times (N-1) \times \cdots \times (m+1) = N! / m!$$
$$|G_2| = m \times (m-1) \times \cdots \times 2 \times 1 = m!$$

the total number of the permutations required for the two-step switching is formulated as below, which gives the VIA character number required for the switch layer in total.

$$|G_1| + |G_2| = N! / m! + m!$$

The methodology above is applicable to switch implementations with more than 2 steps, and it implies that more VIA layers in the switch layer could result in less CP characters.

In our 9-track architecture, the 1-step switching requires $9! = 362,880$ VIA characters and $362,880 \times 9 \times 9 \sim 3.0 \times 10^7$ grids of stencil area, which exceeds the 3.6×10^6 limit. However, if we adopt 2-step switching with $6! = 720$ flexibility in the second VIA layer, the required flexibility for the first VIA layer is $9!/6! = 540$, resulting in the total VIA character number of $720 + 540 = 1260$. It is quite a small number compared to the 1-step case. It enables stencil design within the limited stencil area.

In our design flow, we first generate a table of all N_T-to-N_T permutations σ and their decomposed permutations σ_1 and σ_2. Then we choose VIA placements in each tile from the table according to the port-to-track correspondence given as the node coloring result from process (ii) in Figure 7.

4. EXPERIMENTAL RESULTS

4.1 Reference Designs and VSB Performance

We developed an SRA design flow using Synopsys IC Compiler and a few own programs. EB exposure performance by CP method of the SRA layout results were evaluated and compared with VSB exposure performance of their reference designs.

First, we created the reference designs with a typical design flow. HDL codes of four logic circuits in Table 2 were got from OpenCores [13], and they went through the synthesis and the place-and-route steps using Synopsys Design Compiler and Synopsys IC Compiler, respectively. We assumed a low-power 65nm CMOS library from Fujitsu Semiconductor Ltd, but its cell layouts were modified to have a regular port shape (a fixed-length rectangle in MET2 layer) so as to enable the port-to-track connection through the switch layers as discussed in section 3.5. We assumed the layout results (summarized in Table 2) as the reference designs in our EBDW performance evaluation.

EBDW performance was evaluated by the EB shot count required for the whole-wafer exposure of 14nm technology, assuming that the target layout was covering the wafer. We used the formula below for the shot-count calculation based on the effective shot area of 10.0 μm^2 (the target shot area for 100 Giga show per wafer, scaled to 65nm technology). The target range of the EB shot count is 100~173 G-shot/wafer for realistic applications [8].

$$(\text{Gshot/wafer}) = 100 \times \frac{10.0 \times (\text{shot\#})}{(\text{layout area})}$$

VSB shot counts of VIA2 and MET3 of the reference designs are presented at the bottom of Table 2. These layers are most dense layers in the VIA and the metal layers for routing, respectively, and thus they define the overall performance of the EBDW equipments used for the chip production. The estimated shot counts are around 1000 G-shot/wafer, which are far larger than the target shot count range. In addition, it is expected that the VSB exposure requires

Table 2: Reference layout designs of example logic circuits

Design name	8080 Compat. CPU	USB 2.0 Function Core	AES (Rijndael) IP Core	Discrete Cosine Transform
Clock period [ns]	600	300	600	1,000
Layout area [um²]	11,750.40	29,309.40	60,269.76	63,907.20
Cell area [um²]	11,321.28	26,997.48	51,955.92	55,013.40
Cell utilization	96.3%	92.1%	86.2%	86.1%
Logic cell#	4,326	7,708	16,755	13,880
Signal net#	4,346	8,089	17,014	14,295
VIA2#	15,311	30,340	58,513	54,170
VIA3#	6,854	10,957	27,106	19,330
VIA4#	3,366	5,304	15,049	6,613
VIA2 dens [um⁻²]	1.303	1.035	0.971	0.848
VIA3 dens [um⁻²]	0.583	0.374	0.450	0.302
VIA4 dens [um⁻²]	0.286	0.181	0.250	0.103
MET3 segment #	13,174	24,398	49,818	43,732
MET4 segment #	5,934	9,461	24,579	15,507
MET5 segment #	2,852	4,162	12,739	4,858
MET3 dens[um⁻²]	1.121	0.832	0.827	0.684
MET4 dens[um⁻²]	0.505	0.323	0.408	0.243
MET5 dens[um⁻²]	0.243	0.142	0.211	0.076
VSB shot count estimation [G-shot/wafer]				
VIA2	1303.0	1035.2	970.9	847.6
MET3	1121.2	832.4	826.6	684.3

Table 3: Summary of tile netlists and node coloring results

Design name		8080 Compat. CPU	USB 2.0 Function Core	AES (Rijndael) IP Core	Discrete Cosine Transform
Tile cell #		3,479	8,316	16,859	17,597
Tile port #		14,698	28,380	55,366	51,074
Total node#		4,346	8,077	17,014	14,297
Node# per track color	1	829	1,747	4,115	3,623
	2	787	1,554	3,720	3,048
	3	726	1,311	3,122	2,438
	4	665	1,299	2,407	2,068
	5	597	1,114	1,795	1,496
	6	523	745	1,338	1,063
	7	204	261	502	414
	8	15	38	14	133
	9	0	8	1	14
Max. color		8	9	9	9

more EB shots for long wire segments due to the shot size limitation.

4.2 Tile-Cell Netlist Generation and Track Group Assignment

The SRA design flow was processed based on the netlists and the cell placement information of the reference designs.

Table 3 summarizes the results of tile-cell netlist generation and the node coloring (track group assignment) that correspond to the outputs from processes (i) and (ii) in Figure 7. We used a node coloring algorithm by Brelaz [12]. Track group assignment was successful for all example designs, although the node numbers were not uniform among the colors due to the simple coloring algorithm for fast processing.

4.3 Partial Netlist Routing and Superposition

After the track group assignment, detailed routing of the partial netlists was carried out for each track group separately using Synopsys IC Compiler (process (iii)). Figure 11 shows the routing results of two track groups (colors 2 and 7) of "USB 2.0" as examples. Since metal segments are only allowed on the specific tracks, they may look very sparse in each partial design.

The partial routing results were then joined into a single layout shown in Figure 12. The dedicated tracks to the track groups are superposed without conflicts (short circuits) to each other. Then, the resultant layout design could be treated as the final routing result of the whole tile-cell netlist. Figure 12 (b) shows VIA4, MET5, and MET1 (power and ground) layers in the resultant

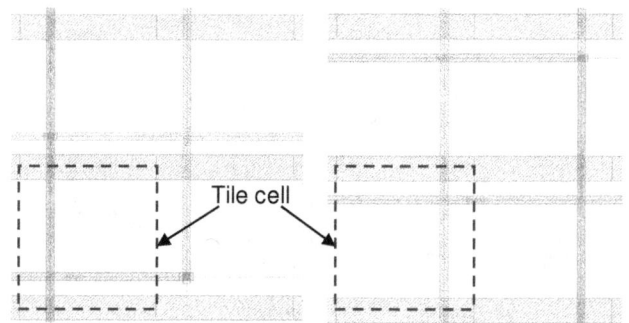

(a) Track group 2   (b) Track group 7

Figure 11: Routing results of partial netlists (USB 2.0)

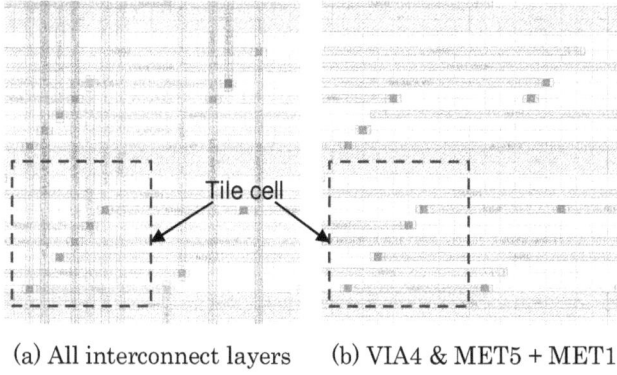

(a) All interconnect layers   (b) VIA4 & MET5 + MET1

Figure 12: Superposed interconnect design (USB 2.0)

layout, where VIAs were placed on diagonal lines as we expected. The MET5 wire segments started or ended at such diagonal VIA grids with the normalized segment lengths by the tile size.

Table 4 compares the total wire lengths in each metal layer in the reference and the SRA layouts of "USB 2.0". Wire routing was done in MET4 layer and above in SRA due to the switch-layer insertion for the port-to-track switching. Despite the coarse grid feature of SRA, the total wire length was shorter than the reference layout. Looking at the other designs' results in Table 5, wire lengths were either shorter or longer, but they differed by less than 20%.

4.4 CP Performance of SRA Layout Results

We evaluated the EBDW performance of VIA and metal layers of the SRA layout results by CP method using the CP stencils discussed in section 2.4. Table 6 shows the results of "USB 2.0". There, "VIA (Segment) utilization" is the ratio of the occupied VIA (or metal-segment) sites to the potential sites. "Shot number" is the required EB shot count for the design.

EB shot counts for the whole wafer ("Giga-shot/wafer") were calculated using the formula presented in section 4.1. The largest shot count values were seen in the bottom layers of the structured routing layers (VIA4 and MET5), and were 138.8 G-shot/wafer in the VIA layers and about 204.6 G-shot/wafer in the metal layers,

Table 4: Wire length comparison by layers of the original and the structured routing architectures (USB 2.0)

Reference		Structured	
Layer	Length [um]	Layer	Length [um]
MET1	1,079.6	<MET4	43.2
MET2	15,879.8	MET4	17,398.8
MET3	37,475.9	MET5	40,914.0
MET4	31,732.7	MET6	34,495.2
MET5	30,313.6	MET7	25,819.2
MET6	18,980.9	MET8	15,049.8
>MET6	9,573.9	>MET8	7,374.6
Total	**145,036.3**	**total**	**141,094.8**
		Structured/Original	**97.3%**

Table 5: Wire length comparison of the original and the structured routing architectures (USB 2.0)

	Original	Structured	**Str/Orig**
8080	68,174.9	80,567.6	**118.2%**
USB	145,036.3	141,094.8	**97.3%**
AES	340,980.5	371,617.2	**109.0%**
DCT	191,027.1	158,561.6	**83.0%**

respectively. In this example, VIA layer exposure could meet the target shot count range (100~173 G-shot/wafer), but metal layer exposure could not.

CP efficiency ("E_{CP}") was also calculated as the average number of exposed VIA or metal figures in one EB shot (section 1.1). It ranged from 1 to 6 in both VIA and metal layers, and showed higher efficiency in the layers with more figures in general.

CP performance estimations of VIA4 and MET5 layers of the example circuits are shown in Table 7. The estimated shot counts ranged 120.3~152.3 G-shot/wafer for VIA4, and 168.8~219.2 G-shot/wafer for MET5. E_{CP} ranged from about 4 to 7.

These results are almost same as what we have seen in the "USB 2.0" results; VIA layers met the target performance but metal layers did not. The gap between the current and the target CP exposure performance in the metal layers should be filled by metal segment reduction by tile size increase. It would also result in increase in the local connections inside the tiles and thus in less global routing wires through the structured routing layers.

Table 8 summarizes the EBDW performance improvement from the reference designs with VSB method to the SRA designs with CP method. In average, the proposed scheme improved the EBDW performance by about 7.5 times in the VIA layers and 4.4 times in the metal layers, respectively.

Table 6: CP performance estimation of VIA and metal layers of "USB 2.0" SRA layout

Layer	VIA#	VIA util.	Shot#	G-shot/ wafer	E_{CP}	Char. efficiency
VIA4	22,288	27.4%	4,068	138.8	5.48	32.2%
VIA5	9,871	12.1%	3,163	107.9	3.12	18.4%
VIA6	5,215	6.4%	2,301	78.5	2.27	13.4%
VIA7	1,996	2.5%	1,223	41.7	1.63	9.6%
VIA8	600	0.7%	444	15.1	1.35	7.9%
VIA9	100	0.1%	91	3.1	1.10	6.5%
Layer	Seg#	Segment util.	Shot#	G-shot/ wafer	E_{CP}	Char. efficiency
MET5	35,482	43.6%	5,998	204.6	5.92	53.8%
MET6	25,836	31.7%	5,410	184.6	4.78	43.5%
MET7	17,600	21.6%	4,461	152.2	3.95	35.9%
MET8	9,480	11.6%	3,168	108.1	2.99	27.2%
MET9	3,746	4.6%	1,440	49.1	2.60	23.6%
MET10	583	0.7%	446	15.2	1.31	11.9%

Table 7: CP performance estimation of VIA and metal layers of the example circuits (VIA4 and MET5 layers)

Design name	VIA4#	VIA4 util.	Shot#	G-shot/ wafer	E_{CP}	Char. efficiency
8080	11,861	36.3%	1,790	152.3	6.63	39.0%
USB	22,288	27.4%	4,068	138.8	5.48	32.2%
AES	46,266	27.6%	8,253	136.9	5.61	33.0%
DCT	29,914	16.9%	7,689	120.3	3.89	22.9%
Design name	MET5 Seg#	MET5 util.	Shot#	G-shot/ Wafer	E_{CP}	Char. efficiency
8080	15,563	47.7%	2,576	219.2	6.04	54.9%
USB	35,482	43.6%	5,998	204.6	5.92	53.8%
AES	64,850	38.7%	11,928	197.9	5.44	49.5%
DCT	46,602	26.3%	10,785	168.8	4.32	39.3%

Table 8: EBDW throughput improvement from VSB (reference) to CP (SRA)

Design name	VIA			Metal		
	G-shot/wafer		VSB /CP	G-shot/wafer		VSB /CP
	VSB (VIA2)	CP (VIA4)		VSB (MET3)	CP (MET5)	
8080	1303.0	152.3	8.55	1121.2	219.2	5.11
USB	1035.2	138.8	7.46	832.4	204.6	4.07
AES	970.9	136.9	7.09	826.6	197.9	4.18
DCT	847.6	120.3	7.05	684.3	168.8	4.05
Average	1039.2	137.1	7.58	866.1	197.6	4.38

5. CONCLUSION

In this study, we focused on throughput improvement of character projection (CP) method for EBDW, especially from the viewpoint of the interconnect design in VIA and metal layers.

We introduced a structured routing architecture (SRA) into the interconnect layers, and presented its routing design flow using a commercial routing tool. In SRA, possible VIA and metal layout patterns are restricted to diagonal arrangements of VIAs or fixed-length metal segments, resulting in less CP character variations compared to ones required for the conventional layout results. We also developed an area-efficient design methodology of character stencils with the VIA and metal patterns above as the CP characters. As their results, CP character stencils with more figures in each character can be realized within the limited area, and great improvement in CP exposure performance is expected compared to the VSB performance of the conventional layouts.

We demonstrated the SRA design flow with some example designs, and evaluated EBDW performance of the CP strategy with the area-efficient CP character stencils. The required EB shots were reduced from 1000 G-shot/wafer range by VSB method on the reference designs to 100~220 G-shot/wafer range by the proposed CP strategy. The typical improvement factors were about 7.5 and 4.4 in VIA and metal layers, respectively. CP efficiency, E_{CP}, was evaluated as the average number of the exposed figures in one EB shot, and it achieved more than 6 figures per shot in the densest layers (VIA4 and MET5).

It was concluded that the exposure throughput of VIA layers almost achieved the target throughput for 14nm technologies while metal layer throughput missed the target range. Further EB shot reduction should be attempted by tile-size optimization that might require a revisit of the cell library architecture like the cell height and the tracks pitch.

6. REFERENCES

[1] Pfeiffer, H. C. 1978. Variable spot shaping for electron-beam lithography. *Journal of Vacuum Science & Technology*, 15 (May 1978), 887-890.

[2] Pfeiffer, H. C. 1979. Recent advances in electron-beam lithography for the high-volume production of VLSI devices. *IEEE Trans. Electron Devices* ED-264 (Apr. 1979), 663-674.

[3] Yamada, A., Yasuda, H., and Yamabe, M. 2008. Electron beams in individual column cells of multicolumn cell system. *Journal of Vacuum Science & Technology B: Microelectronics and Nanometer Structures* 26 (Nov./Dec. 2008), 2025-2031.

[4] Inanami, R., Magoshi, S., Kousai, S., Hamada, M., Takayanagi, T., Sugihara, K., Okumura, K., and Kuroda, T. 2000. Throughput enhancement strategy of maskless electron beam direct writing for logic device. In *Technical Digest of 2000 IEEE International Electron Devices Meeting* (San Francisco, CA, December 11-13, 2000). IEDM '00. IEEE, 833-836.

[5] Sugihara, M., Takata, T., Nakamura, K., Inanami, R., Hayashi, H., Kishimoto, K., Hasebe, T., Kawano, Y., Matsunaga, Y., Murakami, K., and Okumura, K. 2006. Cell library development methodology for throughput enhancement of character projection equipment. *IEICE Trans. Electronics* E89-C, (Mar. 2006), 377-383.

[6] Kosai, S., Inanami, R., Hamada, M., Magoshi, S., and Hatori, F. 2006. Throughput enhancement in electron beam direct writing by multiple-cell shot technique for logic devices. In *Proceedings of the 17th IEEE/SEMI Advanced Semiconductor Manufacturing Conference* (Boston, MA, May 22-24, 2006). ASMC 2006. ACM/IEEE, 253–256.

[7] Kazama, T., Ikeda, M., and Asada, K. 2006. LSI design flow for shot reduction of character projection electron beam direct writing using combined cell stencil. *IEICE Trans. Fundamentals* E89-A (Dec. 2006), 3546-3550.

[8] Maruyama, T., Machida, Y., Sugatani, S., Takita, H., Hoshino, H., Hino, T., Ito, M., Yamada, A., Iizuka, T., Komatsu, S., Ikeda, M., and Asada, K. 2012. CP element based design for 14nm node EBDW high volume manufacturing. In *Proceedings of SPIE 8323: Alternative Lithographic Technologies* IV (San Jose, CA, February 12-16, 2012).

[9] Minh, H. P. D., Iizuka, T., Ikeda, M., and Asada, K. 2008. Minimization of electron beam shots to enhance the throughput of character projection electron beam direct writing. In *Proceedings of SPIE 6921: Emerging Lithographic Technologies* XII (San Jose, CA, February 26-28, 2008).

[10] Du, P., Zhao, W., Weng, S.-H., Cheng, C.-K., and Graham, R. 2012. Character design and stamp algorithms for character projection electron-beam lithography. In *Proceedings of 17th Asia and South Pacific Design Automation Conference* (Sydney, Australia, January 30- February 2, 2012). ASP-DAC 2012. 725-730.

[11] Ikeno, R., Maruyama, T., Komatsu, T., Iizuka, T., Ikeda, M., and Asada, K., High-throughput Electron Beam Direct Writing of VIA Layers by Character Projection using Character Sets Based on One-dimensional VIA Arrays with Area-efficient Stencil Design. In *Proceedings of 18th Asia and South Pacific Design Automation Conference* (Yokohama, Japan, January 22-25, 2013). ASP-DAC 2013. 255-260.

[12] Brelaz, D. 1979. New methods to color the vertices of a graph. *Communications of ACM* 22 (Apr. 1979), 251-256.

[13] http://opencores.org

Simultaneous OPC- and CMP-aware Routing Based on Accurate Closed-Form Modeling *

Shao-Yun Fang[1], Chung-Wei Lin[1,2], Guang-Wan Liao[1], and Yao-Wen Chang[1,3,4]

[1]Graduate Institute of Electronics Engineering, National Taiwan University, Taipei 106, Taiwan
[2]Department of Electrical Engineering and Computer Science, University of California, Berkeley, CA 94720
[3]Department of Electrical Engineering, National Taiwan University, Taipei 106, Taiwan
[4]Research Center for Information Technology Innovation, Academia Sinica, Taipei 115, Taiwan
{yuko703, enorm, gwliao}@eda.ee.ntu.edu.tw; ywchang@cc.ee.ntu.edu.tw

ABSTRACT

As the process technology advances to the nanometer nodes, Optical Proximity Correction (OPC) is the most popular Resolution Enhancement Technique (RET) in industry for subwavelength lithography, and the inter-level dielectric (ILD) thickness variation caused by the planarization step of a Chemical Mechanical Polishing (CMP) process also plays a key role for interconnect yield. Considering the OPC and CMP effects simultaneously during routing can significantly alleviate the width and thickness variations (and thus the whole 3D geometry variations) of post-layout RET and CMP operations. In this paper, we first present an efficient, yet sufficiently accurate closed-form formula for printed width computation and dummy-insertion-aware routing cost derivation. The formula provides an effective cost modeling to guide a router for better post-layout OPC correction and CMP planarization. Compared with the state-of-the-art OPC-friendly router, QL-MGR (which does not consider CMP), the experimental results show that our router can achieve respective 19% and 6% reductions in the maximum and average layout distortions. Compared with the state-of-the-art CMP-aware router, TTR (which does not consider OPC), the experimental results show that our router can achieve respective 19% and 25% reductions in the peak-to-peak thickness and thickness variance. These results demonstrate that our simultaneous OPC- and CMP-aware router can lead to significant improvements for layout integrity.

Categories and Subject Descriptors

B.7.2 [**Integrated Circuits**]: Design Aids

General Terms

Algorithms, Design, Performance

*This work was partially supported by IBM, SpringSoft, TSMC, and NSC of Taiwan under Grant No's. NSC 100-2221-E-002-088-MY3, NSC 99-2221-E-002-207-MY3, NSC 99-2221-E-002-210-MY3, and NSC 98-2221-E-002-119-MY3.

Keywords

Chemical Mechanical Polishing, Optical Proximity Correction, Resolution Enhancement Technique, Routing

1. INTRODUCTION

As IC technology advances to the nanometer territory, the industry faces severe challenges of manufacturing limitations. Optical Proximity Correction (OPC) is the most popular Resolution Enhancement Technique (RET) in industry for subwavelength lithography, and the inter-level dielectric (ILD) thickness variation caused by the planarization step of a Chemical Mechanical Polishing (CMP) process plays a key role for interconnect yield.

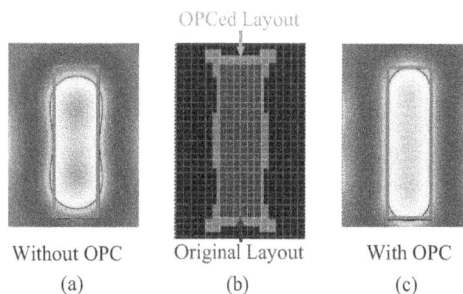

Figure 1: (a) Distortion without OPC. The red rectangle is the desired image. (b) The red area inside the green area is the original layout. The green area is the OPCed layout. (c) Smaller distortion with OPC [7].

The OPC operations may change layout shapes, as shown in Figure 1 [7]. In some situations, the shapes are too close to reserve enough spacing for OPC processing, and thus a larger spacing between design shapes is preferred for OPC operations. However, it increases chip area. Therefore, it is desirable to consider the trade-off between image distortions and area costs.

On the other hand, the thicknesses of dielectric and metal layers after a CMP process strongly depend on layout patterns [16]. Because of the difference in hardness between metal and dielectric materials, a planarization process might generate topography irregularities. Non-uniform distribution of feature density on each layer makes a wafer over polished or under polished and incurs metal dishing and/or dielectric erosion, as illustrated in Figure 2(a). To tackle these problems, dummy fill is often used to enhance the layout pattern uniformity, as illustrated in Figure 2(b).

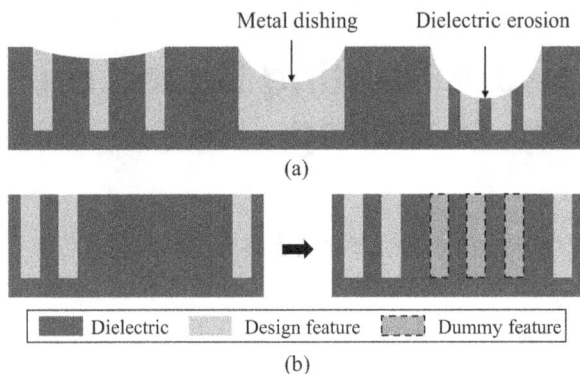

Figure 2: (a) Dishing and erosion in a CMP process due to different metal and dielectric polish rates. (b) Dummy fill for reducing the thickness variation of a metal layer.

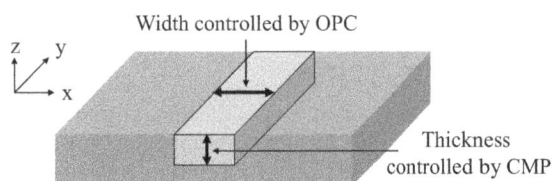

Figure 3: OPC and CMP affect the width and thicknesses of a pattern.

In modern IC process, layout distortion might occur not only in pattern width (*critical dimension*, *CD* for short) and length, but also in pattern height (thickness). Because OPC focuses on the minimization of CD errors and CMP on the minimization of thickness errors, the effects on all dimensions must be considered to minimize the total distortion, as illustrated in Figure 3.

Many researchers tried to minimize the distortions of pre-OPC shapes [4, 13, 18]. Nevertheless, OPC tools often change design shapes significantly and also add new shapes. As a result, it is not sufficient to consider only pre-OPC shapes. Instead, it is of importance to also consider the effects of routing patterns and spacing on OPC. An OPC-aware router should reserve necessary routing resource based on the behavior prediction of an OPC tool. Such a router should consider post-OPC shapes and traditional routing objectives as well. Thus, some researchers tried to focus on post-OPC shapes [4, 6, 8].

Dummy pattern insertion is often used to maximize the pattern density uniformity of a chip layout. However, inserting dummy patterns may hurt timing closure due to the additional capacitances induced by the patterns, and further it is very difficult to extract the full-chip RC of floating dummy patterns. Consequently, previous density-aware routers all try to avoid dummy pattern insertions by improving wire-density uniformity [3, 5, 11, 15], where the wire density was modelled as routing congestion like a traditional routing cost. Li et al. [11] presented the first routing system addressing variations induced by CMP. By setting a desired density in the cost function of a global router, the routing results could have more balanced interconnect distribution. Cho et al. [5] proposed a pioneering work to consider CMP variation during global routing. Chen et al. [3] considered wire-density uniformity for CMP variation control during

routing and achieved excellent results on wire-density uniformity. All these routers try to meet density constraints without dummy pattern insertions.

However, the coupling capacitance induced by dummy features is not the only key factor that affects circuit timing. The variations of interconnection resistance and coupling capacitance due to large ILD thickness variation also significantly affect circuit timing. Consequently, a router achieving maximum wire-density uniformity might not always lead to desired circuit timing. For example, the router in [11] tries to meet density constraints without considering post-routing dummy insertion; after a CMP process, however, the ILD thickness might still have large variation, thus still requiring post-routing dummy insertion. Due to the uniformly distributed wires routed by the router, a dummy insertion tool could incur a large failure rate because there might not be enough space to insert dummy features for metal density control. For the example shown in Figure 4, the two paths both have the same wire length, the number of vias, and metal density, but the path of Figure 4(b) has a better dummy insertion controllability to control the ILD thickness variation.

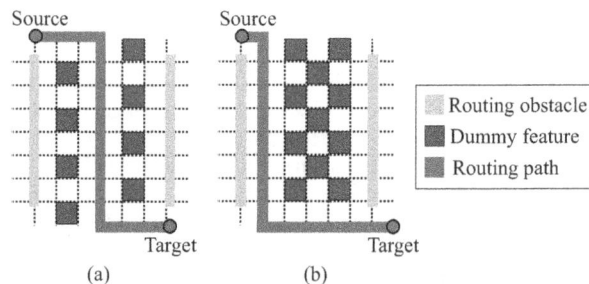

Figure 4: Different routing paths are generated by different objectives. (a) A density-driven route. (b) Another route with a larger dummy metal insertion rate.

Because wire density distribution significantly affects dummy feature filling, it is desirable to control wire density distribution properly during routing. In addition to the traditional objectives of a CMP-driven router, maximizing dummy fillable area during routing could also generate better dummy insertion controllability and thus layout density controllability. To tackle the coupling capacitance induced by dummy patterns, a proper buffer space, a minimum spacing between dummy and design features, can be added to significantly reduce the coupling capacitance between dummy and design features. Further, timing-critical nets can be assigned larger buffer spaces than non-critical ones.

In this paper, we first present an efficient, yet sufficiently accurate closed-form formula for printed width computation and dummy-insertion-aware routing cost derivation. The formula provides an effective cost modeling for post-layout OPC and CMP optimization during routing. Incorporating the OPC and CMP costs, a router can be guided to optimize the effects of layout correction and planarization. Compared with the state-of-the-art OPC-friendly router, QL-MGR [4] (which does not consider CMP), the experimental results show that our router can achieve respective 19% and 6% reductions in the maximum and average layout distortions. For fair comparison, the TSMC VCMP tool [17] was used to extract the resulting thickness variations. Compared with the state-of-the-art CMP-aware router, TTR [3] (which does not consider OPC), the experimental results show that our router can achieve respective 19% and 25% reductions in the peak-to-peak thickness and thickness variance. These results show that our simultaneous OPC- and CMP-aware router achieves significant improvements for layout integrity.

The rest of this paper is organized as follows. Section 2 gives the OPC cost model. Section 3 presents the CMP cost model. Section 4 presents our router that can optimize OPC and CMP simultaneously. Experimental results are reported in Section 5. Finally, we conclude our work in Section 6.

2. OPC-AWARE ROUTING COST MODELING

It is typically too complicated to enumerate a 2D pattern library for practical applications, and thus lithography engineers usually resort to 1D patterns to measure the process quality. In this section, we focus on 1D patterns first, and then propose a method to decompose a 2D layout into 1D patterns. By this method, we can efficiently deal with a 2D layout without using a complicated 2D pattern library.

2.1 Litho-Prediction for 1D Pattern

A 1D pattern is a set of wires with the same spacing and width. A 1D pattern might have distortion along the direction of its width (x-axis). For a 1D pattern, it can be characterized by the spacing s and the width w. Assume that the pitch between two wires is $p = s + w$. The normalized electric field of a 1D pattern is

$$E(x) = \begin{cases} 1, & np - \frac{w}{2} \leq x \leq np + \frac{w}{2}, n = 0, \pm 1, \pm 2, \cdots, \\ 0, & \text{otherwise,} \end{cases} \quad (1)$$

where x denotes the x-coordinate. After the light propagation, the electric field is transformed with Fourier transform with 1D patterns. The electric field on a lens L is

$$E_L(x) = \sum_{n=-\infty}^{\infty} \left(\delta(x - \frac{n}{p}) \frac{\sin(\pi x w)}{\pi x w} \right), \quad (2)$$

where

$$\delta(x) = \begin{cases} 1, & x = 0, \\ 0, & \text{otherwise.} \end{cases} \quad (3)$$

Due to the size limitation of a lens, only the electric field between $-1 \leq n \leq 1$ will be caught, and thus the electric field on a lens L can be calculated by

$$E_L'(x) = \left(\delta(x - \frac{1}{p}) + \delta(x) + \delta(x + \frac{1}{p}) \right) \frac{\sin(\pi x w)}{\pi x w}. \quad (4)$$

Off-axis illumination (OAI) is often used in advanced processes. With OAI, this printing can be formulated as two-beam imaging [12] and approximated as follows:

$$E_{OAI}(x) = \left(\delta(x - \frac{1}{p}) + 2\delta(x) + \delta(x + \frac{1}{p}) \right) \frac{\sin(\pi x w)}{\pi x w}. \quad (5)$$

Assuming $y = w/p$, after the light passes through the lens, the electric field on a wafer W is given with inverse Fourier transform as follows:

$$E_W(x) = 2 + \frac{1}{\pi y^2} \left(e^{i\frac{\pi x}{py}} + e^{-i\frac{\pi x}{py}} \right) \quad (6)$$

$$= 2 + \frac{1}{\pi y^2} \cos \left(\frac{\pi x}{py} \right). \quad (7)$$

Letting $\theta = \pi x/py$, the light intensity on a wafer W is then given by

$$I_W(x) = (E_W(x))^2 \quad (8)$$

$$= \left(2 + \frac{1}{\pi y^2} \cos \theta \right)^2 \quad (9)$$

$$= 4 + \frac{4}{\pi y^2} \cos \theta + \frac{1}{\pi^2 y^4} \cos^2 \theta. \quad (10)$$

If we want a design pattern to be printed, the intensity of the pattern on its wafer must be larger than the threshold of photoresist, I_t. The width of a printed pattern can be computed by the range of x when $0 \leq x \leq p$ and $I_W(x) \geq I_t$. Consequently, the width can be obtained by solving the following equations:

$$I_t = 4 + \frac{4}{\pi y^2} \cos \theta + \frac{1}{\pi^2 y^4} \cos^2 \theta \quad (11)$$

$$\cos^2 \theta + 4\pi y^2 \cos \theta + \pi^2 y^4 (4 - I_t) = 0 \quad (12)$$

If Equation (12) is solvable, $\cos \theta = \cos \left(\frac{\pi x}{py} \right) = -\pi y^2 (2 \pm \sqrt{I_t}$. The result of x can thus be computed by

$$x = \frac{w}{\pi} \cos^{-1} \left(\frac{\pi w^2}{(w+s)^2} \left(-2 \pm \sqrt{I_t} \right) \right). \quad (13)$$

The width of the printed pattern can be computed by using Equation (13). This closed-form solution can help engineers quickly compute the printed width of a 1D pattern, and thus the lithography cost can be set as the difference between the original wire width and the printed width computed by the above formulae. Figure 5 shows an example. There are three 1D wires with different pitches and spacings. To derive the lithography cost of the middle wire, the pitch and spacing values are used in Equation (13) to compute the wire width difference. Thus, the two edges of the middle wire, e_1 and e_2, have different lithography costs due to the spatial relationships between the edges and their neighboring wires.

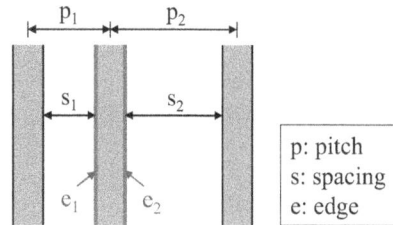

Figure 5: 1D pattern lithography cost computation.

Note that it is often infeasible for layout design with an over complicated (thus over time-consuming) cost model, and there is always inevitable trade-off between the efficiency and accuracy for different cost models. So our goal here is to derive an efficient, yet sufficiently accurate closed-form formula to guide our routing optimization. To achieve a better trade-off, we resort to the OAI modeling for higher accuracy than the state-of-the-art OPC model for routing used in [4] and the simpler two-beam imaging approximation and 1D patterns for better efficiency. It will be clear from the experimental results that our scenario leads to a better quality and efficiency trade-off than the work in [4]. Further, Equation (5) was derived from Equation (4) with a linear combination; for more accurate (yet more time-consuming) modeling, diffraction grating [14] can be used.

2.2 Litho-Prediction for 2D Pattern with Layout Decomposition

Most real designs consist of 2D patterns, but it is much harder to model the process window of a 2D pattern. Consequently, it might not be accurate to evaluate the quality of layout printing by using an incomplete 2D pattern library. Furthermore, 2D pattern modeling is a time-consuming process. As a result, most lithography engineers handle 2D patterns by decomposing them into 1D patterns because process parameters can be extracted easily with a complete 1D pattern library.

Focusing on the routing stage, we shall derive an effective cost function to guide the routing in an un-routed area. This cost function shall model the lithography effect of the un-routed area and guide a router to find a lithography-friendly wiring path.

We first decompose a layout into a set of subregions so that all the points in a subregion are closer to a target wire edge. Because the optical effect comes from all neighboring patterns, we classify all un-routed areas by nearest design patterns. A nearest design pattern is more critical than other patterns because the pattern with a larger spacing can be corrected more easily by an OPC tool. As a result, this decomposition is related to the operation of a Voronoi diagram. The input of a traditional Voronoi diagram is a set of *points*, while the input of the pattern decomposition is a set of *edges*. So we shall generalize the Voronoi diagram for pattern decomposition, where any point in a generalized Voronoi diagram has only one nearest input edge. See Figure 6 for an example of the generalized Voronoi diagram.

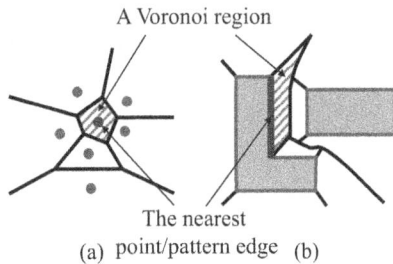

Figure 6: Layout decomposition by Voronoi diagram. (a) Traditional Voronoi diagram. (b) Generalized Voronoi diagram.

2.3 OPC Cost Calculation

For an un-routed net, the algorithm summarized in Figure 7 is used to compute the lithography cost to guide the routing for a net. This cost calculation can be used by a traditional routing algorithm because this lithography cost is associated with the edges of maze grids. The overall edge cost can be defined by a linear combination of the wire length and lithography cost.

3. CMP-AWARE ROUTING COST MODELING

In this section, we derive a cost model for CMP-aware routing.

Algorithm: *Lithography-aware Routing*
Input: S – a layout with a set of design patterns
Output: P – the path with a minimal lithography cost
1. Feed the original layout S into a maze router and construct the data structure of maze grids;
2. Decompose S into several regions by generalized Voronoi diagram;
3. Use the 1D cost function to calculate the lithography cost of all wire edges and routing grids;
4. Perform maze routing with the lithography cost to find a path P with a minimal lithography cost;
5. **return** P.

Figure 7: A simple algorithm for one net routing

3.1 Dummy Insertion

In order to improve the CMP quality, foundries often impose recommended layout density rules and insert dummy features into layouts to minimize the variation on each layer. Dummy features may either be connected with power/ground (tied fills) or left floating (floating fills) [10]. The tied fill has predictable but higher capacitance, while the floating fill has lower but unpredictable one due to the floating nature. However, electrical impacts of dummy fills could be negligible if there is enough buffer space for each wire to reduce the coupling capacitances.

In this paper, we employ a simple, yet popular Boolean-based dummy insertion algorithm in the post-routing stage; see Figure 8 for an example. This algorithm is relatively timing-consuming, yet effective. It was also used to explore the effect of dummy features [9].

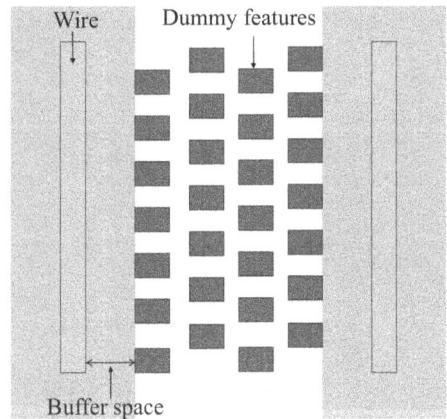

Figure 8: Illustration of dummy insertion. The buffer space (BS) is a user-defined variable for the adjustment of coupling capacitance.

The original objective of dummy insertion is to minimize thickness variation, and a dummy insertion tool chooses a fillable region to place fill patterns. However, the target density might not be the same everywhere in one layer. One key reason is that different positions of a layer could have different multi-layer accumulative effects—the positions of upper layers could be affected by those of lower layers. See Figure 9 for an illustration. Consequently, the target density might not be the same in different layers, and thus a dummy insertion tool should carefully consider the accumulative effect.

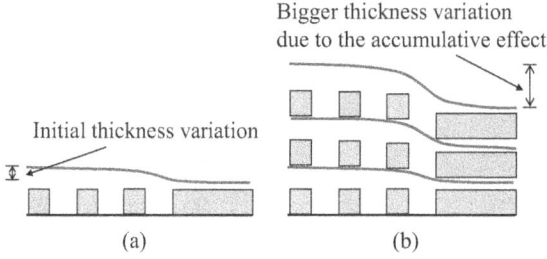

Figure 9: **Example of multi-layer accumulative effect. The thickness variation is accumulated.**

3.2 Dual Problem

Due to multiple unknown parameters, like the buffer space (BS) and available sizes of fill patterns, it is often very difficult to predict post-layout dummy insertion during routing. The original objective of dummy insertion is to minimize the thickness variation. A dummy insertion tool must control metal density to adjust the ILD thickness, which is an important capability for such a tool. Thus, we convert the original problem, the thickness variation minimization problem, into a density controllability maximization problem.

Because the insertion tool uses a fillable area to place patterns, a larger fillable area is more friendly/desirable for dummy insertion. The fillable area can be treated as resources to control density because the density can be affected by inserting dummy features into this area. As shown in Figure 10, due to the larger dummy fillable area in Figure 10(a), density controllability of the layout in Figure 10(a) is better than that of the layout in Figure 10(b). Thus, the density controllability and the size of the fillable area are positively correlated; that is, maximizing a fillable area is equivalent to maximizing density controllability.

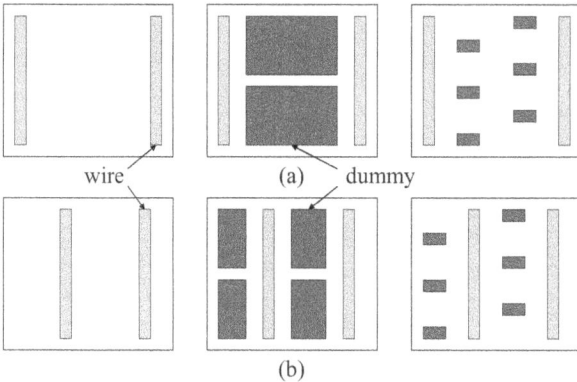

Figure 10: **Density controllability of the layout in (a) is better than that of the layout in (b) due to the larger dummy fillable area.**

The free (un-routed) area is not all fillable because some free area is so close to a routed area that it violates the BS constraint. The fillable area $A_{fillable}$ can be computed as follows:

$$A_{fillable} = A_{total} - \sum_i A_{wire,i} - \bigcup_i (A_{BS,i} \cup A_{S,i}) \quad (14)$$

where A_{total} is the total area, $A_{wire,i}$ is the area of wire i, $A_{BS,i}$ and $A_{S,i}$ are the respective areas of the buffer space

and the minimum space induced by wire i. In this work, we assume that buffer space (BS) is larger than the minimum spacing (MS), which is the case for practical applications.

To maximize dummy fillable areas, some area of buffer space can be overlapped. Figure 11 illustrate an example. When two wires are placed more closely, the non-fillable area would be overlapped and the fillable area could become larger. A larger fillable area can provide better density controllability. Thus, we shall guide CMP-aware routing by solving the fillable area maximization problem, which can be converted into a non-fillable area minimization problem. Note that the non-fillable area minimization problem is the dual problem of the density controllability maximization problem.

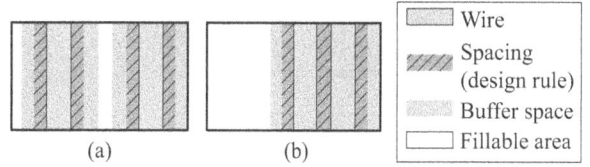

Figure 11: **Different wire positions induce fillable area change. When two wires are placed more closely, the non-fillable area could be overlapped and the fillable area becomes larger.**

3.3 CMP cost

In this section, we propose a three-step method to estimate the non-fillable area. Because the full-chip calculation is too timing-consuming and most routers sequentially process the un-routed nets, we calculate only the cost of routing path candidates. The CMP cost of a position is defined by *the increase of non-fillable areas* if a new wire is placed at this position. Our cost function can help a router find a proper routing path to place the wire, which would minimize the increase of non-fillable area.

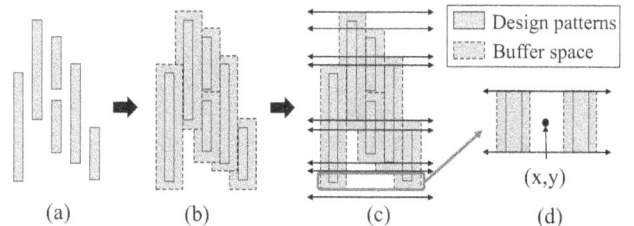

Figure 12: **Illustration of the CMP cost estimation. (a) Original pattern shapes. (b) Pattern expansion. (c) Trapezoidal decomposition. (d) Cost calculation of a point (x, y) in a region.**

The basic idea for the CMP cost modeling is to decompose the whole area into sub-regions and find a closed-form equation to calculate the change of the non-fillable area. Our cost estimation consists of three steps: (1) layout expansion, (2) trapezoidal decomposition, and (3) closed-form calculation. Figure 12 illustrates our cost calculation. First, we expand all boundaries of the features outward by BS along both the vertical and horizontal directions (see Figure 12(b)). Given a set of expanded rectangles, the trapezoidal decomposition is obtained by slicing along the horizontal boundaries of the rectangles (see Figure 12(c)). With this decomposition, the

Table 1: Statistics of the MCNC benchmarks.

Circuit	Size (μm^2)	#L.	#Nets	#Conn.	#Pins	Width (nm)	Spacing (nm)
Mcc1	162.0×140.4	4	802	1693	3101	90	72
Mcc2	548.6×548.6	4	7118	7541	25024	90	72
Struct	735.5×735.6	3	1920	3551	5471	90	180
Primary1	1128.3×748.2	3	904	2037	2941	90	180
Primary2	1565.7×973.2	3	3029	8197	11226	90	180
S5378	108.8×59.8	3	1694	3124	4818	90	90
S9234	101.0×56.3	3	1486	2774	4260	90	90
S13207	165.0×91.3	3	3781	6995	10776	90	90
S15850	176.3×97.3	3	4472	8321	12793	90	90
S38417	286.0×154.8	3	11309	21035	32344	90	90
S38584	323.8×168.0	3	14754	28177	42931	90	90

Table 2: Comparison between QL-MGR and OPC-MR by total wirelength, memory usage, routing time, maximum EPE, and average EPE.

Circuit	QL-MGR [4]					OPC-MR (Ours)				
	WL (mm)	Memory	CPU (s)	EPE_{max}	EPE_{avg}	WL (mm)	Memory	CPU (s)	EPE_{max}	EPE_{avg}
Mcc1	102	39	107	21	7.1	100	38	103	13	6.8
Mcc2	1463	302	2719	18	7.7	1456	292	2608	13	7.5
Struct	127	12	5	17	7.5	126	12	5	12	7.0
Primary1	154	10	7	16	7.4	154	10	7	12	6.9
Primary2	626	21	37	16	7.4	625	20	35	12	6.7
S5378	18	12	9	11	7.3	19	12	9	10	6.9
S9234	14	11	9	12	7.4	14	11	9	10	6.8
S13207	42	24	31	10	7.5	42	24	30	10	7.0
S15850	53	34	38	13	7.4	52	34	37	12	7.1
S38417	114	76	95	12	7.4	113	75	93	11	7.0
S38584	160	733	369	15	7.3	158	720	354	11	7.0
Comparison	1	1	1	1	1	0.99	0.98	0.98	0.81	0.94

CMP cost of a point in a fillable region can easily be computed based on the length and width of this fillable area, as illustrated in Figure 12(d).

For a point $p = (x, y)$ in a free space, we define $D_L(x, y)$ to be the minimum distance between x and the right boundary of the rightmost feature on the left side of p and $D_R(x, y)$ to be the minimum distance between x and the left boundary of the leftmost feature on the right side of p. Then, the cost of p can be derived by computing the following equation:

$$C(x, y) = C_L(x, y) + C_R(x, y), \qquad (15)$$

where $C_L(x, y)$ and $C_R(x, y)$ are the increased non-fillable areas on the left and right sides of p, respectively. Thus, $C_L(x, y)$ can be computed as follows:

$$C_L(x,y) = \begin{cases} \Phi \times l, & D_L(x,y) \geq 2\Phi, \\ (D_L(x,y) - \Phi) \times l, & \Phi \leq D_L(x,y) \leq 2\Phi, \\ 0, & otherwise, \end{cases} \qquad (16)$$

where l is the pattern length and Φ is the buffer space. Note that $C_R(x, y)$ can be computed in a similar way.

4. SIMULTANEOUS OPC- AND CMP-AWARE ROUTING

To demonstrate the effectiveness of our cost modeling in reducing the distortion of wire width and the variation of wire thickness, we can incorporate the OPC and CMP costs computed by the aforementioned modelings into MR [2]. Note that MR is a popular academic router with publicly available source codes [2]. The OPC and CMP costs during routing are defined as the 1D OPC cost and non-fillable area, respectively. Both costs are first normalized and then integrated together by equal weights (of course, we can also define user-specified parameters to weigh the two factors).

The integrated cost function can then be used to guide MR for the wire width and ILD thickness variation minimizations.

5. EXPERIMENTAL RESULTS

We integrated the proposed OPC and CMP cost functions into MR in the C++ language. The experiments were performed on a 1.2 GHz SUN Blade-2000 workstation with 8 GB memory. We used the commonly used MCNC benchmarks because their detailed-routing information is available to the public. To investigate the effects of OPC and CMP, we modified the original benchmarks with the 65 nm technology. The wire width of each benchmark was shrunk to 90 nm. The layout dimensions and design rules were also shrunk accordingly.

Table 1 gives the statistics of the benchmarks. In this table, "Circuit" gives the names of the circuits, "Size (μm^2)" gives the layout dimensions in μm^2, "#L." denotes the number of used routing layers, "#Nets" gives the total number of nets, "#Conn." gives the number of two-pin connections after net decomposition, "#Pins" gives the number of pins, "Width (μm)" gives the design rules for wire/via width, and "Spacing (μm)" gives the design rules for wire/via spacing.

First, we compared QL-MGR [4] and MR with our lithography cost (called *OPC-MR*). QL-MGR considers OPC behavior to model the OPC demand for edge-placement error (EPE) minimization in the routing stage. The program of QL-MGR was provided by the authors of [4] and was run on the same machine as ours. For fair comparison, OPC-MR used the same setting as QL-MGR for the size of routing tiles in all benchmarks. Note that as reported in [4], QL-MGR achieves better solutions than previous works, and thus we shall directly compare QL-MGR and OPC-MR.

The geometries of the routing results of QL-MGR and OPC-MR were extracted and fed into Mentor Graphics Calibre-OPC [1] to add sub-resolution assistant features (SRAF's), perform OPC operations, and estimate EPE. The optical parameters of Calibre-OPC were $\lambda = 193nm$, $NA = 0.85$, and $defocus = 0/0.1\mu m$. The illumination system was off-axis illumination (OAI) with $\sigma = 0.94/0.74$. The thicknesses of photoresist and the bottom anti-reflective coating (BARC) were 75 nm and 80 nm, respectively.

Table 2 gives the comparison of QL-MGR and OPC-MR based on the total wirelength, the memory usage, the routing time, the maximum EPE, and the average EPE. In this table, "WL" gives the total wirelength in μm, "Memory" gives the memory usage in MB, "CPU" gives the running times for routing in second, "EPE_{max}" gives the maximal EPE in nm, and "EPE_{avg}" gives the average EPE in nm. Note that all benchmarks were 100% routed. From the table, the memory requirement and running time of OPC-MR are smaller than QL-MGR for all the MCNC circuits; further, OPC-MR achieves 19% improvement for the maximum EPE and 6% improvement for the average EPE. The results show that our lithography cost function is effective and efficient.

We then compared MR and MR with our CMP cost (called *CMP-MR*) by traditional routing objectives, like the routing completion rate, the total wirelength, the routing running time, and the number of vias. Table 3 gives the comparison of MR and CMP-MR. In this table, "WL (μm)" gives the total wirelength in μm, and "$\#via$" gives the number of vias. From the table, CMP-MR achieves the same routability and almost the same total wirelength and number of vias as MR, with very small overhead in the running time. The results show that CMP-MR can optimize the CMP cost while maintaining comparable quality for those traditional routing cost metrics.

We further compared our CMP-MR with TTR [3]. TTR considers wire-density uniformity for CMP variation control during routing, and TTR can achieve very excellent results on wire-density uniformity. The results of TTR was provided by the authors of [3], and the design rules used by CMP-MR were the same as those used by TTR. For fair comparison, CMP-MR used the same setting as TTR for the size of routing tiles in all benchmarks. Note that as reported in [3], TTR achieves the best results compared with the previous works, and thus we shall directly compare with TTR. The geometry of the routing results of TTR and CMP-MR were extracted and fed into a dummy insertion tool to perform the insertion. The inserted layouts were then fed into TSMC VCMP [17] to extract the post-CMP thickness variation. Both TTR and CMP-MR obtained 100% routing completion on the MCNC benchmarks. As shown in Table 4, CMP-MR averagely reduces the post-CMP peak-to-peak thickness (T_{PP}) by 19% and the post-CMP thickness variance (T_{var}) by 25%, compared with TTR.

We further compared the following four routers: (1) MR, (2) MR with integrated OPC and CMP costs (called *DFM-MR*) with the weights for the OPC and CMP costs both being 0.5, (3) QL-MGR, and (4) TTR. Specifically, we compared with MR on the wirelength and routing time, QL-MGR on EPE_{max} and EPE_{avg}, and TTR on T_{pp} and T_{var}, which are the cost metrics for the corresponding routers. See Table 5 for the results. From the table, our DFM-MR has only 2% and 13% overheads in wirelength and running time, respectively. The results show that our DFM-MR can achieve comparable results while optimizing OPC and CMP

costs simultaneously. For the OPC comparison, the geometries of the routing results of DFM-MR were extracted and fed into Mentor Graphics Calibre-OPC to estimate the EPE under the same optical parameters as QL-MGR. As shown in Table 5, our DFM-MR can averagely achieve respective 13% and 5% improvements on the maximum and average EPE's over QL-MGR. For the CMP comparison, our DFM-MR can averagely achieve 18% and 16% improvements on the post-CMP peak-to-peak thickness and post-CMP thickness variance over TTR, respectively. The results show that our DFM-MR can simultaneously optimize OPC and CMP costs effectively while maintaining good solution quality for traditional routing completion rates, wirelength, and running time. Figure 13 shows the routing solutions of "Mcc2" generated from MR, CMP-MR, and TTR.

Table 3: Traditional objective comparison between MR and CMP-MR; both are 100% routed.

Circuit	MR			CMP-MR (Ours)		
	WL (nm)	CPU (s)	$\#via$	WL (nm)	CPU (s)	$\#via$
Mcc1	100499	72	5487	100507	92	5509
Mcc2	1457462	1924	32402	1457133	2982	32402
Struct	127178	6	9984	127181	6	10009
Prim1	154150	5	6658	154151	6	6663
Prim2	627662	34	27291	627690	35	27389
S5378	17889	13	7246	17894	14	7286
S9234	13158	9	6129	13157	9	6126
S13207	42467	41	15926	42469	43	16007
S15850	52884	60	18950	52886	63	19064
S38417	114770	171	47032	114750	1778	47202
S38584	159664	376	63698	159678	391	63875
Comp.	1	1	1	1.00	1.10	1.00

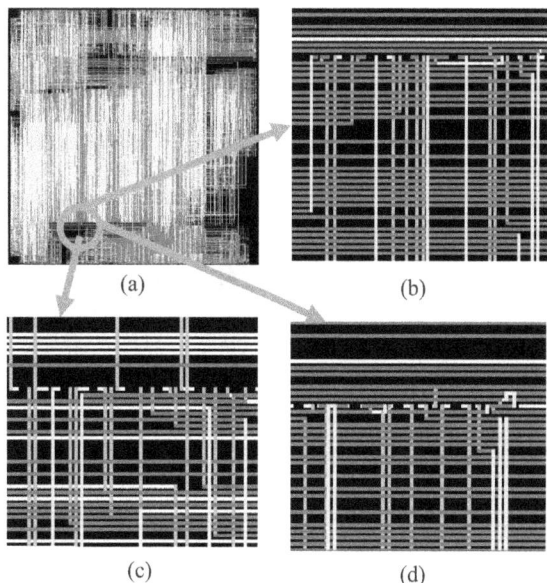

(a)   (b)

(c)   (d)

Figure 13: Routing solutions for "Mcc2" obtained from MR, CMP-MR and TTR. (a) Full-chip solution generated by CMP-MR. (b) Partial solution generated by CMP-MR. (c) Partial solution generated by TTR. (d) Partial solution generated by MR. Wires in Metal 1, Metal 2, Metal 3, and Metal 4 are colored in white, blue, red, and yellow, respectively.

Table 4: Comparison between TTR and CMP-MR. All results are normalized to the post-dummy-fill-insertion results of TTR.

Circuit	Ratios of the CMP-MR results vs. TTR's							
	Metal 1		Metal 2		Metal 3		Metal 4	
	T_{PP}	T_{var}	T_{PP}	T_{var}	T_{PP}	T_{var}	T_{PP}	T_{var}
Mcc1	0.83	0.63	0.72	0.65	0.81	0.73	0.70	0.78
Mcc2	0.68	0.46	0.70	0.52	0.78	0.49	0.87	0.53
struct	1.03	0.78	1.04	0.71	1.05	1.11	-	-
Prim1	1.64	1.23	0.40	0.63	1.13	1.05	-	-
Prim2	0.86	0.63	0.93	1.09	1.16	1.14	-	-
s5378	0.75	1.05	0.90	0.75	0.95	0.83	-	-
s9234	0.72	0.72	0.55	0.57	0.54	0.50	-	-
s13207	1.03	1.09	0.92	0.75	0.99	0.84	-	-
s15850	0.66	0.56	0.69	0.71	0.70	0.71	-	-
s38417	0.72	0.74	0.92	0.83	0.91	0.79	-	-
s38584	0.61	0.83	0.85	0.94	0.78	0.83	-	-
Average	0.87	0.79	0.78	0.74	0.82	0.82	0.79	0.65

Table 5: Normalized comparisons of DFM-MR with the three routers. The wirelength and routing time are compared with MR, EPE_{max} and EPE_{avg} are compared with QL-MGR, and T_{pp} and T_{var} are compared with TTR.

Circuit	Comp. Rates	WL	Routing Time	EPE_{max}	EPE_{avg}	T_{PP}	T_{var}
Mcc1	1.00	1.06	1.25	0.80	0.97	0.81	0.72
Mcc2	1.00	1.07	1.20	0.87	0.98	0.78	0.75
Struct	1.00	1.00	1.06	0.82	0.93	0.95	0.84
Primary1	1.00	1.01	1.07	0.81	0.95	0.90	0.88
Primary2	1.00	1.03	1.04	0.76	0.94	0.91	0.95
S5378	1.00	1.01	1.06	0.92	0.95	0.81	0.93
S9234	1.00	1.00	1.08	0.92	0.93	0.82	0.98
S13207	1.00	1.02	1.13	0.89	0.94	0.72	0.81
S15850	1.00	1.01	1.18	1.01	0.97	0.70	0.81
S38417	1.00	1.02	1.17	0.94	0.96	0.79	0.79
S38584	1.00	1.03	1.20	0.84	0.96	0.76	0.80
Average	1.00	1.02	1.13	0.87	0.95	0.82	0.84

6. CONCLUSIONS

We have presented an efficient, yet sufficiently accurate closed-form formula for printed width computation and dummy-insertion-aware thickness estimation. The formula provides a cost modeling for post-layout OPC and CMP optimization during routing so that the router can be guided to optimize the effects of layout correction and planarization. Experimental results have shown that our simultaneous OPC- and CMP-aware router has a significant improvement for layout integrity. Future work lies in even better trade-off with more accurate modeling by diffraction grating [14] and higher efficiency by advanced approximation.

7. REFERENCES

[1] Calibre OPC, Mentor Graphics Corporation.
[2] Y.-W. Chang and S.-P. Lin. MR: A new framework for multilevel full-chip routing. *IEEE Trans. Computer-Aided Design*, 23(5):793–800, May 2004.
[3] H.-Y. Chen, S.-J. Chou, S.-L. Wang, and Y.-W. Chang. A novel wire-density-driven full-chip routing system for CMP variation control. *IEEE Trans. Computer-Aided Design*, 28(2):193–206, Feb. 2009.
[4] T.-C. Chen, G.-W. Liao, and Y.-W. Chang. Predictive formulae for OPC with applications to lithography-friendly routing. *IEEE Trans. Computer-Aided Design*, 29(1):40–50, January 2010.
[5] M. Cho, D. Pan, H. Xiang, and R. Puri. Wire density driven global routing for CMP variation and timing. *Proc. ICCAD*, pages 487–492, Nov. 2006.
[6] M. Cho, K. Yuan, Y. Ban, and D. Z. Pan. ELIAD: efficient lithography aware detailed routing algorihm with compact and macro post-OPC printability prediction. *IEEE Trans. Computer-Aided Design*, 28(7):1006–1016, July 2009.

[7] S. Chow, A. B. Kahng, and M. Sarrafzadeh. Modern physical design: algorithm technology methodology. *ICCAD Tutorial*, June 2000.
[8] D. Ding, J.-R. Gao, K. Yuan, and D. Z. Pan. AENEID: A generic lithography-friendly detailed router based on post-RET data learning and hotspot detection. *Proc. DAC*, pages 795–800, 2011.
[9] K.-H. Lee, J.-K. Park, Y.-N. Yoon, D.-H. Jung, J.-P. Shin, Y.-K. Park, and J.-T. Kong. Analyzing the effects of floating dummy-fills: from feature scale analysis to full-chip rc extraction. *Proc. IEDM*, 2001.
[10] K.-S. Leung. Spider: simultaneous post-layout IR-drop and metal density enhancement with redundant fill. *Proc. ICCAD*, pages 33–38, 6-10 Nov. 2005.
[11] K. S.-M. Li, Y.-W. Chang, C.-L. Lee, C.-C. Su, and J. E. Chen. Multilevel full-chip routing with testability and yield enhancement. *IEEE Trans. Computer-Aided Design*, 26(9):1625–1636, September 2007.
[12] B. Lin. *Lecture 9 in Lecture Notes on Microlithography Theory and Practice.* 2006.
[13] J. Mitra, P. Yu, and D. Z. Pan. RADAR: RET-aware detailed routing using fast lithography simulations. In *Proc. DAC*, pages 369–372, Anaheim, CA, June 2005.
[14] C. Palmer. *Diffraction Grating Handbook.* 6th Ed., Newport Corporation, 2005.
[15] Y. Shen, Q. Zhou, Y. Cai, and X. Hong. ECP- and CMP-aware detailed routing algorithm for dfm. *IEEE Trans. Very Large Scale Integration (VLSI) Systems*, 18(1):153–157, January 2010.
[16] R. Tian, D. Wong, and R. Boone. Model-based dummy feature placement for oxide chemical-mechanical polishing manufacturability. *Proc. DAC*, pages 667–670, 2000.
[17] T. S. M. C. (TSMC). Virtual chemical mechanical polishing (VCMP).
[18] Y.-R. Wu, M.-C. Tsai, and T.-C. Wang. Maze routing with OPC consideration. In *Proc. ASP-DAC*, pages 18–21, Shanghai, China, January 2005.

Planning for Local Net Congestion in Global Routing[*]

Hamid Shojaei, Azadeh Davoodi
Department of
Electrical & Computer Engineering

Jeffrey T. Linderoth
Department of
Industrial & Systems Engineering

University of Wisconsin - Madison WI 53706
{shojaei,adavoodi,linderoth}@wisc.edu

ABSTRACT

Local nets are a major contributing factor to mismatch between the global routing (GR) and detailed routing (DR) stages. A local net has all its terminals inside one global cell (gcell) and is traditionally ignored during global routing. This work offers two contributions in order to estimate and manage the local nets at the GR stage. First, a procedure is given to generate gcells of non-uniform size in order to *reduce* the number of local nets and thus the cumulative error associated with ignoring or approximating them. Second, we *approximate* the resource usage of local nets at the GR stage by introducing a capacity for each gcell in the GR graph. With these two complementary approaches, we offer a mathematical model for the congestion-aware GR problem that captures local congestion with non-uniform gcells along with other complicating factors of modern designs including variable wire sizes, routing blockages, and virtual pins. A practical routing procedure is presented based on the mathematical model that can solve large industry instances. This procedure is integrated with the CGRIP congestion analysis tool. In the experiments, we evaluate our techniques in planning for local nets during GR while accounting for other sources of congestion using the ISPD11 benchmarks.

Categories and Subject Descriptors

B.7.2 [**Integrated Circuits**]: Design Aids—*layout*

Keywords

congestion, detailed routing, global routing

1. INTRODUCTION

Many factors complicate the routing process for modern designs. Variation in the the wire sizes and spacings of the metal layers, routing blockages, and *virtual* pins in higher metal layers all contribute to make the routing process more difficult. Additionally, modern designs often have high pin density and vias, which further contribute to congestion.

[*]This research is supported by National Science Foundation under award CCF-0914981.

To improve a design's routability, one set of techniques focus on *routability-driven placement* procedures that rely on mechanisms for predicting routing congestion. There are many successful efforts in this area, including [7, 8, 9, 10, 11, 14, 16].

Another avenue for improving routability is to adjust the global routing (GR) procedure to account for modern complicating factors. Along this line, CGRIP [13] is a flexible GR tool that can handle arbitrary routing blockages and virtual pins, and varying wire sizes and spacings. However, CGRIP does not account for local congestion effects. Moreover, CGRIP was primarily tuned to provide a rapid estimation of congestion by imposing a small runtime budget—much smaller than typically spent during GR *on unroutable designs* which contain routing congestion.

The focus of this work is to improve routability by modifying the GR procedure to account for congestion caused by local nets, in addition to the complicating factors already handled by CGRIP. A local net has all its terminals in one global cell (gcell). The routing of a local net is determined during the detailed routing stage, and local nets are typically ignored during GR. In many instances, a significant portion of the nets to be routed are local. For example, for the ISPD11 benchmarks [15], using the winning placement solutions, on average 31.20% of the nets are local. The aim of this work is to account for these local nets during GR which translates into significant reduction in the effort right after one iteration of ripup-and-reroute during detailed routing.

This work offers two complementary techniques to manage the impact of local nets during GR. First, we propose to *reduce* the number of local nets by using gcells of non-uniform size. A procedure is presented to transform a given GR instance into one with fewer local nets. The procedure results in an increase in the number of global nets however to have a control on the runtime complexity at the GR stage, it keeps the number of gcells intact. So the effort required at the GR stage may increase but a significantly better solution in terms of the induced overflow is observed, just after one iteration of ripup-and-route at the detailed routing stage. Second, we *approximate* the area required to route the local nets and adjust the areas of the affected gcells before GR.

Using these two complementary techniques to approximate and reduce the local nets, a graph model and a mathematical formulation of GR with non-uniform gcells is presented which includes vertex capacity in addition to the conventional edge capacity. Our model also captures other factors contributing to congestion such as varying wire size and spacing, routing blockages, and virtual pins.

The recent work [17] proposes several methods to approximate the routing usage of local nets inside the gcells, all of which are translated into a reduction in the capacities of related edges in the graph model of GR. In this work, we show using a motivational example and experiments that adding a vertex capacity during GR to reflect this local usage, results in GR solutions which provide a better starting point to perform detailed routing in terms of the induced (detailed routing) overflow, compared to solely reducing the edge capacities. The focus of this work is not on computation of the vertex capacity; any model of the usage of routing resources within a gcell, such as the ones explored in [17], can directly be used as the vertex capacities in our framework. Overall, we show to reflect local congestion during GR, the use of varying vertex capacities together with (unreduced) edge capacities avoid cutting the size of the feasible search space which may otherwise happen if the edge capacities are solely reduced and vertex capacities are not used.

Our experiments are conducted using the recently-released ISPD11 benchmarks [15] which capture a large set of the factors contributing to routing congestion in modern designs. To the best of our knowledge, there is no work in the open literature *on global routing* which have used this challenging set of benchmarks to evaluate routing congestion. In fact we show in an experiment that ignoring these factors during GR and just considering local congestion is not effective in reducing the actual routing congestion.

Our ideas are implemented into a practical routing framework which handles large industry-sized instances. The practical realization is based on integration with the CGRIP congestion analysis tool [13], revising its functionality to suit the goals of this research effort.

2. NON-UNIFORM GCELLS

To reduce the error caused by the modeling approximation of local nets, we propose to define the gcells in a non-uniform manner in order to reduce the number of local nets. Reducing the number of local nets increases the number of global nets (since the total number of nets is constant) and thus creates a more difficult global routing (GR) instance. However, the effort required for detailed routing can be significantly reduced. We will give computational results in Section 5 indicating that the extra effort applied during GR often pays off, generating improved overall designs with less total computing time.

In this section, we first provide mathematical notation for describing local nets in the presence of gcells of unequal size. We then present our non-uniform gcell generation procedure.

2.1 Notation and the Binning Problem

To more accurately account for local effects in GR, we revisit how instances for GR arise out of the design process. At the most fundamental level, we are given a 3-dimensional *placement grid*

$$P = \{0, 1, \ldots X\} \times \{0, 1, \ldots, Y\} \times \{1, \ldots L\}.$$

In most instances, like the ISPD11 benchmark, there are $L = 9$ layers in the grid.

The design problem also consists of a set of nets $\mathcal{N} = \{T_1, T_2, \ldots, T_{|\mathcal{N}|}\}$, where each net $T_n \in \mathcal{N}$ consists of a set of *pin locations* on the placement grid P. As a first step, each net with multiple pins is decomposed into two-terminal subnets based on a minimum spanning tree connecting the pins.

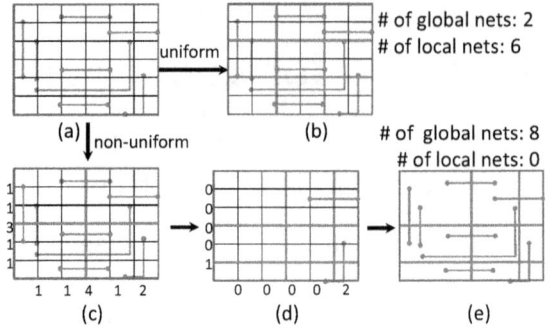

Figure 1: Binning to create non-uniform gcells.

Therefore, after decomposition, each net consists of two *pin locations*: ($T_n = \{p_1^n, p_2^n\}$). The pin locations are coordinates on the placement grid ($p_k^n = (x_k^n, y_k^n, \ell_k^n) \in P$), and typically all pin locations are in the first layer ($\ell_k^n = 1$). However virtual pins outside of the first layer are also possible. For example, in the ISPD11 benchmarks, virtual pins are located at layer 4.

To create an instance that can be solved by GR software, this very detailed placement grid P is replaced with a less-refined version, where coordinates of the placement grid are aggregated into global cells.

Each layer $\ell \in \{1, \ldots, L\}$, consists of a number of *global cells* (gcells),

$$\mathcal{C}_\ell = \{C_{ij\ell}\}_{i=1,\ldots,N_x^\ell, j=1,\ldots,N_y^\ell},$$

where each gcell $C_{ij\ell}$ is a rectangular region on a fixed level of the placement grid P. In our work (and in all benchmark instances), we will assume that a gcell $C_{ij\ell}$ is characterized as a pair of intervals

$$C_{ij\ell} = ([\alpha_i^\ell, \alpha_{i+1}^\ell], [\beta_j^\ell, \beta_{j+1}^\ell]), \qquad (1)$$

and the intervals have the property that they partition individual coordinate axes of the placement grid, i.e.

$$0 = \alpha_1^\ell < \alpha_2^\ell < \ldots < \alpha_{N_x+1}^\ell = X \qquad (2)$$

$$0 = \beta_1^\ell < \beta_2^\ell < \ldots < \beta_{N_y+1}^\ell = Y \qquad (3)$$

We define the selection of the intervals ($[\alpha_i^\ell, \alpha_{i+1}^\ell], [\beta_j^\ell, \beta_{j+1}^\ell]$) for the gcells as the *binning problem*. In most GR instances, the gcell intervals have a uniform size. For example in the ISPD11 benchmark instances, all gcell intervals are of length 40. In Section 2.2 we discuss advantages of creating instances whose gcells are of different sizes.

2.2 Binning Procedure

When selecting intervals that define gcells, we do not wish to change the total *number* of gcells, just their individual sizes. Specifically, N_x^ℓ and N_y^ℓ will remain unchanged in Equations (2) and (3), but the cell starting locations $\alpha_i^\ell, \beta_j^\ell$ in Equation (1) will be adjusted by the binning procedure. For purposes of making a 2D-projection of the instance straightforward, the binning procedure will keep all gcells to be of uniform size between layers, i.e. $\alpha_i^{\ell_1} = \alpha_i^{\ell_2}$ and $\beta_j^{\ell_1} = \beta_j^{\ell_2}$ $\forall i, j, \ell_1, \ell_2$.

We explain the procedure for gcell definition using the example shown in Figure 1. We assume that each multi-terminal net is decomposed into two-terminal subnets based on its Minimum Spanning Tree (MST). The MSTs are shown on the placement grid in (a). The standard instance has 9 equal-sized gcells specified by 3x3 grid shown in (b). This gcell configuration results in 2 global nets and 6 local nets.

Our procedure for defining gcells is based on an iterative bi-partitioning. At each iteration, both a horizontal and vertical cut are made to maximize the number of nets that are cut. For example, in (c) the maximum number of nets cut in the horizontal and vertical directions are 3 and 4, respectively, and the corresponding cuts are shown. After each iteration, the nets that are cut are removed from consideration (as shown in (d)). The process completes when the number of intervals N_x^ℓ and N_y^ℓ are of the requisite size. By allowing for gcells of non-uniform size, the number of local nets is reduced to 0, but the number of global nets is increased to 8, as shown in (e).

The binning procedure reduces the number of local nets by maximizing the number of nets that pass a vertical or horizontal cut at each step of the algorithm. It is also computationally useful to consider a parameterized version of this algorithm that controls the tradeoff between the increase in global nets and the decrease in local nets. In the parameterized version of the procedure, when deciding the location of a vertical or horizontal cut at each iteration, we select the cut location that results in the number of cut nets to be closest to ηN_{max}, where N_{max} is the maximum number of nets that can be cut at that iteration. The user-specified parameter η is between 0 and 1 and allows for more fine-tuned control of the eventual number of local nets.

Post-Processing for Local Congestion Balancing: Another important consideration in the binning problem is to generate gcells with "balanced" local congestion. This means to reduce the deviation in local congestion among neighboring gcells. This consideration is helpful in detailed routing because local nets are typically routed inside the corresponding gcell. Reducing this deviation makes it easier to route the local nets during the detailed routing stage. Therefore, after generating non-uniform gcells, we post-process the intervals, perturbing the boundaries of the gcells in order to balance the local congestions among the gcells. For each gcell C_{ijl}, we define the local congestion ratio LC_{ijl} as

$$LC_{ijl} = \frac{R_{ijl}}{A_{ijl}}, \qquad (4)$$

where R_{ijl} denotes the routing resources consumed by local nets inside C_{ijl} and A_{ijl} is the area of gcell C_{ijl}. (We discuss a method for estimation of R_{ijl} in the next section.)

The post-processing step is a greedy heuristic with the objective to decrease the deviation of the congestion ratios among the gcells, while ensuring that the number of local and global nets remains the same. The procedure sequentially considers the impact of adjusting each cut line (e.g., up and down for horizontal cuts), computing the updated local congestion ratios and number of global and local nets if the interval boundary was changed. The new cut line location is chosen to be the one that (1) does not change the number of local and global nets; and (2) results in the maximum decrease in total deviation of the gcells' congestion ratios from the average. The latter helps to reduce the deviations in the local congestion ratios among the gcells as a mean to balance the local congestion.

The effect of post-processing is depicted in Figure 2. In Figure 2(a), the congestion ratios are shown for each gcell, calculated assuming the dimensions of each gcell is 4×4, and local congestion is computed using the bounding box of each local net (which is 2 for all local nets). Figures (b) to (d) show the steps of post-processing for the horizontal cuts.

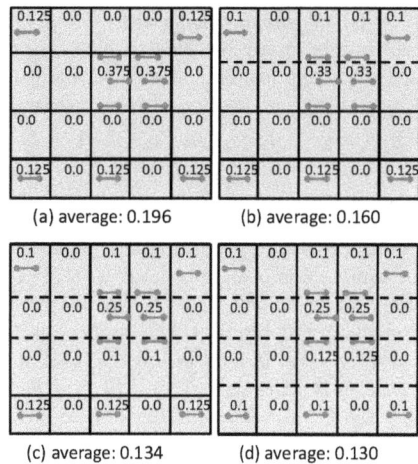

Figure 2: Post-processing example.

In Figure (b), the top cut line (shown by the dashed line) is slightly moved down and as a result some of the local nets are moved from the bottom gcells to their neighboring top gcells. The new congestion ratios corresponding to this step are shown in the gcells and the total deviations in the congestion ratios are reduced from 3.26 to 2.17, compared to the average in each case. Similarly, in Figures 2(c) and (d) the remaining horizontal cut lines are visited and in the end the total deviation in the congestion ratios is reduced from 1.67 to 1.63 in Figures (c) and (d).

An important consideration when applying binning is the amount of routing blockage in the design. For example, in the sb10 benchmark instance, over 67% of the first four metal layers are defined as routing blockages. In such a case, generating bins of non-uniform size can significantly complicate the global routing procedure and add to its complexity. Therefore, in our framework, we only apply the binning procedure if the amount of routing blockage compared to the total chip area is beyond 50%. In such a case, the less-invasive post-processing step is solely applied.

In Section 5, we show that the above binning strategy is effective for the ISPD11 benchmarks. So far, these benchmarks are the only instances which capture some of the factors that challenge the routability in modern designs and are openly available to the research community. In cases that are not captured by these benchmarks such as designs with more complex routing blockages and/or containing various-sized macros, it would be interesting to investigate our binning strategy when it is selectively applied to "appropriate" regions of the layout.

3. LOCAL CONGESTION MODELING

In this section we describe mathematical models for approximating the routing resources consumed by local nets while considering factors such as non-uniform wire size, wire spacing, and routing blockages. These models are embedded into an Integer Programming (IP) formulation to describe a congestion-aware global routing (GR) problem.

3.1 Local Congestion and Graph Models

To create a GR instance, the gcell information is used to construct a graph $G = (V, E)$, where

$$V = \cup_{\ell \in \mathcal{L}} \mathcal{C}_\ell,$$

with one vertex $v = (i, j, \ell) \in V$ for each gcell $C_{ij\ell}$. Edges $e \in E$ are created between adjacent gcells.

In a given layer, all wiring tracks are oriented in one direction. Additionally there are *vias* to allow for routes to move between adjacent layers. Therefore, each edge $e = ((i_1, j_1, \ell_1), (i_2, j_2, \ell_2)) \in E$ is of one of the following types:

- (EW): if $\ell_1 = \ell_2 = \ell$, $j_1 = j_2$, and $i_2 = i_1 + 1$ or $i_2 = i_1 - 1$,
- (NS): if $\ell_1 = \ell_2 = \ell$, $i_1 = i_2$, and $j_2 = j_1 + 1$ or $j_2 = j_1 - 1$,
- (VIA): if $i_1 = i_2$, $j_1 = j_2$ and $\ell_2 = \ell_1 - 1$ or $\ell_2 = \ell_1 + 1$.

We assume without loss of generality if ℓ is odd, edges are of (EW) type, and if ℓ is even, edges are of (NS) type.

Pin locations for each net $T_n \in \mathcal{N}$ are mapped to gcell locations. Specifically, if pin $p_1^n \in C_{ij\ell}$ and $p_2^n \notin C_{ij\ell}$, then vertex $v = (i, j, \ell) \in V$ is a terminal of T_n that must be connected in the GR instance. A net T_n is a *local net* for gcell $C_{ij\ell}$ if $p_1^n, p_2^n \in C_{ij\ell}$. As explained in the introduction, local nets are not explicitly considered by the GR instance, which may result in difficulty at the detailed routing stage.

In the GR instance, (EW) edges and (NS) edges $e = (i_1, j_1, \ell), (i_2, j_2, \ell) \in E$ have a normalized capacity u_e that depends on the length of the interval defining the gcell:

$$u_e = \begin{cases} \dfrac{\beta_{j_2}^\ell - \beta_{j_1}^\ell}{w^\ell + s^\ell} & \text{if } e \text{ is a EW edge} \\ \dfrac{\alpha_{i_2}^\ell - \alpha_{i_1}^\ell}{w^\ell + s^\ell} & \text{if } e \text{ is a NS edge} \end{cases} \tag{5}$$

where w^ℓ and s^ℓ denote the wire size and spacing in layer ℓ. By dividing the interval with $w^\ell + s^\ell$, the normalized capacity u_e reflects the number of wiring tracks that can pass between the boundary of adjacent gcells. The first layer $\ell = 1$ is not available for GR in ISPD11 benchmarks, so the capacity is set to 0 for these edges. VIA edges are given a capacity $u_e = \gamma$ that is determined by the technology. For the ISPD11 benchmark instances, $\gamma = \infty$.

To account for local effects, we also consider that each vertex $v = (i, j, \ell) \in V$ has a normalized *vertex capacity* r_v. The vertex capacity is designed to limit the total number of routes that can pass through a vertex (gcell). To capture local effects in a GR instance, the vertex capacity should be reduced based on the area required to route local nets contained in gcell $C_{ij\ell}$. We estimate the area required to route local net $T_n = \{(x_1^n, y_1^n, \ell), (x_2^n, y_2^n, \ell)\}$ in level ℓ by multiplying the wire width by a length equal to the net's half-perimeter bounding box in the lowest possible levels:

$$d_n^\ell = \begin{cases} w^\ell |x_1^n - x_2^n| & \text{if } \ell = 3 \\ w^\ell |y_1^n - y_2^n| & \text{if } \ell = 2. \end{cases}$$

Let $\mathcal{A}_{ij} \subset \mathcal{N}$ be the set of nets local to cell C_{ij1}. The routing resources consumed by local nets at level ℓ is then

$$R_{ij\ell} = \begin{cases} \sum_{n \in \mathcal{A}_{ij}} d_n^\ell & \text{if } \ell = 2, 3 \\ 0 & \text{otherwise.} \end{cases}$$

Note that we make a practical assumption that the local nets of gcell C_{ij1} only impact the capacities corresponding to gcells C_{ij2} and C_{ij3}. *Other models for approximating the routing usage of local nets such as [17] can be incorporated in the above equations.* If there are routing blockages inside a gcell C_{ijl}, the blockage area should be added to $R_{ij\ell}$.

The capacity of vertex $v = (i, j, \ell)$ is the area of the corresponding gcell reduced by the estimated area required for routing local nets. The capacity is normalized by the length of the appropriate gcell interval (depending on whether v is in an (EW) layer or a (NS) layer), so r_v becomes a measure of the number of routes that may cross the gcell. We let

$$r_v = \frac{A_{ij\ell} - R_{ij\ell}}{\lambda_{ij\ell}}, \tag{6}$$

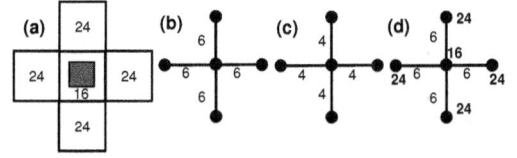

Figure 3: Various global routing graph models.

where $A_{ij\ell} = (\beta_{j+1} - \beta_j)(\alpha_{i+1} - \alpha_i)$ and $\lambda_{ijl} = (\alpha_{i+1}^\ell - \alpha_i^\ell)$ if ℓ is even, and $\lambda_{ijl} = (\beta_{j+1}^\ell - \beta_j^\ell)$ if ℓ is odd.

3.2 Integer Program Model

Input to the integer programming model consists of the grid-graph $G = (V, E)$ and a set of multi-terminal *global nets* denoted by $\mathcal{N} = \{T_1, T_2, \ldots, T_N\}$ with $T_i \subset V$, created as described in Section 3.1. Let $\mathcal{T}_n$ be the *candidate global routes* for net T_n—the collection of all Steiner trees connecting the terminals of T_n. We define the parameters $a_{te} = 1$ if Steiner tree t contains edge $e \in E$, and $a_{te} = 0$ otherwise. Similarly, we define parameters $b_{vt} = 1$ if Steiner tree t contains vertex $v \in V$, and $b_{vt} = 0$ otherwise. The model contains binary decision variables x_t that are equal to 1 if and only if tree $t \in \mathcal{T}_n$ is used to route net T_n. An Integer Program (IP) describing the GR with the normalized vertex capacity is given as follows:

$$\min_{x,o,s} \sum_{n \in \mathcal{N}} \sum_{t \in T_n} c_t x_t + M_1 \sum_{e \in E} o_e + M_2 \sum_{v \in V} s_v \quad \text{(IP-LC)}$$

$$\sum_{t \in \mathcal{T}_n} x_t \geq 1 \qquad \forall n \in \mathcal{N} \quad (\lambda_n)$$

$$\sum_{n \in \mathcal{N}} \sum_{t \in \mathcal{T}_n} a_{et} x_t - o_e \leq u_e \qquad \forall e \in E \quad (\pi_e)$$

$$\sum_{n \in \mathcal{N}} \sum_{t \in \mathcal{T}_n} b_{vt} x_t - s_v \leq r_v \qquad \forall v \in V \quad (\mu_v)$$

$$x_t \in \{0, 1\} \quad \forall n \in \mathcal{N}, \forall t \in \mathcal{T}_n$$

$$o_e, s_v \geq 0 \quad \forall e \in E, \forall v \in V$$

The parameter c_t is the cost of global route t, computed as the *lengths* of its corresponding edges: $c_t = \sum_{e \in E(T)} c_e$. For the non-uniform binning case, the length of each edge is computed as the distance between the centers of the corresponding gcells. In formulation (IP-LC), the first set of inequalities enforces the routing of each net using one of its candidate trees. In the second set of inequalities, o_e is a variable that measures the overflow of the normalized capacity u_e on edge e. In the third set of inequalities, s_v is a variable that measures the overflow of the normalized vertex capacity r_v. The objective is a linear combination of total global routed length, total edge overflow, and total vertex overflow. The two parameters M_1 and M_2 are large numerical constants, so that edge and vertex overflow are minimized.

Motivational Example: Figure 3 depicts a GR instance that demonstrates the utility of vertex capacities. Subfigure (a) shows five gcells with the middle one having high local congestion. The number inside each gcell is an approximation of total number of routes that can cross its four boundaries without causing overflow. A conventional graph model would ignore the local congestion, resulting in the instance depicted in Subfigure (b). In this case, each edge is given a capacity of 6, reflecting the maximum number of routes that can pass the boundary of the two corresponding gcells. Vertex capacities are not included in this model.

Subfigure (c) shows a model that accounts for the local congestion inside the middle gcell by reducing the the number of routes that can pass from each of its boundaries to 4. This case accounts for the local congestion better than (b), but it does so by reducing edge capacities. However, accounting for local congestion by reducing edge capacities may be too restrictive. For example, a solution to the instance depicted in Subfigure (a) can have 6 routes crossing the top boundary, and this solution is excluded by reducing the edge capacities in (c). Subfigure (d) depicts the alternative presented in this work, which uses the same edge capacities as (a). In addition, a vertex capacity of 16 is placed on the middle gcell. This case most accurately models the true instance by accounting for local congestion without unduly restricting the search space.

4. ROUTING FRAMEWORK

In this section we describe our routing framework, which is an extension of CGRIP [13]. By extending from CGRIP, our framework can also account for the same complicating factors— varying wire size and spacing at different metal layers, routing blockages, and virtual pins—as CGRIP. We start by giving a brief overview of CGRIP and then discuss specific extensions to plan for local net congestion.

Overview of CGRIP: The routing procedure in CGRIP starts with a 2D-projection of the instance and the creation of an initial 2D-solution. A rip-up and reroute (RRR) procedure is iteratively applied until a time limit is reached, or no improvement in overflow is possible, similar to other routing procedures [2, 3, 5, 6, 7, 12, 19, 18]. A congestion-aware layer assignment is applied at the last step to account for wire size and spacing, virtual pins, and routing blockages. CGRIP's procedure assumes gcells have uniform size.

CGRIP relies heavily on an IP formulation similar to (IP-LC). A reduced size linear program (RLP), consisting of a subset of the variables of the IP corresponding to critical edges and nets are selected and relaxed to take continuous values. The RLP is solved to determine the amount of utilization on the critical edges which are subsequently fed into the current step of the RRR procedure. The iterative use of these two procedures (i.e., forming an RLP and integrating its solution with the current RRR iteration) leads to a tremendous reduction in total overflow.

Note that CGRIP was designed for rapid congestion analysis. It relies on an input *resolution parameter*. In this extension, we set this parameter to be of maximum resolution. For details about CGRIP, please refer to [13].

Extensions: We now discuss how individual steps in CGRIP are revised to handle the new IP formulation (IP-LC) and the local congestion models with non-uniform gcell sizes.
1) 2D Projection: The 2D projection is done similar to CGRIP. It is done in a straightforward manner using our mathematical notations (presented in Section 3.1) and because even with non-uniform gcell sizes in each layer, we require uniformity across the layers in our gcell generation procedure. In our 2D projection, cells $C_{ij\ell}$, for a fixed location (i, j), whether of uniform or non-uniform size, are aggregated so that there is one gcell at each location (i, j) representing all the layers. Specifically, the 2D graph is $G_{2D} = (V_{2D}, E_{2D})$, where $V_{2D} = \{(i, j)\}_{i=1,...,N_x, j=1,...,N_y}$, and there is an edge $e = ((i_1, j_1), (i_2, j_2)) \in E$ if $i_2 = i_1 + 1$, $i_2 = i_1 - 1$, $j_2 = j_1 + 1$ or $j_2 = j_1 - 1$.

Normalized capacity for edges $e = ((i_1, j_1), (i_2, j_2)) \in E_{2D}$ are computed by summing capacity defined in Eq. (5) over the metal layers:

$$u_e = \sum_{\ell=1}^{\mathcal{L}} u_{(i_1,j_1,\ell),(i_2,j_2,\ell)}.$$

Similarly, vertex capacity for $v = (i, j) \in V_{2D}$ is obtained by summing vertex capacity (Equation 6) over the metal layers:

$$r_v = \sum_{\ell=1}^{\mathcal{L}} r_{i,j,\ell}. \tag{7}$$

2) Initial Solution: To create an initial solution, multi-terminal nets are decomposed into two-terminal subnets according to their Minimum Spanning Tree (MSTs), as in [4, 12, 13]. The same decomposition is used during the procedure to generate non-uniform gcells which was explained in Section 2.

To generate an initial solution, the reduced-sized linear program (RLP) for the modified IP given by formulation (IP-LC) is solved to identify an initial solution from a set of pre-defined candidate routes obtained using maze and pattern routing. This procedure is similar to CGRIP except in the way the RLP is formed which will be explained next.
3) Reduced-Sized Linear Program (RLP): We form a reduced-sized version of (IP-LC) by writing the formulation only for a subset of optimization variables $o_{\hat{e}}$, $s_{\hat{v}}$ and $x_{\hat{t}}$ and the corresponding constraints that include these variables. Variables are identified by first identifying a set of *critical edges*, based on current utilization estimates. (The procedure for identification of critical edges is the same as CGRIP.) For each identified critical edge $\hat{e} = (\hat{v}_1, \hat{v}_2)$, the variable $o_{\hat{e}}$ and two variables $s_{\hat{v}_1}$ and $s_{\hat{v}_2}$ are added. This is because vertex capacities are considered in formulation (IP-LC).

If a critical edge is used in the current solution to route a net T_n, then this net is a *critical net*, and a sufficiently large subset of routes $\hat{t} \in \mathcal{S}_n$ corresponding to candidate routes ($\mathcal{S}_n \subset \mathcal{T}_n$) will have their decision variables $x_{\hat{t}}$ included in the RLP. We note, this process to define the reduced version of formulation (IP-LC) depends on whether gcells are of uniform size. This is because both vertex and edge capacities are computed based on the gcell dimensions. We note however that our non-uniform gcell generation procedure ensures the size of the grid-graph (in term of number edges and vertices) remain the same as the uniform case.
4) Rip-up and Re-Route (RRR): The RRR procedure rips up nets with high overflow and re-routes them. The nets to rip-up are based on the computed overflow in the current solution. The re-route step solves a weighted shortest path problem. In contrast to CGRIP, the shortest path problem in our extension also has weights on vertices. This is because we consider capacity of edges and vertices.

Specifically, in CGRIP, each edge e has a weight of $1 + f(\frac{g_e}{u_e})$, where f is an exponential function of an estimate of utilization of edge e, denoted by g_e, and the normalized capacity of e. Similar to CGRIP, if e is a critical edge, then the utilization g_e is the dual value corresponding to the edge capacity constraint for e (π_e shown in (IP-LC)). Otherwise, g_e is computed as the number of global routes that use edge e in the most current routing solution.

In our framework, each edge $e \in E_{2D}$ has an edge weight of $l_e + f(\frac{g_e}{u_e})$, where l_e is the length of edge e, computed as the distance between the centers of the two gcells that edge e connects. The role of l_e in this edge weight expression is to account for the used wirelength associated with each edge.

In contrast, CGRIP uses $l_e = 1$ in its edge weight expression. This is because the cell sizes are equal in CGRIP and the relative contributions of the edges in terms of wirelength are equal to each other.

In our extension, during the re-routing, we also have weights for each vertex $v = (i, j) \in V_{2D}$. The vertex weights are $f(\frac{h_v}{r_v})$, where f is the same function used for the edge weight, h_v is an estimate of the vertex utilization, and r_v is the normalized vertex capacity (7). If v is an endpoint of a critical edge, then its utilization h_v is taken from the dual value corresponding to the vertex capacity constraint for v, denoted by μ_v in formulation (IP-LC). Otherwise, the utilization h_v is taken to be the sum of the edge utilizations for all edges incident to vertex v: $h_v = \sum_{e \in \delta(\{v\})} g_e$.

5) Congestion-Aware Layer Assignment: The RLP and RRR procedures are iterated for a pre-specified time limit, or until no additional overflow improvement is identified. The routes in G_{2D} are then converted into routes on G using a congestion-aware layer assignment procedure. CGRIP uses a greedy procedure for layer assignment which accounts for virtual pins, routing blockages and varying wire sizes and spacing. Here we extend the CGRIP procedure in two ways: (1) we use updated edge capacities which account for local congestion using our model which assumes the local nets are routed at the two lowest layers; and (2) computation of routing resource utilization in the greedy procedure of CGRIP is extended to account for non-uniform gcell dimensions. Specifically, the routing resource of an edge e in G is computed using an estimated length l_e and the corresponding wire size for that layer. The length l_e is computed as the distance between the centers of the two gcells corresponding to the two vertices of edge e.

5. SIMULATION RESULTS

The routing framework described in Section 4 and the binning procedure of Section 2 to create global routing (GR) instances with non-uniform cells were both implemented in C++ and integrated with the CGRIP congestion analysis tool [13]. For the binning procedure, the parameter η was set to 0.9, given significant weight to reducing the number of local nets in the instance. For solving linear programs, CPLEX 12.0 was used. ISPD11 benchmarks were used to validate our framework, and GR instances were created from the winning placement solutions of the ISPD11 contest [1]. For each benchmark, Table 1 shows the placement solution used, the grid size, and the number of nets before and after terminal decomposition. These benchmarks consider non-uniform wire size and spacing, routing blockages, and virtual pins. They are specifically designed to be challenging instances for routability.

GR Variations: We implemented four GR variations:

1. **U-E** (Uniform-Edge): Uniform grid with edge capacity only, without any adjustment for local congestion;

2. **U-AE** (Uniform-Adjusted-Edge): Uniform grid with adjusted edge capacity to capture the impact of local nets without any vertex capacity;

3. **U-AV** (Uniform-Adjusted-Vertex): Uniform grid with unadjusted edge but adjusted vertex capacity to reflect local congestion;

4. **NU-AV** (Nonuniform-Adjusted-Vertex): Non-uniform grid with unadjusted edge capacity and adjusted vertex capacity to reflect local congestion.

Table 1: ISPD 2011 benchmark info

Bench	Placer	Grid Size	#Nets	#2T-Nets
sb1	SimPLR	704x516	822744	2038444
sb2	Ripple	770x1114	990899	2237446
sb4	Ripple	467x415	567607	1316401
sb5	Ripple	774x713	786999	1713307
sb10	RADIANT	638x968	1085737	2579974
sb12	SimPLR	444x518	1293436	3480633
sb15	Ripple	399x495	1080409	2736271
sb18	mPL11	381x404	468918	1395388

The method U-E is identical to CGRIP [13]. Methods U-AV and NU-AV are the ones proposed in this work. They both consider vertex capacity based on local congestion as well as (unadjusted) edge capacity. The only difference between them is whether or not gcells are of uniform size. The method U-AE adjusts the edge capacity u_e to account for local nets similar to Figure 3(c). Specifically, for each edge $e = (i, j)$ of type EW and NS, the edge capacity is reduced to reflect a smaller number of routes that can pass the corresponding boundary of two neighboring gcells. The local congestion is computed using vertex capacities r_i and r_j (given by Equation (6)) for the two end endpoints of an edge. The adjusted edge capacity is then computed by replacing the numerator in Equation (5) by $\min(r_i, r_j)$ indicating that the reduction in the edge capacity is dominated by the gcell with more local congestion. The strategy of reducing the edge capacity to reflect the resource usage due to local congestion is also used in [17].

The above variations are used to create four different GR solutions. The termination criterion in all cases was set to be when no additional improvement in overflow was obtained during the rip-up and reroute (RRR) phase of the algorithm.

Detailed Routing (DR) Emulation: To measure the impact of different GR solutions on the detailed routing stage, we (obviously) require a mechanism for performing detailed routing (DR). In this work, we had to implement our own *detailed routing emulator* to perform this evaluation. We would prefer to use an actual DR tool, but at the time of this writing, there were no DR tools, having an interface that takes a GR solution as input, available to us.

We did obtain a binary of the DR tool `RegularRoute`, kindly shared by the authors of [20]. `RegularRoute` is one of the most recent, competitive academic DR tools. However, at this phase, and similar to other DR tools, `RegularRoute` does not account for the complicating factors introduced in the ISPD11 benchmarks, such as zero metal-1 capacity, virtual pins, and non-uniform wire sizes and spacings. Our attempts to use `RegularRoute` after simplifying the benchmarks (by removing these factors) also failed.

We also carefully evaluated the use of Cadence's `wroute` to perform DR. Unfortunately for our purposes, `wroute`, similar to other commercial detailed routers that we considered, does not have an interface that takes a GR solution as input. Rather, it accepts a *placement instance* as input. This same issue is also mentioned in the paper [20].

Thus, we created our own DR emulator, designed to illustrate the impact of considering local congestion right after the first iteration of detailed routing. We will show that the impact is significantly different among the GR variations that were tested. The goal of our DR emulator is to illustrate the impact of the generated GR solution *immediately after one iteration* in the detailed routing stage, as a relatively-accurate *surrogate measure* to reflect the difficulty of the corresponding detailed routing instance.

Table 2: Comparison of wirelength and total overflow at various stages

	U-E (CGRIP)			U-AE			U-AV			NU-AV		
	GR-OF	DR-OF	GR-WL	GR-OF	DR-OF	GR-WL	GR-OF	DR-OF	GR-WL	GR-OF	DR-OF	GR-WL
sb1	0	23142	153.36	0	23020	154.20	0	12740	154.65	0	806	154.67
sb2	3168	18880	335.80	14496	18506	335.36	10526	13154	344.46	7780	9180	350.31
sb4	228	28696	114.40	2024	27476	115.12	880	13296	119.20	418	876	119.88
sb5	0	10878	184.74	322	9256	187.45	0	2588	187.78	450	1036	188.82
sb10	124	84842	270.32	4502	73862	282.65	872	66780	281.01	766	65232	281.23
sb12	0	44556	256.58	274	44416	264.11	302	36414	259.03	12120	15732	261.46
sb15	0	29982	191.81	1022	29800	192.32	846	18886	192.01	2630	7678	193.50
sb18	0	11406	105.47	0	11184	106.90	0	558	106.70	0	444	107.80
average	1.0X	1.0X	1.00X	6.4X	0.9X	1.00X	3.8X	0.7X	1.00X	6.9X	0.4X	1.01X

Our emulator uses a projected DR instance that can be used to estimate the overflow occurring during DR. Specifically, our DR emulator works on a two-layer grid, with one layer containing NS edges and one containing EW edges connected with VIAs. Each route in a given GR solution is then projected to this two-layer grid. Only the original (uniform) gcells are used at this stage to define the gcells where each gcell in the projected instance has a 40x40 detailed routing grid DRG superimposed upon it. This 40x40 granularity gives the same resolution as the placement grid P in the ISPD11 instances, ensuring that pins of all nets (global and local) are vertices of the DR grid. The capacity of each NS/EW edge in the DRG is equal to the number of the NS/EW layers above that edge that do not contain an obstacle at that location.

Global cells are individually routed over the DRG in a sequential, breadth-first manner, starting from bottom left. When visiting gcell c, first the local nets are routed, and then a track assignment is made for all global routes mapped to c in the current GR solution. This process is similar to other published detailed routing algorithms such as [20]. Some track assignments are imposed by neighboring (previously-visited) gcells. For global routes that connect to c through a VIA edge in the GR model, a utilization of one unit of the DRG grid edge inside c is used to reflect this VIA usage. Once the track assignment is made for c, rip-up and reroute (RRR) is applied to route the remaining subnets of each global route and all local nets inside c. In our emulator, we do only one iteration of RRR, so that our emulator shows the immediate impact of the translation of a GR solution into a DR solution. When doing RRR, the same net ordering as in the GR procedure is used. Each net inside c is routed using its shortest path after updating the routing resource usage by the previously-routed nets, similar to the GR framework. Each net, however, is restricted to be routed within a bounding box of its terminals.

Evaluation Metrics: For each generated GR solution, the overflow (denoted by GR-OF) is measured using the (unadjusted) edge capacities that are used in the U-E method. The wirelength of each case (denoted by GR-WL) is also measured. In NU-AV, the wirelength is computed while accounting for non-uniform gcells for fair comparison. For example, an edge in NU-AV which is twice than an edge in U-E due to non-uniform gcells is counted as 2 units of wirelength. For each GR solution, the total overflow computed by the DR emulator (denoted by DR-OF) is computed.

Comparison of Evaluation Metrics: Table 2 shows comparison of GR-OF, DR-OF, and GR-WL for each tool variation. The results confirm the following:

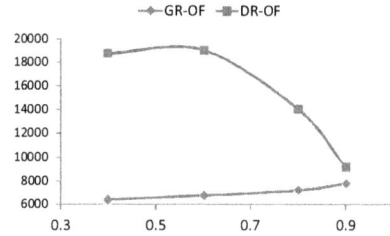

Figure 4: Tradeoff in DR-OF and GR-OF with η.

- Methods U-AE, U-AV, and NU-AV, which account for local nets, result in a reduced DR-OF of 0.9X, 0.7X, and 0.4X, respectively compared to U-E (1.0X).

- Methods U-AE, U-AV, NU-AV all have a significantly higher GR-OF than U-E.

- The GR-WL of methods U-AE, U-AV, NU-AV are up to 1% larger than U-E. Wirelength is increased due to detours as a result of reduced vertex or edge capacities in these methods.

The increase in GR-OF in methods U-AE, U-AV, and NU-AV is to be expected, since solutions are generated for instances with a reduced edge capacity. The resulting solution when evaluated with the original unadjusted edge capacities may have a high overflow compared to U-E. However, in most cases, this increase in the GR-OF is *more than offset* by a decrease in the DR-OF.

Consideration of vertex capacity (U-AV) results in more improvement in DR-OF compared to only reducing the edge capacity (U-AE). It allows reaching higher quality solutions which are excluded from the model of U-AE. Similarly, non-uniform binning results in more improvement in DR-OF.

Impact of Non-Uniform Binning in NU-AV: Our binning procedure is parameterized by a value η that controls the reduction of local nets when generating non-uniform gcells. Figure 4 demonstrate the impact of varying η for the instance sb2. (Similar behavior was observed in the other instances.) In the figure, we see that increasing η (which decreases the number of local nets), increases the GR-OF, but decreases the DR-OF. The runtimes have a reverse trade-off: lower η (higher local nets) results in lower GR runtime but higher DR runtime. Note that our binning procedure keeps the total number of gcells the same in all the cases.

When using the binning procedure in NU-AV, we note that for all the benchmarks, DR-OF was improved both with and without the post-processing step. However, post-processing provided additional reduction in these metrics. For example in sb2 for $\eta = 0.9$, the DR-OF with and without post-processing were 11378 and 9180 respectively —both smaller than DR-OF of the other methods.

Table 3: Comparison of runtime (min) and local nets

Bench	U	NU	U-E		U-AE		U-AV		NU-AV	
	%LC	%LC	GR	DR	GR	DR	GR	DR	GR	DR
sb1	30.8	14.1	3	**28**	7	**21**	5	**18**	7	**7**
sb2	28.9	13.5	352	**22**	321	**17**	303	**17**	389	**17**
sb4	35.2	16.8	180	**39**	60	**25**	201	**25**	60	**8**
sb5	29.4	12.2	135	**42**	184	**33**	164	**24**	221	**5**
sb10	34.1	34.0	251	**62**	341	**51**	329	**32**	342	**33**
sb12	28.6	14.6	238	**41**	360	**42**	309	**37**	306	**22**
sb15	34.4	15.8	212	**34**	269	**24**	259	**19**	233	**10**
sb18	28.2	15.0	10	**32**	20	**20**	16	**15**	10	**9**
ave	31.2%	17.0%	1.0X	**1.0X**	1.1X	**0.8X**	1.1X	**0.6X**	1.1X	**0.4X**

Comparison of Runtimes: Table 3 gives a runtime comparison for the different methods. The table shows that the CPU time of the DR emulation is reduced on average by 0.8X, 0.6X, and 0.4X for U-AE, U-AV, NU-AV, respectively, compared to U-E. The GR runtime on average is increased around 10% compared to U-E. Columns 2 and 3 report the percentage of local nets for uniform and non-uniform cases.

The termination condition of CGRIP can be set by the user based on the design flow—whether the user wants a "quick-and-dirty" solution or a solution that spends higher effort to produce a high-quality GR solution. In this work, we set the parameters of CGRIP to reduce the overflow of the GR-solution as much as possible. Specifically, the termination condition for each method is when no improvement in its objective is made for two consecutive RRR iterations. With this termination criterion, the GR runtimes of U-E are 3min and 10min for sb1 and sb18, respectively. However, the runtimes are multiple hours for the other instances. This longer runtime could be reduced at the expense of larger GR-OF values. For example, running CGRIP (the method U-E) for one hour results in GR-OF of 13568 for sb2, and overflow exists in 6 of the 8 benchmarks. A comparison of these values to column 2 of Table 2 indicates that the additional effort can significantly reduce GR-OF.

Impact of Local Congestion versus Wire Sizes: Local nets and non-uniform wire sizes are two factors that contribute to congestion, but are ignored or not mentioned in the GR published works. To evaluate the individual impact of these two factors, we conducted a small experiment comparing three GR methods: 1) when local congestion is ignored but non-uniform wire sizes are considered; 2) when non-uniform wire sizes are ignored but local congestion is considered; and 3) when both non-uniform wire size and local congestion are considered. In all cases, the gcells are of uniform size. Case 1 is same as U-E (i.e., CGRIP). Case 3 is U-AV with adjusted vertex capacity. In case 2, we assumed the same wire size and spacing in all layers to be equal to layer 1, which was done by changing the benchmark header line describing the per layer wire size and spacing values. (The routing procedures remain intact.) As a result, if layer 4 had a wire size of 2 units, after getting normalized to wire size of 1 unit in layer 1, then it passes double the number of wires on each edge. This transformation resulted in an increase in the normalized capacity of the edges in the higher layers, since these layers allow for more routing tracks to pass an edge once their wire sizes are reduced to match layer 1. This behavior is the same as the ISPD08 benchmarks. After generating GR and DR solutions, we evaluated each solution using the original wire size using our DR emulator, similar to the previous experiments.

In all cases, the DR-OF, which we are using as a surrogate measure for the goodness of the true DR solution, was significantly reduced by considering both additional complicating factors. For example, for benchmark sb4, the DR-OF for cases 1, 2, 3 were 28696, 3348396, and 13296, respectively. Note that case 2 had *significantly* higher DR-OF than the other cases. We conclude that non-uniform wire sizes and local nets are both crucial factors for routability. Our routing models and framework can account for *both* of these factors.

6. CONCLUSIONS

We proposed two techniques for considering local effects during global routing. First, we introduced a technique for constructing GR instances with gcells of non-uniform size that can decrease the number of local nets while controlling the complexity of the GR procedure. Second, we proposed a model to approximate the congestion induced by local nets and incorporated this mathematical model as a vertex capacity constraint into a congestion-aware Integer Programming (IP) formulation. The IP formulation also accounts for non-uniform wire sizes, routing blockages, and virtual pins.

7. REFERENCES

[1] ISPD 2011 routability-driven placement contest [online] http://www.ispd.cc/contests/11/ispd2011_contest.html.
[2] Y.-J. Chang, Y.-T. Lee, T.-C. Wang. NTHU-Route 2.0: A fast and stable global router. In *ICCAD*, pages 338–343, 2008.
[3] Y.-J. Chang, T.-H. Lee, T.-C. Wang GLADE: A modern global router considering layer directives. In *ICCAD*, pages 319–323, 2010.
[4] H.-Y. Chen, C.-H. Hsu, and Y.-W. Chang. High-performance global routing with fast overflow reduction. In *ASP-DAC*, 2009.
[5] M. Cho, K. Lu, K. Yuan, D.Z. Pan. BoxRouter 2.0: A hybrid and robust global router with layer assignment for routability. *TODAES*, 14(2), 2009.
[6] K.-R. Dai, W.-H. Liu, Y.-L. Li. NCTU-GR: Efficient simulated evolution-based rerouting and congestion-relaxed layer assignment on 3-D global routing. *TVLSI*, 20(3):459 –472, 2012.
[7] M. Hsu, S. Chou, T.-H. Lin, Y.-W. Chang. Routability-driven analytical placement for mixed-size circuit Designs. In *ICCAD'* pages 80–84, 2011.
[8] X. Hu, T. Huang, L. Xiao, H. Tian, G. Cui, E.F.Y. Young Ripple: an effective routability-driven placer by iterative cell movement. In *ICCAD*, pages 74–79, 2011.
[9] M.-C. Kim, J. Hu, D. Lee, Igor L. Markov A SimPLR method for routability-driven placement. In *ICCAD*, pages 67–73, 2011.
[10] M. Pan and C. Chu. IPR: an integrated placement and routing algorithm. In *DAC*, pages 59–62, 2007.
[11] J. A. Roy, N. Viswanathan, G.-J. Nam, C.J. Alpert, I.L. Markov CRISP: congestion reduction by iterated spreading during placement. In *ICCAD*, pages 357–362, 2009.
[12] J. A. Roy and I. L. Markov. High-performance routing at the nanometer scale. *IEEE TCAD*, 27(6):1066–1077, 2008.
[13] H. Shojaei, A. Davoodi, J.T. Linderoth Congestion analysis for global routing via Integer Programming. In *ICCAD*, pages 256–262, 2011.
[14] P. Spindler and F. M. Johannes. Fast and accurate routing demand estimation for efficient routability-driven placement. In *DATE*, pages 1226–1231, 2007.
[15] N. Viswanathan, C.J. Alpert, C.C. N. Sze, Z. Li, G.-J. Nam, J.A. Roy. The ISPD-2011 routability-driven placement contest and benchmark suite. In *ISPD*, 2011.
[16] M. Wang, X. Yang, K. Eguro, M. Sarrafzadeh. Multi-center congestion estimation and minimization during placement. In *ISPD*, pages 147–152, 2000.
[17] Y. Wei, C.C.N. Sze, N. Viswanathan, Z. Li, C.J. Alpert, L.N. Reddy, A.D. Huber, G.E. Téllez, D. Keller, S.S. Sapatnekar. Glare: global and local wiring aware routability evaluation. In *DAC*, pages 768–773, 2012.
[18] Y. Xu and C. Chu. MGR: Multi-level global router. In *ICCAD'* pages 250 –255, 2011.
[19] Y. Xu, Y. Zhang, C. Chu. FastRoute 4.0: global router with efficient via minimization. In *ASPDAC*, pages 576–581, 2009.
[20] Y. Zhang and C. Chu. RegularRoute: an efficient detailed router with regular routing patterns. In *ISPD*, 2011.

Escape Routing of Mixed-Pattern Signals Based on Staggered-Pin-Array PCBs *

Kan Wang,　Huaxi Wang,　Sheqin Dong

Tsinghua National Laboratory for Information Science and Technology(TNList)
Department of Computer Science & Technology
Tsinghua University, Beijing, China 100084
wangkan09@mails.tsinghua.edu.cn, thomas.kasim@gmail.com,
dongsq@mail.tsinghua.edu.cn

ABSTRACT

Escape routing has become a critical issue in high-speed PCB routing. Most of previous work paid attention to either differential-pair escape routing or single-signal escape routing but few considered them together. In this paper, a unified ILP model is used to formulate the problem of escape routing of differential pairs together with single signals (mixed-pattern signals) on staggered pin array. A mixed-pattern escape routing algorithm is proposed to solve the problem and a slice-based heuristic method is presented to speed up the algorithm. Experimental results show that the proposed method is very efficient. It can solve all the test cases in short time and improve wire length and chip area by 13.8% and 13.4% respectively compared to traditional pin array. At the same time, the method can increase routability by 16.3% and reduce the wire length by 9.3% compared to a two-stage method.

Categories and Subject Descriptors

B.7.2 [**Integrated Circuits**]: Design Aids

General Terms

Algorithm, Design, Performance

Keywords

PCB Routing, Escape Routing, Differential Pair, Staggered Pin Array, Mixed-Pattern Signals

1. INTRODUCTION

High speed printed circuit board (PCB) routing has become more and more difficult for manual design due to increased pin count and dwindling routing resource. To ad-

*This paper is supported by MOST of China project 2011DFA60290 and NSFC 61176022

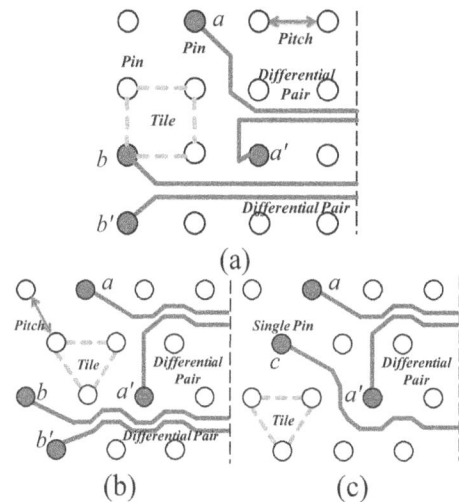

Figure 1: Escape routing pattern: (a) single-pattern escape routing on grid pin array (b) single-pattern escape routing on staggered pin array (c) mixed-pattern escape routing on staggered pin array

dress the problem, many methods were proposed, including PCB pin array structure [1]-[2] and escape routing algorithms [3]-[12].

For pin array structure, although *rid pin array*(GPA)[3] has been widely used, it cannot satisfy the demands of the ever-increasing pin number on PCB. Then another pin array structure, *sta ered pin array*(SPA) was formed [1]. Compared to GPA, the SPA can increase pin density greatly under the similar number of pins and same area constraints [2]. Due to the different routing resources and constrains, the routing on SPA will be much different.

For escape routing, differential-pair escape routing (two nets for each signal) [4]-[7] and single-signal escape routing (one net for each signal) [8]-[12] have been developed on GPA or SPA in recent years. However, there are still some disadvantages. For GPA, the previous works focused on only differential-pair escape routing or only single-signal escape routing(as shown in Fig.1(a)) and did not consider them together. For SPA, only single-signal escape routing was developed and there is no work on differential-pair escape routing on SPA, which is illustrated in Fig.1(b). Furthermore, to the best of our knowledge, there is still no work to consider the escape routing of both differential pairs and single sig-

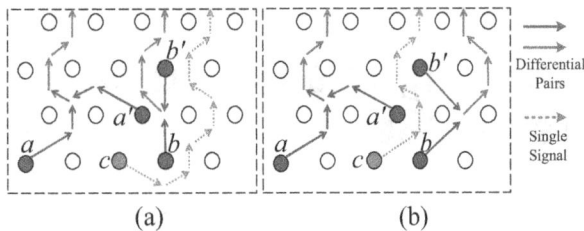

Figure 2: Routing examples (a)single-pattern escape routing will greedily optimize the routing of pin a and a' which can lead to congestion of single signal c while (b)mixed-pattern escape routing will optimize both signals

nals simultaneously on neither GPA nor SPA. Because of the high noise immunity and low electromagnetic interference, differential pairs are always used for the high-speed signal transmission on a PCB. However, due to the limitation of resources, not all signals will be transmitted by differential pairs. As a result, the signals of differential pairs and single signals will coexist on board and hence the research on escape routing of both differential-pair and single signals simultaneously will be quite valuable. For convenience, the escape routing only for differential pairs or only for single signals is referred as *sin le-pattern escape routin* and the escape routing of both differential-pair and single signals simultaneously is referred as *mixed-pattern escape routin* . Note that the *sin le-pattern escape routin* is just a special case of *mixed-pattern escape routin* . In this paper, the mixed-pattern escape routing problem on SPA is considered, as shown in Fig.1(c).

Although network flow algorithm [4]-[7] has been successfully applied to single-pattern escape routing on GPA, however, it is not suitable for mixed-pattern escape routing. Because on one hand, the two kinds of signals take different routing resource: differential pair takes two units of resource while single signal takes one, and on the other hand, they have different constraints. For example, differential pairs should satisfy the length-matching constraint [5] while single signals need not.

One feasible solution for mixed-pattern escape routing is to solve differential-pair routing and single-signal routing separately. However, the routing result cannot be optimized as the routing of the two kinds of signals will influence each other. As a result, the routing of one signal will probably fail if the routing resource is occupied by the other. This is hard to avoid even though congestion-aware method is used. Fig.2(a) shows an example where the optimization of differential pairs will lead to congestion of routing resource and longer routing path of single-signals. On this occasion, a mixed-pattern escape routing, which can optimize both differential-pair routing and single-signal routing as shown in Fig.2(b), is required.

Another feasible solution is to use unified model such as network flows to solve the problem. However, there are some challenges. First, the two kinds of signals are required to satisfy different constraints, which should be formulated in the unified model. Second, as the two signals are independent sources and take different routing resources, the mixed-pattern escape routing problem becomes multi-commodity flow problem which has been proven to be NP-hard. To solve

the problem, some efficient methods are required. Besides, some rules such as non-crossing rule should also be satisfied. Since it is more likely to obtain the best solution, in this paper, the unified modeling method based on ILP formulation is adopted, and a slice-based algorithm is presented to solve it.

The major contributions of this paper are summarized as follows:

1. A mixed-pattern escape routing algorithm on staggered pin array is proposed. To the best of our knowledge, this is the first work for mixed-pattern escape routing. Experimental results show that it is very efficient to solve both mixed-pattern routing and single-pattern routing.

2. A unified ILP model is formulated for mixed-pattern escape routing problem. Design rules such as non-crossing rule and length-matching rule[5] are also considered.

3. A slice-based heuristic method is presented to prune the variables of ILP and speed up the solving. The proposed approach can achieve 100% routability for all of ten test cases in reasonable running time.

The rest of the paper is organized as follows. Section II introduces the related work and Section III introduces the routing style in SPA and gives the formulation of mixed-pattern escape routing. In Section IV, the detailed routing algorithm is described and experimental results are shown in Section V. Conclusions are provided in Section VI.

2. RELATED WORK

The related work can be classified into two parts: one is SPA based modeling and escape routing algorithms and the other is GPA based escape routing including differential-pair escape routing and single-signal escape routing.

On one hand, [1] analyzed the properties of the SPA and proposed three escape routing strategies for SPA. [2] proposed an escape routing algorithm based on SPA and showed the advantages of SPA compared to GPA. However, they did not consider the differential-pair routing on SPA.

On the other hand, [8]-[12] proposed network flow based escape routing algorithms on GPA. However, they only focused on single-signal escape routing problem. A work on chip-package-board co-design [7] considered differential pairs, but it paid more attention to co-design instead of the optimization of differential pair escape routing. [4] proposed a negotiated congestion-based differential-pair routing but it did not take length-matching rule into account. A recent work [5] proposed a five-stage algorithm for escape routing of differential pairs considering length matching. However, all of them are based on GPA and cannot be applied to SPA directly. Furthermore, all previous work is for single-pattern escape routing and an algorithm for mixed-pattern escape routing is required.

3. PROBLEM DEFINITION

3.1 Preliminaries

For SPA, triangular tiles are usually modeled to stand for routing resource, as shown in Fig.3. Similarly to [5], in this paper, tile node is used as intermediary and wires are routed via tile nodes. Before we present the definition of the problem, some definitions for routing network are introduced first.

DEFINITION 1 (STAGGERED PIN ARRAY). *A $m \times n$ sta -*

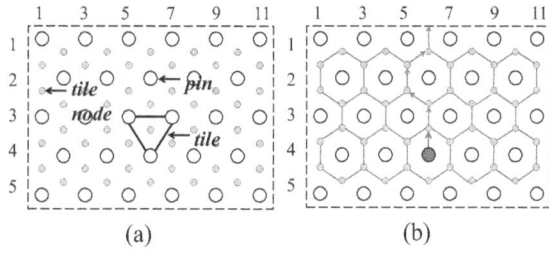

Figure 3: Tile network (a) staggered pin array (b) example in tile routing

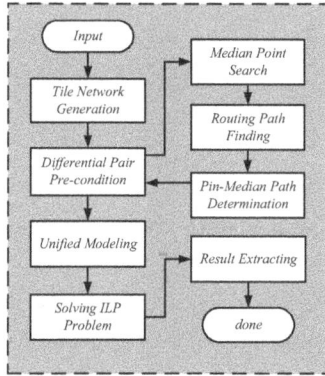

Figure 4: Overview flow of MPERA

ered pin array (S A) is composed o n rows, and in each row there are m (in odd rows) or m − 1 pins (in even rows). A trian ular tile is composed o three adjacent pins and in each tile there is a tile node, as shown in i .3(a).

DEFINITION 2 (TILE NETWORK). *The tile network is enerated by connectin trian ular tiles with each other in the orm o hexa ons, as shown in i .3(b). The ed es o tile hexa ons are the channels or escape routin and the an le between the routin channels is 120-de ree. An example o routin path is also shown in i .3(b).*

3.2 Problem Formulation

Based on the definitions, the problem of mixed-pattern escape routing (MPER) can be defined as follows: Given (1) an $m \times n$ staggered pin array; (2) a differential pairs and b single signals to be routed to the boundary; (3) design rules including wire length matching of differential pairs, non-crossing rules and (4) constraints such as routing resource and wire width constraint, the objective is to escape all marked pins to the array boundary with minimized total wire length via the tile network. At the same time, no design rule is violated and 100% routability is guaranteed. In the following section, a mixed-pattern escape routing algorithm (MPERA) is proposed to solve the problem.

4. THE MIXED-PATTERN ESCAPE ROUTING ALGORITHM

As mentioned in Section 1, MPER can be solved by a unified model. However, it is very difficult to take the wire length matching rule of differential pairs into unified model since it is only for differential pairs. To address it, the wire

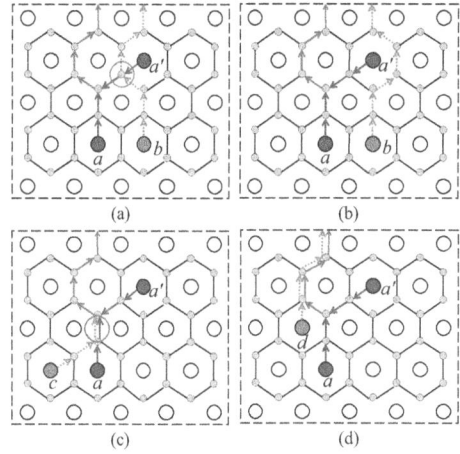

Figure 5: Routing Constraints in MPER: (a) is not allowed due to the non-crossing rule; (c) is not allowed due to differential pair protection constraint; while (b) and (d) are both legal examples

length matching rule will be solved separately. Consider that the wire length matching rule only work at the beginning of differential-pair escape routing, which is before two signals of differential pair meet with each other at median point[5], the escape routing process can be divided into two stages: pin-median-point routing of differential pairs and the mixed-pattern escape routing containing both differential pairs and single signals. The overview flow of MPERA is shown in Fig.4.

Given the routing requirement, tile network is generated first. Then the differential pairs are preconditioned to reduce the complexity of the problem. Median point searching and pin-median-point path determination algorithms are used for differential-pair escape routing and some strategies guarantee the differential pairs to satisfy the length-matching rule. After that, a unified ILP modeling is proposed to formulate the mixed-pattern escape routing problem and a heuristic algorithm is performed to solve the ILP in reasonable time. If the problem cannot be solved as a result of congestion, then go back to differential-pair precondition. The details of MPERA are described in the following sections.

4.1 Differential Pair Pre-conditioning

Different from single signals, the pins of differential pairs have more constraints. First, the routing of differential pairs should satisfy the length-matching rule. Without considering this, large signal skews will be created, which can lead to degradation of performance. Second, in order to avoid signal crosstalk, before the two signals of differential pair meet with each other, no other signal is allowed to be close to. That is, the routing path between differential-pair pins and median points do not allow other signals to cross or go along, as shown in Fig.5. In this paper, the constraint is named as *differential pair protection constraint*.

It is hard to take these constraints into a unified modeling. To solve it, the pin-median-point routing is handled separately. The median point routing, which routes median point to boundary, is solved with single-signal escape routing together through a unified modeling. Since the pin-median-

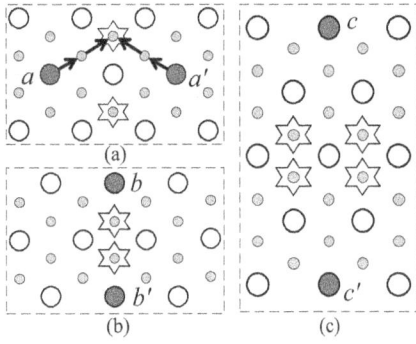

Figure 6: Simple cases for median points searching

point paths occupies routing resource, it will directly impact solution quality of mixed-pattern escape routing. Therefore, it is quite necessary to reduce the total length of pin-median-point paths. In this section, a two-stage pin-median-point routing is proposed, including median point searching and pin-median-point path determination. The detailed algorithms will be introduced as follows.

4.1.1 Median point searching algorithm

An effective method was proposed[5] to find median point candidates for each pair. However, it is based on GPA. For SPA, it would be different and the previous method cannot be adopted directly due to different structure and different constraints. In this section, a median point searching method for staggered pin array is proposed.

Let $(x_a{}^p, y_a{}^p)$ and $(x_b{}^p, y_b{}^p)$ be the coordinates of two pins of differential pair. Based on the features of the staggered pin array, the median point searching can be solved according to the following six cases:

Case 1: $y_a{}^p = y_b{}^p$: there are two min-cost median point candidates, which lie on the midperpendicular between two pins, as shown in Fig.6(a).

Case 2: $x_a{}^p = x_b{}^p$ and $|y_a{}^p = y_b{}^p|$ is not multiple of 4: There are two min-cost median point candidates, which lie on the line between two pins, as shown in Fig.6(b).

Case 3: $x_a{}^p = x_b{}^p$ and $|y_a{}^p = y_b{}^p|$ is multiple of 4: There are four min-cost median point candidates, which lie around the middle pin of the two pins, as shown in Fig.6(c).

Before we formally investigate the following three cases, the *pin hexa on* and *minimum intersectin hexa on* are first defined:

DEFINITION 3 (PIN HEXAGON). *ne pin hexa on is composed o certain pins which have the same distance to this pin. The pin in the center is called as host pin while the pins around are called as local pin, as shown in i .7(a). The size o pin hexa on is determined by the distance between local pin and host pin.*

DEFINITION 4 (THE MINIMUM INTERSECTING HEXAGON). *Given a differential pair, the intersectin hexa ons o the pair are two intersectin or adjacent hexa ons with the same size and the minimum intersectin hexa ons are those with the minimum size, as shown in i .7(b).*

Based on the definitions, the following three cases can be formulated:

Case 4: $x_a{}^p \neq x_b{}^p$, $y_a{}^p \neq y_b{}^p$ and minimum intersection hexagons are adjacent: the candidates lie in each hexagon

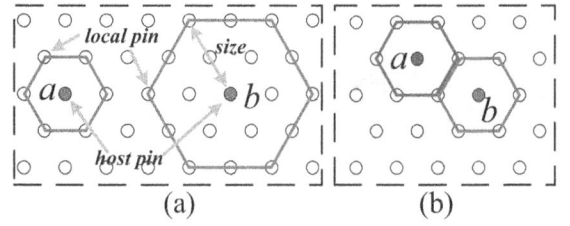

Figure 7: Pin hexagon (a) examples of pin hexagon (b) minimum intersection hexagons

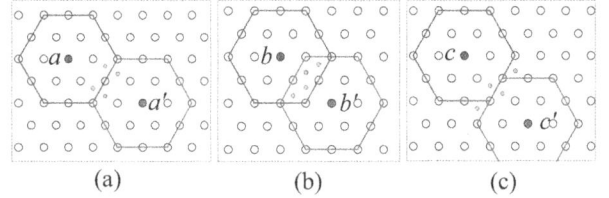

Figure 8: Complicated cases for median points searching

and beside the adjacent line of the hexagons, as shown in Fig.8(a).

Case 5: $x_a{}^p \neq x_b{}^p$, $y_a{}^p \neq y_b{}^p$ and minimum intersection hexagons are intersectional to each other: The candidates lie in the intersection region of minimum intersection hexagons of two pins, as shown in Fig.8(b).

Case 6: $x_a{}^p \neq x_b{}^p$, $y_a{}^p \neq y_b{}^p$ and minimum intersection hexagons are partly adjacent: The candidates lie around the middle point of two pins and beside the adjacent line of the hexagons, as shown in Fig.8(c).

Based on the above cases, several median point candidates can be obtained, and then the corresponding pin-median paths can be found according to the relative positions between pins and median points. A dynamic programming algorithm can easily solve the pin-median paths finding for each differential pair with length-matching constraint. Then the obtained paths will be treated as path candidates and the following sections will introduce how to select the final paths from all candidates.

4.1.2 Acute-angle Avoidance for Differential Pairs

Though a lot of path candidates are found in the previous section, some of them are not legal. It's easy to find that based on SPA, the acute-angle(60-degree) paths are possibly generated, especially for the differential pairs whose pins are close to each other. The acute-angle routing results will reduce the strength of signals and even cause undercutting or over etching of the circuitry. Therefore, it is necessary to reduce the numbers of acute-angle routing. In this paper, we bring in *path priority* for the selection of path candidates. The paths without acute-angles will be assigned high priorities. And the higher priority a path possesses, the higher possibility it will be selected as a routing solution. This method can remove the acute-angles by more than 50%. For the rest unavoidable cases, the 60-degree angle can be split into two 120-degree angles by adding an additional segment with a little wire length sacrificed. We don't discuss this in detail as the acute-angle issues are not emphasis of this paper.

4.1.3 Simultaneously Median Point and Shortest Pin-median-point Path Determination

After the acute-angles removal, the paths with high priority will be considered as the final candidates. Then differential pairs can be classified into K groups according to the crossing possibility of path candidates while K is the maximum value that makes the paths in groups without crossing with each other. A similar method as [5] can be used to solve it.

After that an ILP formulation is used to determine median points and pin-median-point paths for all differential pairs in each group.

Assume that each group G_k contains a_k differential pairs. For each pair i, there are n_{ki} path candidates. l_p denotes the sum of the length of path p from differential pair pin to median point and the distance of median point to the nearest boundary.

For group G_k, binary decision variables x_{ip} ($1 \leq i \leq a_k$, $1 \leq p \leq n_{ki}$) are defined such that x_{ip} is 1 if path p is selected for differential pair i, and x_{ip} is 0 otherwise. For each differential pair, one single path is assigned, so:

$$\sum_{p=1}^{n_{ki}} x_{ip} = 1 \; s.t. \; x_{ip} = 0 \; or \; 1, \forall 1 \leq i \leq N \qquad (1)$$

Let PCC_k stand for the path crossing cluster for group G_k. For each crossing pair of $path_{ip}$ and $path_{jq}$ in G_k, x_{ip} and x_{jp} should satisfy:

$$x_{ip} + x_{jq} \leq 1, (ip, jq) \in PCC_k \qquad (2)$$

Based on Eq.(1)-Eq.(2), the objective can be written by:

$$Min \sum_{i=1}^{a_k} \sum_{p=1}^{n_{ki}} x_{ip} \cdot l_p \qquad (3)$$

4.2 Unified Modeling for MPER

By solving the previous problem, the median points for differential pairs and the shortest paths for pins to median points can be determined. However, the paths of median points to boundary cannot be guaranteed as all pairs compete for the limited routing resource, especially with single signals considered. Therefore, it is quite necessary to consider them together in a unified model.

The network flow based ILP formulation method [2] has succeeded on the problem of single-pattern routing problem. However, when considering mixed-pattern routing, the situation will be different. First, there are two kinds of input sources including both differential pairs and single signals. Second, as the two kinds of signals take different network resources, more constraints will be brought in to distinguish them. As a result, the problem will become a multi-commodity problem which has been proven to be NP-hard. In this section, we focus on mixed-pattern escape routing problem and propose a unified ILP modeling to formulate it. A heuristic algorithm is also performed to solve the ILP in reasonable time. Some notations are defined for the routing network, as shown in Table 1.

Thereafter, the problem can be formulated as the following ILP objective function:

$$Min \; \alpha \times \sum_{e_{i,j} \in E} l(e_{i,j}) \times f_d(e_{i,j}) + \sum_{e_{i,j} \in E} l(e_{i,j}) \times f_s(e_{i,j}) \quad (4)$$

Table 1: notations of MPER

arameters o nodes on network raph	
T_D	The set of determined median point of differential pairs
P_S	The set of escaped single signals
T_S	The set of tiles of escaped single signal pins
$Tile$	The set of routing tiles
S_0	The source node of total network
S_D	The source node for differential pair
S_S	The source node for single signal pins
S_t	The sink node of total network
$Path_{DP}$	The set of tiles which has been occupied by pin-median routing of Differential Pair

arameters o ed es on network raph	
E	The set of all edges
e_{S_0,S_D}	The directed edge from S_0 to S_D
e_{S_0,S_S}	The directed edge from S_0 to S_S
e_{S_D,t_i}	Directed edge from S_D to a median point t_i
e_{S_S,p_i}	Directed edge from S_S to a single signal pin p_i
e_{p_i,t_i}	Directed edge from signal pin p_i to nearby tile t_i
e_{t_i,t_j}	Directed edge from an tile node t_i to another t_j
e_{t_i,S_t}	Directed edge from boundary tile t_i to sink node

arameters o constraints	
$l(e_{i,j})$	The length of edge $e_{i,j}$
$f_d(e_{i,j})$	The flow of differential pair on edge $e_{i,j}$
$f_s(e_{i,j})$	The flow of single signals on edge $e_{i,j}$
$f(e_{i,j})$	The total flow on edge $e_{i,j}$
$c(e_{i,j})$	The capacity of edge $e_{i,j}$
$Integer$	The set of non-negative integers

subject to:

$$\sum_{e_{i,j} \in E} f(e_{i,j}) = 1, \; \forall i \in T_D \bigcup P_S \qquad (5)$$

$$\sum_{e_{i,j} \in E} f(e_{i,j}) = |T_D|, \; \forall i \in S_D \qquad (6)$$

$$\sum_{e_{i,j} \in E} f(e_{i,j}) = |P_S|, \; \forall i \in S_S \qquad (7)$$

$$\sum_{e_{i,j} \in E} f(e_{i,j}) = |T_D| + |P_S|, \; \forall i \in S_0 \qquad (8)$$

$$\sum_{e_{i,j} \in E} f(e_{i,j}) = |T_D| + |P_S|, \; \forall j \in S_t \qquad (9)$$

$$\sum_{e_{i,j} \in E} f_d(e_{i,j}) = \sum_{e_{j,k} \in E} f_d(e_{j,k}), \; \forall j \in Tile \qquad (10)$$

$$\sum_{e_{i,j} \in E} f_s(e_{i,j}) = \sum_{e_{j,k} \in E} f_s(e_{j,k}), \; \forall j \in T_S \bigcup Tile \qquad (11)$$

$$2 \times f_d(e_{i,j}) + f_s(e_{i,j}) \leq c(e_{i,j}) \qquad (12)$$

$$\sum_{e_{i,j} \in Path_{DP}} f_s(e_{i,j}) \leq 0 \qquad (13)$$

$$f_d(e_{i,j}) \geq 0, \ f_s(e_{i,j}) \geq 0 \qquad (14)$$

$$f_d(e_{i,j}), \ f_s(e_{i,j}) \in Integer \qquad (15)$$

The objective of Eq.(4) is to minimize the total wire-length including both differential pairs and single signals. As we regard the wire-length as criterion of delay, The wire length of differential pairs is not doubled one because we regard the wire-length as criterion of signal delay instead of the cost of routing resource. α is used to adjust the relative weighting between differential pairs and single signals, and in this paper, α is set to 1. Three source nodes are defined to drive the flows. The S_0 node is used for source of all flows and two sub-source nodes S_D, S_S are used for differential pairs and single signals respectively. For each edge, there are two kinds of flows, $f_d(e_{i,j})$ for differential pairs and $f_s(e_{i,j})$ for single signals. Constraints (5)-(9) ensure the 100% routability of each flow by forcing all escaped pins to be routed to the sink and constraints (10)-(11) ensure that each flow satisfies the flow conservation constraint. Constraint (12)-(13) guarantees the capacity constraint and differential pair protection constraint. Constraint (14)-(15) makes the flow on all edges non-negative integers.

4.3 Slice-based MPER Algorithms

However, directly solving the ILP problem is difficult due to the nature of NP-hard. To address the problem, some heuristic strategies are proposed. Considering that the multicommodity problem can be modified to single commodity problem by adding some constraints, the problem can be divided to three cases and solved respectively as follows.

4.3.1 Modified MPERA for single-pattern escape routing

Considering the single-pattern escape routing for either differential pairs or single signals, the problem will be transformed to the LP problem in [2] and can be solved in polynomial time.

4.3.2 MPERA considering crosstalk between single signals

Considering the crosstalk issues between single signals, the single signals cannot share the routing resources as the close distance will lead to crosstalk. While on the other hand, differential pairs can enjoy the same routing resource as the signals of differential pair will be protected by themselves. Especially for the tile capacity of 2, the constraint of (12) will be transformed into:

$$f_d(e_{i,j}) + f_s(e_{i,j}) \leq 1 \qquad (16)$$

Then the problem will satisfy the unimodularity property [13] and can be transformed to LP formulation without integer constraints (15) as it is guaranteed to be an integer solution according to unimodularity property. The LP problem can be solved in polynomial time.

4.3.3 Slice-based MPERA without considering the crosstalk between single signals

As complement to the cases of 4.3.1 and 4.3.2, if the crosstalk issues between single signals are not considered, the problem will not satisfy the unimodularity property any more. Then a slice-based heuristic method is proposed to speed-up the algorithm.

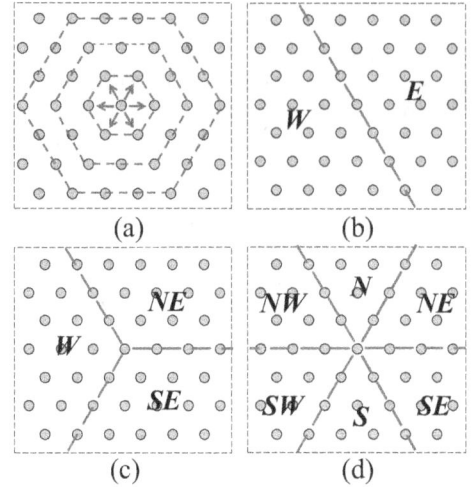

Figure 9: Region division: according to (a) the symmetrical characteristic of SPA, the chip can be partitioned into (b) two regions; (c) three regions; (d) six regions

As it is well known, when a water-drop falls down to the ground, the droplets will spread to all around and if the ground is smooth, the spread will be even enough. Inspired by this physical phenomenon, if the pins of differential pairs and single signals are uniform distributed, the routing paths will be evenly sent out to the boundary, that is, the escape routing satisfies divergence, which can be observed from existing work such as [2]-[5]. In this paper, we refer the property as *diver ence property* of escape routing. Though it cannot satisfy all cases, it is still suitable for most cases. For the special cases, we just need some special treatments to handle them.

Algorithm 1 Variable Pruning for ILP
___

Require: *width, height, sinpins, dpairs, apartNum;*
1: *generatePinArray();*
2: *generateTileArray();*
3: *createRegions*(apartNum);
4: *isdone = false;*
5: **while** *isdone* **do**
6:    **for** each $i \in [1, apartNum]$ **do**
7:       *initalization(i);*
8:       *generateDpair(i);*
9:       *generaterSinpins(i);*
10:      *IDCoorMapping(i);*
11:      *createEdges(i);*
12:      *ILPEscapeRouting(i);*
13:      *resultsExtracting(i);*
14:    **end for**
15:    **if** ILP is optimized solved for all regions **then**
16:       *isdone = true;*
17:    **else**
18:       *redistributeFailurePoints();*
19:    **end if**
20: **end while**
21: *mergeResults();*
22: *outputAllPaths();*
___

Table 2: Effect of escape routing algorithm on staggered pin array

Benchmark information			[5]			Ours			
Benchmark	Dpair num	Array size	Avg. len.	Run-time (s)	Area	Pin Array len.	Avg. len.	Run-time (s)	Area
case1	10	11x8	2.7	<1	70	2.67	2.31	<1	60.62
case2	18	22x7	4.0	<1	126	4.00	3.46	<1	109.12
case3	18	18x12	4.8	<1	187	4.74	4.10	<1	161.95
case4	8	11x3	2.0	<1	20	1.33	1.15	<1	17.32
case5	11	14x3	2.6	<1	26	1.76	1.52	<1	22.52
case6	11	17x6	4.1	<1	80	4.18	3.62	<1	69.28
case7	18	17x6	3.0	<1	80	3.04	2.63	<1	69.28
case8	20	9x16	3.5	<1	120	3.50	3.03	<1	103.92
case9	20	8x15	3.3	<1	98	3.30	2.86	<1	84.87
case10	60	35x35	12.2	<1	1156	13.50	11.68	2	1001.13
Ratio			1	0.91	1	0.996	0.862	1	0.866

Considering *diver ence property*, the problem can be divided into several subproblems and solved individually. Specifically, for the case of SPA, the whole region can be divided into 2, 3 or 6 sub-regions according to the characteristics of hexagons, as illustrated in Fig.9. Due to this, for each subproblem, the variables of ILP can be reduced by at least 50% to 87%.

The detailed algorithm is shown in Algorithm 1. Initially, the pin array and tile network are generated according to the input information(line 1-2). The chip is partitioned into $apartNum$ regions(line 3), each of which will solve the ILP independently. Some initialization is done to store the information of pin array and tile array in each region(line 7). Then differential pairs and single signals are also classified into each region according to coordinates of median point and pin respectively(line 8-9). Correspondingly, the IDs of pins are also re-mapped to a new set(line 10). After that, edges are generated for the network flow based ILP formulation(line 11). Then an ILP solver is performed to solve the problem(line 12). If the ILP is optimally solved for all regions, then go to end. However, the region is not always equally partitioned. As a result, some regions may be congested and not able to get a feasible solution. To address it, the failed signals in congested regions will be redistributed heuristically into nearby region which has the most routing resource, until the problem is solved successfully(line 15-19). Finally, the results in each region are merged and final routing results are stored(line 21-22).

5. EXPERIMENTAL RESULTS

In this section, two experiments are conducted to show the efficiency of the proposed algorithms on benchmarks generated according to [5]. All experiments are implemented in C++ on a workstation with Intel Xeon 2.40GHz CPU and 12GB physical memory. The capacity of edge on tile network is set to 2 and only one routing layer is considered. lp_solve[14] is used for the ILP and LP solving. The wire lengths of experimental results are all normalized values.

5.1 Effect of MPERA on Staggered Pin Array

In this section, we show the effect of the proposed escape routing algorithm on single-pattern escape routing based on staggered pin array.

In Table 2, *Dpair num* denotes the number of differential pairs, while *Sin.num* denotes the number of single signals,

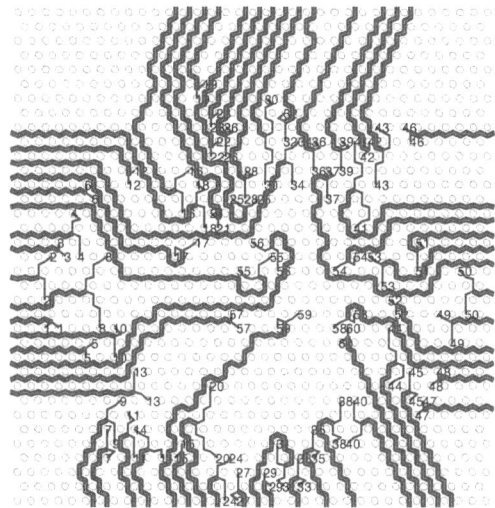

Figure 10: Escape routing results of case10

Array size shows the size of the staggered pin array and *Avg.len* shows the average number of tiles from two pins to the boundary. To make the comparison more fair, we create ten similar grid pin arrays according to the test cases in [5] and generate the ten benchmarks by shifting specific columns of the grid pin arrays. Table 2 shows that the error of our generated grid pin array compared to [5] is only 0.4%. *Run-time(s)* gives the running time for routing in second and *Area* gives the total chip area based on the pin array pattern.

The proposed method is very efficient on differential-pair escape routing. It can solve the routing problem in quite short time. Furthermore, compared to traditional pin array, the staggered pin array can reduce wire length by about 13.8% on average and chip area by 13.4%. Fig.10 shows the routing result of case 10.

5.2 Effect of MPERA on Mixed-Pattern Escape Routing

In this section, the effect of the proposed mixed-pattern escape routing is shown. To make the comparison more clear, we also implement a two-stage network flow based routing algorithm named *TS* to be a competitor to show

Table 3: Effect of the proposed mixed-pattern escape routing

Benchmark information			Two-stage method				Ours			
Benchmark	Dpair num	Sin. num	array size	DPair length	Sin. length	Routability (%)	Runtime (s)	DPair length	Sin. length	Routability (%)	Runtime (s)
case1-2	5	5	11x8	8.8	8.5	80%	<1	8.8	8.2	100%	<1
case2-2	9	9	22x7	5.78	6.1	88%	<1	5.78	5.67	100%	<1
case3-2	9	9	18x12	10.44	7.5	88%	<1	10.44	7.33	100%	<1
case4-2	5	5	11x3	2.4	2.6	90%	<1	2.4	2.2	100%	<1
case5-2	6	5	14x3	3.67	2	100%	<1	3.67	2	100%	<1
case6-2	6	7	17x6	7.17	5.67	85%	<1	7.67	4.43	100%	<1
case7-2	9	9	17x6	4.67	4.88	88%	<1	4.78	4.55	100%	<1
case8-2	10	10	9x16	6.4	7	90%	<1	6.4	6.5	100%	<1
case9-2	10	10	8x15	5.7	6.63	80%	<1	5.8	6.1	100%	<1
case10-2	30	36	35x35	18.33	16.12	88%	2	18.4	14.52	100%	3
case11-2	50	66	35x35	15.92	12.61	69%	2	16.0	11.36	100%	13
Ratio				1	1.093	1	-	1.010	1	1.163	-

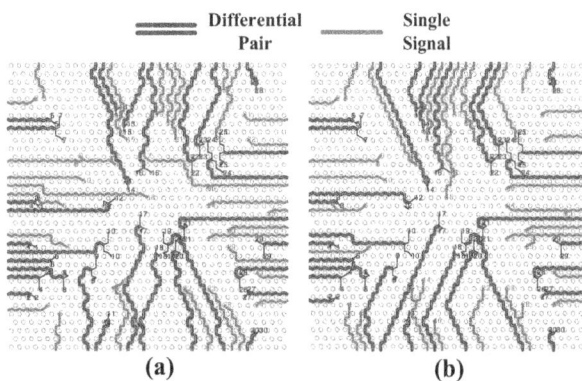

Figure 11: Escape routing results of case 10-2 (a) two-stage network algorithm (b) the proposed algorithm

the advantage of our method. Specifically, TS is to solve differential-pair routing and single-signal routing respectively.

The results are shown as Table 3, where we can see that TS cannot solve all test cases, while the proposed method can solve all single-signal escape routing. The routability is increased by about 16.3%. Furthermore, for the routed pins, the proposed method can also reduce the average wire length of single signals by 9.3% on average and 22.0% at most. At the same time, compared to the pure ILP solving which cannot converge even during thousands of seconds for large cases, the proposed method can simultaneously solve both differential pairs and single signals in short time.

Fig. 10 shows the escape routing result of case 10-2 using two-stage network algorithm(Fig.11 (1)) and the proposed algorithm based on 3 sub-regions(Fig.11 (2)).

6. CONCLUSION

Staggered pin array has become more and more popular for printed circuit board (PCB), as it has higher pin density than grid pin array. At the same time, differential-pair is a good method to increase noise immunity of signal for high-speed signal transmission on. In this paper, an algorithm for escape routing of simultaneously differential-pair and single signals is proposed on staggered pin array based

PCB. Experimental results show that the proposed method can solve both single-pattern and mixed-pattern effectively.

7. REFERENCES

[1] R. Shi and C.-K. Cheng. Efficient escape routing for hexagonal array of high density I/Os. A , 2006, pp. 1003-1008.

[2] Y. Ho, H. Lee, and Y. Chang. Escape Routing for Staggered-Pin-Array PCBs. A , 2011, pp. 306-309.

[3] T. Yan and M. D.-F. Wong. Recent research development in PCB layout. I A , 2010, pp. 398-403.

[4] T. Yan, Pei-Ci Wu, Qiang Ma, Martin D. F. Wong. On the Escape Routing of Differential Pairs. I A ,2010, pp. 614-620.

[5] T. Li, W. Chen, X. Cai, T. Chen. Escape Routing of Differential Pairs Considering Length Matching. AS - A , 2012, pp 139-144.

[6] T. Yan, Martin D. F. Wong. A Correct Network Flow Model for Escape Routing. A ,2009, pp. 332-335.

[7] J.-W. Fang, K.-H. Ho, and Y.-W. Chang. Routing for chip-package-board co-design considering differential pairs. I A ,2008, pp. 512-517.

[8] J.-W. Fang, I.-J. Lin, Y.-W. Chang, and J.-H. Wang. A network-flow-based RDL routing algorithm for flip-chip design. IEEE Trans. omputer-Aided esi n, vol. 26, no. 8, Aug. 2007.

[9] J.-W. Fang and Y.-W. Chang. Area-I/O flip-chip routing for chip-package co-design. I A , pp. 518-522, 2008.

[10] M.-F. Yu, J. Darnauer, and W. W.-M. Dai. Interchangeable pin routing with application to package layout. I A , 1996, pp. 668ÍC673.

[11] R. Wang, R. Shi, and C.-K. Cheng. Layer minimization of escape routing in area array packaging. I A , 2006, pp. 815ÍC819.

[12] M.-F. Yu and W. W.-M. Dai. Single-layer fanout routing and routability analysis for ball grid arrays. I A , 1995, pp. 581ÍC586.

[13] R. J. Vanderbei. Linear Programming: Foundations and Extensions. Sprin er, 2007.

[14] "Lp_solve". http://www.lpsolve.sourceforge.net/5.5/

Delay-Driven Layer Assignment in Global Routing under Multi-tier Interconnect Structure*

Jianchang Ao[1], Sheqin Dong [1], Song Chen [2], Satoshi Goto[3]

[1]Department of Computer Science and Technology, Tsinghua University, Beijing, China 100086
[2]Dept. of Electronic Science and Technology, University of Science and Technology of China, Hefei, China 230027
[3]Graduate School of Information, Production and System,   Waseda University, Kitakyushu, Japan 802-0001
aojianchang@163.com, dongsq@mail.tsinghua.edu.cn, songch@ustc.edu.cn, goto@waseda.jp

ABSTRACT

A multilayer routing system usually adopts multiple interconnect configuration with different wire sizes and thicknesses. Since thicker layers of metal lead to fatter wires with smaller resistance, the layer assignment of nets has a large impact on the interconnect delay. However, such layer dependent characteristics have been ignored by most of the state-of-the-art academic layer assignment methods. To remedy this deficiency, this work studies a more effective layer assignment problem under such multi-tier interconnect structure, which arises during 3D global routing and focuses on minimizing both delays and via count. This work presents a two-stage algorithm to solve the problem, which first minimizes the total delay and via count simultaneously by dynamic programming and negotiation technique, and then further minimizes the maximum delay carefully while not increasing the via count. The experimental results on ICCAD09 benchmarks show that the proposed algorithm can significantly reduce the total delay and maximum delay while still keeping roughly the same via count, compared with the state-of-the-art via count minimization layer assignment method NVM.

Categories and Subject Descriptors

B.7.2 [**Integrated Circuits**]: Design Aids–*Placement and Routing*;
J.6 [**Computer-Aided Engineering**]: Computer-Aided Design

General Terms

Algorithms, Design, Performance

Keywords

Global Routing, Layer Assignment, Delay Optimization, Via Minimization, Multi-tier Interconnect Structure.

1. INTRODUCTION

As VLSI technology enters the nanoscale regime, interconnect delay has become the bottleneck of the circuit timing since interconnects scale much slower than devices. Even with use of

*This paper is supported by MOST of China project 2011DFA60290 and NSFC 61176022.

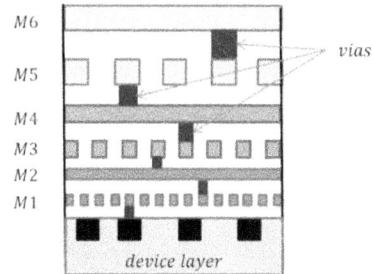

Fig. 1. Three-tier six-layer interconnect structure.

copper wires, low dielectric constant materials, and improved metal aspect ratio, the interconnect delay is still more dominant than the logic delay [1]-[3]. In VLSI designs, a multilayer routing system usually adopts multiple interconnect configuration with diverse specifications of wire sizes and thicknesses for metal layers, and wires usually become thicker and wider on higher metals. Such layer dependent characteristics are becoming exceedingly common in design of 65nm and smaller [2]-[5]. Fig. 1 shows a three-tier six-layer interconnect structure.

Multilayer routing plays critical role for interconnect delay, reliability and manufacturability/yield of a chip, and are generally carried out by global routing, track routing and detailed routing. While holding 3D (multilayer) global routing (GR) contests [6, 7], ISPD07 and ISPD08 have contributed to considerable research on 3D global routing. To reduce the complexity of 3D global routing, one major approach is to first condense the 3D grid graph to a 2D one, then use a 2D global router to obtain a 2D routing result. Finally, layer assignment (LA) assigns each net edge to the corresponding metal layer to obtain the final 3D routing solution. Fig. 2 shows the common design flow adopted by the participants in the ISPD07 and ISPD08 global routing contests.

As a key step of such multi-layer global routing approach, layer assignment should try to avoid creating additional wire congestion, and perform some optimization tasks, such as via optimization [8]-[17], antenna avoidance [16, 17], etc. Since the 3D global routing benchmarks of ISPD07 and ISPD08 contests do not specify different wire sizes or pitches on metal layers, these recently developed layer assignment methods mainly focus on the minimization of via count, and ignore the importance of delay optimization due to the layer dependent characteristics. However, since thicker layers of metal lead to fatter wires with smaller resistance, the assignment of net edges to routing layers has a large impact on the interconnect delay. Therefore, the delays incurred by this assignment also need to be considered.

Timing constrained minimum cost layer assignment or buffer insertion, e.g., [3][18], regards multi-tier interconnect structure, but they are very different from GR LA because they deal primary with single trees rather than tree sets, more importantly, they aim to assign wires to thick metals or insert buffers such that the

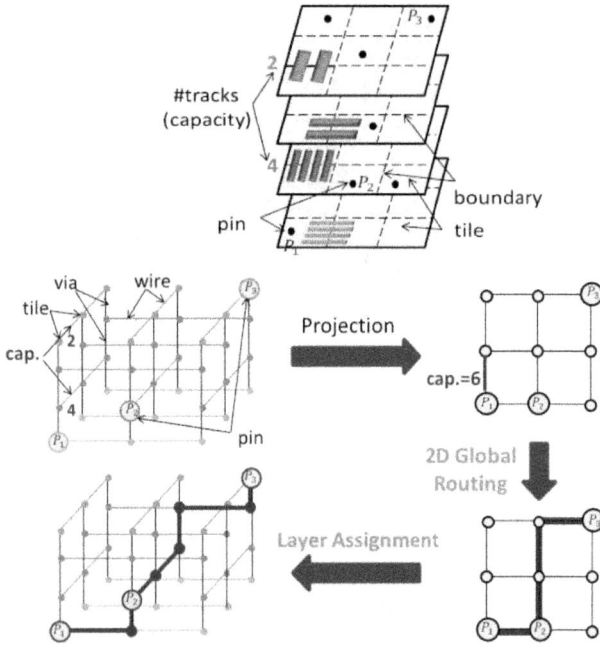

Fig. 2. 3D global routing under multi-tier interconnect structure.

timing constraint of a net is satisfied and the usage of thick metals or buffers is minimized, while GR LA aims to assign nets to metals such that wire congestion constraints of 3D global routing are satisfied and via count (or delay, etc.) of all nets is minimized.

Global routers honoring layer directives, e.g., [19][20], specify candidate routing layers (higher and thicker metals) for the appointed timing-critical nets, but do not come down to the actual calculations of delays. The classical performance driven layer assignment, e.g., [21]-[23], is different from GR LA in the sense that it handles delay optimization in the post-layout stage and does not consider the wire congestion constraints of 3D global routing but the strict constraints of design rules on the layout [10]. Some works regard the timing optimization for coupling capacitance in layer assignment, e.g., [24], but do not consider the multi-tier interconnect structure or the layer dependent characteristics.

This work studies the problem of delay-driven layer assignment under the multi-tier interconnect structure, which arises during 3D global routing. It not only considers traditional routing objective such as via count, but also focuses on delay optimization, which has been ignored by most recent layer assignment methods in the literature. The problem is to transform a 2D global routing solution into a 3 (multilayer) one with delays and via count minimized, subjective to the wire congestion constraints. This work presents a two-stage algorithm to tackle the problem, which first minimizes the total delay and via count simultaneously by dynamic programming and negotiation technique, and then further minimizes the maximum delay carefully while not increasing the via count. Resistances and capacitances of metal wires and vias are considered in the corresponding RC model. The proposed method can be used with a wide variety of interconnect configuration for it does not make any assumption on resistances or capacitances of metal wires or vias. The experimental results on ICCAD09 benchmarks [5] show that the proposed algorithm can significantly reduce the total delay and maximum delay while still keeping roughly the same via count, compared to the state-of-the-art via count minimization layer assignment method NVM [15].

The paper is organized as follows. Section 2 formulates the problem and discusses the delay model. Section 3 presents the two-stage algorithm framework. Section 4 presents the single-net layer assignment algorithm for simultaneous delay and via count minimization (SDLA) by dynamic programming. Experimental results are provided in Section 5, followed by conclusion in Section 6.

2. PROBLEM FORMULATION
2.1 Background: 3D Global Routing under Multi-tier Interconnect Structure

For 3D global routing, the routing region is partitioned into tiles and a 3D routing graph composed of vertexes and edges models the routing regions, where a vertex represents a tile and the edges model the relationship between adjacent tiles. Each edge is associated with a wire capacity to denote the number of available detailed routing tracks on tile boundary, i.e., the maximum number of nets that can pass through. Note that tiles on each routing layer may have *different routing track counts* due to *different wire sizes and pitches*. The wire usage of an edge is defined as the number of nets that actually pass through the tile boundary.

For each boundary edge, the number of wire overflows indicates that the wire usage locally exceeds the wire capacity. The total or maximum wire overflow denotes the summation of the wire overflows or the maximum wire overflow of all boundary edges. There is also a set of nets where each net is composed of a set of pins (with each pin corresponding to a vertex). A solution maps nets to trees that connect all the pins of a net using the edges of the routing graph, and performs some optimization tasks, such as minimizing the total/maximum wire overflow, metal wirelength, and via count, etc. Fig. 2 illustrates a two-tier four-layer global routing instance.

2.2 Problem Formulation

Vias often increase the manufacture cost, degrade the reliability and reduce the manufacturing/yield of a design, so the optimization of via count is also taken into consideration in this work. The problem is formulated as follows.

Delay-Driven Layer Assignment in Global Routing under Multi-tier Interconnect Structure (DLA): Given a 5-tuple (G^k, G, S, P, E), G^k represents a k-layer 3D grid graph, G represents a 2D grid graph compacted from G^k, S represents a 2D global routing result on G, P denotes the electrical parameters of the set of nets, i.e., the driven resistances of sources and the load capacitances of sinks for each net, and E denotes the electrical parameters of the routing system, i.e., the resistances and capacitances of wire segments and vias for each layer. The problem requires to map S onto G^k to obtain a 3D global routing result S^k, such that the delays and via count are minimized:

$$\text{min: } \textstyle\sum_{each\ net\ i} (\lambda \cdot delay_i + \#vias_i) \qquad (1)$$

where λ is the parameter to specify the relative importance of the net delay and the via count, and λ can also be specially specified for selected nets to emphasize their critical role in signal propagation. Further, the problem also requires S^k to satisfy the following two wire congestion constraints [10]:
1) Total wire overflow constraint:
$$\text{TWO}(S^k) \leq \text{TWO}(S) \qquad (2)$$
2) Maximum wire overflow constraint:
$$\text{MWO}(S^k) \leq \lceil \text{MWO}(S)/k \rceil \qquad (3)$$
where TWO and MWO denote the total wire overflow and the maximum wire overflow, respectively. Equation (2) ensures that the total wire overflow after layer assignment do not exceed the total wire overflow in the 2D result. Furthermore, if a region has overflow, Equation (3) guarantees to evenly distribute the congestion to each layer.

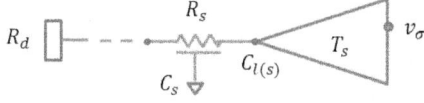

Fig. 3. Contribution of RC segment s to Elmore delay at sink v_σ.

2.3 Delay Model

The distributed Elmore model [25] is adopted to estimate the interconnect delay. Although occasional large errors make Elmore delay unsuitable for critical nets, it has a role in global routing for its fidelity and simplicity, and is a reasonable model for the routing in global stage is coarse and the set of nets is very large.

For each 2D routed net, assume its routing topology is a *tree*; otherwise, remove all its cycles and transform the topology into a tree. A net tree has *one source* and *multiple sinks*, with the *resistance of the driver driving the source* and the *load capacitance* at each *sink*. For an arbitrary net tree, each *wire segment* or *via segment* is viewed as an *individual RC conductor segment*. Under the Elmore distributed RC delay model [25], the signal transmission line is seen as a series circuit composed by series of these RC conductor segments. The delay at any sink v_σ is the sum of delay contributions from each of its ancestors (Fig. 3). Formally,

$$delay(v_\sigma) = \sum_{s \in ans(v_\sigma)} delay(s)$$
$$= \sum_{s \in ans(v_\sigma)} R_s \cdot \left(C_s/2 + C_{l(s)}\right) \quad (4)$$

where R_s, C_s, and $C_{l(s)}$ refer to the *segment resistance*, *segment capacitance*, and *load capacitance*, respectively. Elmore delays are incorporated at multiple sinks by attaching priority a_σ to the delay $delay(v_\sigma)$ at sink v_σ. Without loss of generality, assume $\sum_{\sigma=1}^{m} a_\sigma = 1$, where m is the number of sinks. Then the delay of net T is

$$delay(T) = \sum_{\sigma=1}^{m} [a_\sigma \cdot delay(v_\sigma)]$$
$$= \sum_{s \in T} [wt_s \cdot R_s \cdot (C_s/2 + C_{l(s)})] \quad (5)$$

where wt_s is the *delay weight* of segment s and its value is the summation of the priorities of all sinks in the downstream of segment s. Obviously, the delay weights of *wire* segments for a net tree are independent of the result of layer assignment, but only rely on its 2D topology, so can be pre-computed. In the following, $wt_{sink(v)}$ is also used to denote the priority of sink v.

3. ALGORITHM FRAMEWORK

This section presents the two-stage algorithm framework, which first minimizes the total delay and via count simultaneously, and then further minimizes the maximum delay of the set of nets while avoiding the increase of via count as much as possible.

3.1 Stage 1: Layer Assignment for Simultaneous Delay and Via Count Minimization

Previous work [15] shows that the congestion negotiation technique [26] can achieve better results than traditional net-ordering technique for via count driven layer assignment. Therefore, the negotiation method is also adopted in this stage for the layer assignment to minimize the delay and via count simultaneously.

There are three steps in this stage: initial layer assignment (ILA), rip-up-and-re-assign (RRA) and greedy improvement (GI). First, ILA identifies a minimal cost assignment solution for each net. Second, based on the solution of ILA, RRA iterates one-at-a-time layer assignment passes for nets that passing through 3D edges with overflows. The iteration process of RRA is kept repeating until the total overflow is no more than a pre-defined threshold or a pre-defined number of iterations is reached. Finally, GI rips up and re-assigns each net to further reduce the delay and via count, under the wire congestion constraints.

At each step, a single-net layer assignment algorithm for simul-

taneous delay and via count optimization (*SDLA*, which is detailed in Section 4) performs layer assignment for each net repeatedly until all nets are processed. The processing order of nets is the same at each step, for it only slightly influences the quality of results. SDLA tries to find a layer assignment result of a 2D routed net T (for a 3D routed net that is to be re-assigned, first compress it to 2D) with the minimum total cost of delay, via count and wire congestion; formally, the cost of a layer assignment result of T is

$$cost(T) = \lambda \cdot delay(T) + \#via(T) +$$
$$\sum_{e \in T} congestion\_cost_e \quad (6)$$

At each step, the congestion cost of boundary edge e is computed as follows. In ILA and RRA,

$$congestion\_cost_e^i = p_e \cdot h_e^i \quad (7)$$

where p_e is the present congestion penalty and h_e^i is a historic cost at ith iteration. The congestion penalty p_e is expressed as

$$p_e = 1 + k_1/[1 + \exp(k_2 \cdot (c_e - u_e))] \quad (8)$$

where c_e and u_e are the wire capacity and usage of edge e respectively, and k_1 and k_2 are set to 0.5 and 0.75 respectively. For every iteration, h_e^i is increased by h_{inc} if e overflows, where h_{inc} represents the step of the cost increment and set to 0.3. Obviously, the edges that tend to be congested make their congestion costs increase gradually during iterations of RRA. This helps to distribute the routing demand to other less congested edges. In GI, $congestion\_cost_e$ is set to *zero* for non-congested edges, but *infinite* for congested edges to *forbid* the usage of the edges and violation of wire congestion constraints. Such setting of GI can enforce its solution to obey the two wire congestion constraints (Equation (2) and (3)).

3.2 Stage 2: Minimization of Maximum Delay

This subsection presents a heuristic to carefully reduce the maximum delay of the net set while avoiding the increase of via count as much as possible. It continues to reduce the delay of the net with the maximum delay currently until no improvement can be achieved, which makes the maximum delay of the net set decrease monotonically along with iterations.

The heuristic is detailed in Algorithm 1. Assume net T has the maximum delay D_T for the net set at the current iteration (line 3). The algorithm for delay minimization (*DM*) of net T is composed of three steps (line 4-16). *Step 1* rip-up-and-re-assigns T by SDLA with large λ (larger λ induces smaller delay but more via count) and without considering the wire congestion (line 4). If D_T is not decreased, DM fails (line 5). Otherwise, *step 2* traverses all 3D edges of the new path of T: if a 3D edge causes congestion violation, select the net with the minimum delay from the net set that passes through this 3D edge and push it into a set S (line 6). Finally, *step 3* rip-up-and-re-assigns the nets of S under the wire congestion constraint to eliminate the congestion violation induced by step 1 (line 7-16). In Algorithm 1, "*without* consideration of wire congestion" (line 4) means fixing congestion cost as *zero* for *all* 3D edges, while "*under* wire congestion constraints" (line 9, 12) means setting congestion cost as zero for non-congested edges, but *infinite* for congested edges to *forbid* the usage of edges and violation of congestion constraints.

Overall speaking, Stage 1 focuses on the overall optimization of the total delay and via count of all nets, it can get relative good result for all nets for delay and via count, but may ignore the bad delays of some long nets. On the other hand, Stage 2 focuses on minimizing the maximum delay of the set of nets, but cannot handle the overall optimization of delay and via count of all nets. Therefore, the algorithm framework integrates Stage 1 with Stage 2 to achieve the relative good results for total delay and via count of all nets, while the maximum delay still can be well optimized.

Algorithm 1: Minimization of the Maximum Delay

1. initialize PriorityQueue Q by nets' delay and set Flag←*true*
2. **while** Flag **do**
3.   get net T with maximum delay D_T in Q and its 3D pathP_T
4.   rip-up-and-re-assign T by SDLA with large λ and *without* consideration of the wire congestion, get the new delay D_T'
5.   **if** $D_T' \geq D_T$ **then** recover old path P_T to T and *break* **end if**
6.   traverse the new path of T and get the candidate net set S
7.   **foreach** net $T1 \in S$ in the decreased order of net delay **do**
8.   backup old path P_{T1} of $T1$
9.   rip-up-and-re-assign $T1$ by SDLA with large λ *under* the wire congestion constraint, get the new delay D_{T1}'
10.   **if** $D_{T1}' \geq D_T$ **then**
11.   recover old path P_{T1} to $T1$
12.   rip-up-and-re-assign T by SDLA with large λ *under* the wire congestion constraints, get the new delay D_T''
13.   **if** $D_T'' \geq D_T$ **then** Flag←*false* **end if**
14.   break
15.   **end if**
16.   **end foreach**
17.   update Q by T and $T1$'s with corresponding new delays
18. **end while**

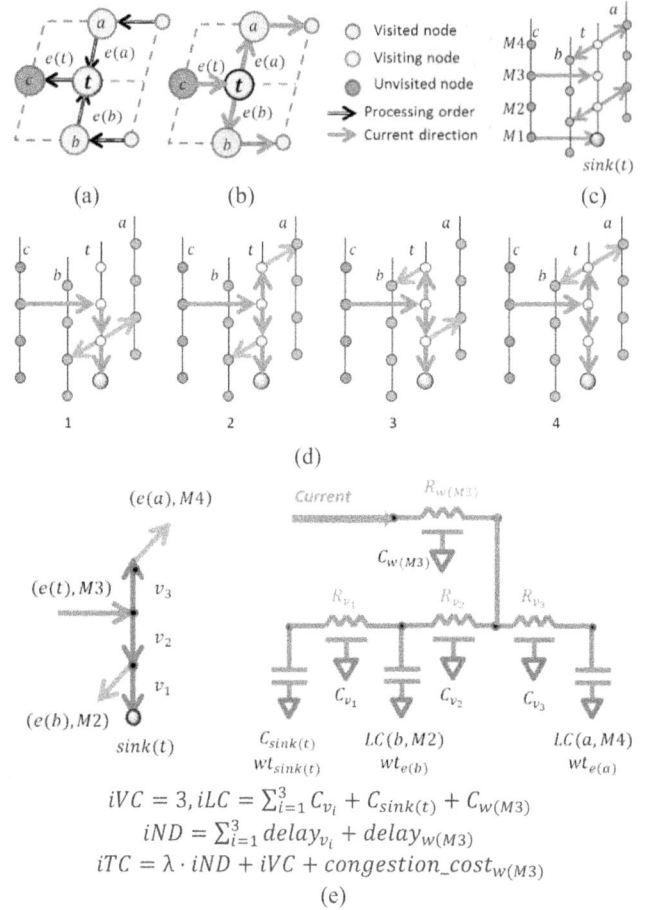

$iVC = 3, iLC = \sum_{i=1}^{3} C_{v_i} + C_{sink(t)} + C_{w(M3)}$

$iND = \sum_{i=1}^{3} delay_{v_i} + delay_{w(M3)}$

$iTC = \lambda \cdot iND + iVC + congestion\_cost_{w(M3)}$

(e)

Fig. 4. The calculation of cost increase.

4. LAYER ASSIGNEMEMNT OF SINGLE NET FOR SIMULTANEOUS DELAY AND VIA COUNT MINIMIZATION

This section presents the Layer Assignment algorithm of a single net for simultaneous Delay and via count minimization (SDLA), which is based on a dynamic programming technique. SDLA tries to find a layer assignment (3D) result of a 2D routed net with the minimum total cost of delay, via count and wire congestion (Equation (6)).

Note that timing constrained minimum cost layer assignment, e.g., [18], is very different from the problem here. Although both of them focus on single-net layer assignment, they have different optimization objectives and constraints: [18] etc. focus on minimizing the usage of thick metal routing resources and meanwhile satisfying the timing constraint of the net, while SDLA focuses on minimizing the delay and via count of the net and meanwhile satisfying the wire congestion constraints of global routing.

4.1 SDLA

Suppose that net tree T is currently under consideration. SDLA takes the *source* of T as the *root*, and processes each tree edge in a bottom-up manner from sinks to the source. SDLA partitions stages by the tree edges, then assigns one 2D edge at a time, and places vias after edge assignment. At each stage, SDLA records the minimum total costs and the corresponding load capacitances of the current stage; then propagates the results to the next stage (the total cost includes the costs of net delay, via count and wire congestion; the load capacitance is used for the calculation of a segment's delay for the next stage, see Equation (5)). Finally, after the root has been handled, the layer assignment solution with the minimum total cost is the required solution. SDLA is detailed as follows.

For convenience, define three terms for each vertex u of T: $par(u)$, $ch(u)$, and $e(u)$. The $par(u)$ is the parent of u, $ch(u)$ is the set of children of u, $e(u)$ is the edge $(u, par(u))$. For example, for vertex t in Fig. 4 (a), $par(t)$ is c, $ch(t)$ is the vertex set $\{a, b\}$, and $e(t)$ is the edge (t, c). Further, define $TC(u, r)$ and $LC(u, r)$ to be the *minimum total cost* and the corresponding *load capacitance* among all possible layer assignment for the sub-tree rooted at u, subject to the constraint

that the edge $e(u)$ is assigned to layer r. $TC(u, r)$ and $LC(u, r)$ can be computed by considering all possible combinations of $TC(u_j, r_j)$'s and $LC(u_j, r_j)$'s for all $u_j \in ch(u)$. Fig. 4 is an example to show how to compute $TC(u, r)$ and $LC(u, r)$.

Fig. 4 (a) and Fig. 4 (b) show a part of a 2D routed net, and Fig. 4 (c) is the corresponding k-layer routing graph, where $k=4$. Note that the current flow is in the opposite direction of the processing order of 2D edges. Assume $TC(a, r_a)$, $LC(a, r_a)$, $TC(b, r_b)$, and $LC(b, r_b)$ for all possible values of r_a and r_b have been computed, where r_a and r_b can be layer $M2$ or $M4$ here. Suppose that $TC(t, M3)$ and $LC(t, M3)$ need to be calculated. Fig. 4 (d) shows all possible combinations of r_a and r_b, with different circuit topologies. For each combination, SDLA needs to place vias to connect the three associated 3D edges and the 3D pins which are projected to the 2D pin t, and then computes the associated amounts of cost increase.

Take Fig. 4 (e) for example, which shows the combination case 2 of Fig. 4 (d). SDLA needs to calculate the increase amount of load capacitance and total cost due to vias v_1, v_2, v_3, and wire $w(M3)$. Let $iND(t)$, $iLC(t)$, $iVC(t)$, and $iTC(t)$ denote the respective increase of net delay, load capacitance, via count, and total cost due to vias v_1, v_2, v_3, and wire $w(M3)$ temporarily. Furthermore, SDLA also needs to consider the *wire congestion constraints*, i.e., the *congestion cost* $congestion\_cost_{w(M3)}$ of edge $w(M3)$. So $iTC = \lambda \cdot iND + iVC + congestion\_cost_{w(M3)}$. The amount of load capacitance and total cost for the combination are computed as $LC(a, M4) + LC(b, M2) + iLC(t)$ and $TC(a, M4) + TC(b, M2) + iTC(t)$, respectively. The increase amount of load capacitance and total cost for each of the other

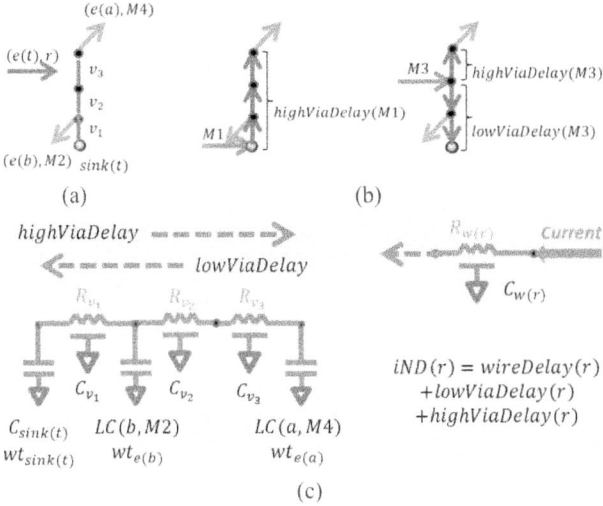

Fig. 5. Fast calculation for delay increase.

Table 1. ICCAD09 benchmarks [5]

circuit	nets	tiles	pins	layers
adaptec1	219794	324*324	942705	6
adaptec2	260159	424*424	1063632	6
adaptec3	466295	774*779	1874576	6
adaptec4	515304	774*779	1911773	6
adaptec5	867441	465*468	3492790	6
bigblue1	282974	227*227	282974	6
bigblue2	576816	468*471	2121863	6
bigblue3	1122340	555*557	3832388	8
bigblue4	2228903	403*405	8899095	8
newblue1	331663	399*399	1237104	6
newblue2	463213	557*463	1771849	6
newblue4	636195	455*458	2498322	6
newblue5	1257555	637*640	4931147	6
newblue6	1286452	463*464	5305603	6
newblue7	2635625	488*490	10103725	8

combinations shown in Fig. 4 (d) are calculated similarly. Among all these combinations, the one with minimum amount of total cost is chosen as the value $TC(t, M3)$.

For each combination of Fig. 4 (d), $iND(u, r)$ and $iLC(u, r)$ can be computed in linear time, as shown in Fig. 4 (e). To further reduce the redundant calculation, for each sub-combination of all children of u, $iND(u, r)$'s and $iLC(u, r)$'s for all layer r's are computed in linear time. As Fig. 5 shows, the increments of delay for the current stage can be partitioned into three parts: the delay of wire segment on layer r, $wireDelay(r)$, and the delays of vias below/above layer r, $lowViaDelay(r)/highViaDelay(r)$, therefore, $iND(r) = wireDelay(r) + lowViaDelay(r) + highViaDelay(r)$, $(r = M1, M2, \dots, M4)$. All $lowViaDelay(r)$'s and $highViaDelay(r)$'s values can be computed on one trip scanning, from the highest layer to the lowest one, or vise versa. The corresponding value of $iLC(u, r)$ is calculated along with $iND(u, r)$.

The above idea of computing $TC(u, r)$ and $LC(u, r)$ is for the case where u is an internal node with sinks. If u has no sinks (but still is an internal node), SDLA does not need to consider the sink capacitances and their connections with the 3D edges. If u is the root, since u has no parent, SDLA does not need to place vias to connect the associated 3D edge assigned for the "virtual" edge $(u, par(u))$, instead, it needs to consider the *delay contribution of the driven resistance of the source*. If u is a leaf, since u has no children, all $TC(u_j, r_j)$'s and $LC(u_j, r_j)$'s are treated as zero.

After the root is reached, the bottom-up computation of all $TC(u, r)$'s and $LC(u, r)$'s has been done and SDLA constructs the layer assignment result of T in the top-down manner. At last, when the method completes layer assignment for a net, it needs to update the wire usage values, the wire overflow values, and the congestion cost for the 3D edges involved in this layer assignment.

Note that although dynamic programming has been used by some recent single-net layer assignment methods, these uses focus on either via optimization [10]-[12][14]-[17] or antenna avoidance [16, 17]; and in this work, SDLA extends such approach to optimize both the net delay and via count.

4.2 Time Complexity

The time complexity of SDLA is analyzed as follows. Given a 2D routed net tree $T = (V, E)$, it first uses a graph traversal algorithm (e.g., depth-first search) to obtain an edge order; this step requires $O(|V| + |E|) = O(|V|)$ time computation due to $|E| = O(|V|)$. Second, for each vertex u of T, the steps of

finding $TC(u, r)$ for a given r require each possible combination of $TC(u_j, r_j)$'s for each $u_j \in ch(u)$ and each possible value of r. Since $|ch(u)|$ is at most four, r_j has $O(k)$ possible values, all $TC(u, r)$'s for u can be done in $O(k^4)$ time. To compute all $TC(u, r)$'s, it requires $O(k^4 \cdot |V|)$ time. Finally, the time spent on constructing T is $O(|V|)$. Thus, the overall time complexity of SDLA is $O(|V| + k^4 \cdot |V| + |V|) = O(k^4 \cdot |V|)$.

5. EXPERIMENTAL RESULTS

The proposed layer assignment algorithm is implemented in C++, and ICCAD09 3D (multilayer) global routing benchmarks [5] are used as test cases. All experiments are performed on a Linux PC with 2.27GHz CPU and 8GB memory. The 3D global routing solutions generated by NTHU-Route 2.0 [27] are projected back to 2D routing results on which the layer assignment algorithm is run.

5.1 Experimental Setup

The ICCAD09 routing benchmarks are detailed in Table 1. The ICCAD09 benchmarks are modified from the ISPD07 and ISPD08 3D (multilayer) global routing benchmarks, and re-adjust the routing capacities of metal layers of the ISPD07 and ISPD08 benchmarks. After the adjustment, high metal layers have less routing tracks than lower metal layers, which shows the fact that wires become thicker and wider on higher metal layers [19, 20]. However, the specific sizes and thicknesses of wires on each routing layer are not specified for these benchmarks. Therefore, this work assumes a set of typical routing specifications of the 45nm library [28].

Without loss of generally, three-tier metal sizes are assumed. For six-layer circuits, the wire width/spacing from the bottom metal M1 to the top metal M6 are set to 0.07, 0.07, 0.14, 0.14, 0.4, and 0.4 um, respectively. For eight-layer circuits, the wire width/spacing from M1 to M8 are set to 0.07, 0.07, 0.07, 0.07, 0.14, 0.14, 0.4, and 0.4 um, respectively. The wire thickness is twice the wire width. The resistance of a driver is set to 100ohm, and the sink load capacitance is set to 1fF. The priority of each sink for a net is set to 1/m, where m is the number of sinks of the net.

5.2 Relationship of Delay and Via Count

The algorithm of stage 1 (See Section 3.1) is used to study the relationship of delay and via count in this subsection. Fig. 6 shows the percent changes of total delay, maximum delay and via count by parameter λ (See Section 2.2) for one benchmark, the span of range for λ is empirically determined. Generally, with larger λ, total delay and maximum delay decrease, but via count increases.

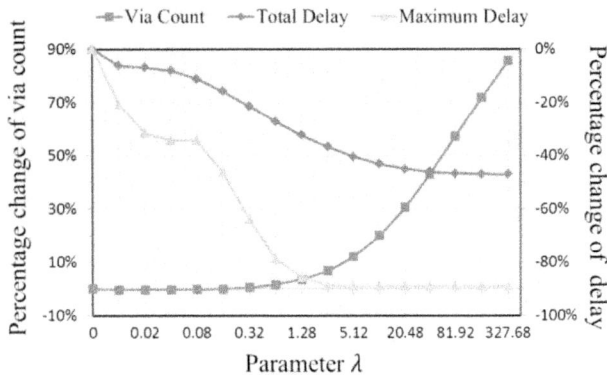

Fig. 6. Percent changes of total delay, maximum delay and via count by parameter λ for adaptec2.

Particularly, when λ is small, as λ increases, the total delay and maximum delay decrease quickly while via count still keeps roughly the same. Two reasons account for such phenomenon: 1) different segments (of the same net or different nets) have *a wide range of delay weights and load capacitances*, some wire segments can generate much smaller delay contributions when assigned to proper layers, which leads to big delay improvement; 2) given a net, even for multiple LA results with the same via count, different via positions lead to *diverse circuit topologies* of the net tree, and then induce diverse delays. Experiments show that the relative good trade-off occurs around 0.15 for all the tested benchmarks, where delay decreases significantly while via count can still keep roughly the same, so $\lambda = 0.15$ is adopted for the following experiments.

5.3 Algorithms Comparison

For the multi-tier interconnect structure, there is no previous work for the delay optimization in layer assignment of 3D global routing. Therefore, this work compares the proposed method (Delay-driven Layer Assignment, DLA) with the state-of-the-art via count minimization layer assignment method NVM [15]. This work also implements a straightforward greedy algorithm for simultaneous delay and via count minimization for comparison. Basically, this greedy approach applies the same negotiation-based algorithm framework as Stage 1 of DLA, so it will produce identical congestion quality to that of DLA. For the single-net layer assignment, the greedy algorithm takes the same processing order of edges as SDLA; but for each edge being processed, it directly chooses the layer that induces the minimum cost currently.

Table 2 shows that DLA can generate a solution with much smaller total delay and maximum delay but roughly the same via count for each test case, compared to NVM. Without Stage 2, DLA decreases the total delay and maximum delay by 22.7% and 82.2% on average while keeping the same via count; with Stage 2, DLA decreases the total delay and maximum delay by 22.8% and 90.8%, respectively but only slightly increases the via count by 0.2%. On the other hand, although the greedy algorithm also reduce the total delay (by 16.3%) and maximum delay (by 51.3%), it results in much more via count (by 40.9%), compared with NVM. In addition, the total delay, maximum delay and via count of the greedy algorithm are obviously much worse than that of DLA; such results are not surprising since the greedy algorithm is just a simple heuristic. As for run time, Table 3 shows that DLA is 1.7X slower than NVM because it needs more time for the calculation of delays. The run time of the greedy algorithm is between that of NVM and DLA. Stage 2 of DLA is fast and only consumes few seconds for each case. Additionally, the resulting total wire

overflow and maximum wire overflow[1] of both DLA and Greedy are the same as NVM and satisfy the two wire congestion constraints, so they are omitted from Table 2.

6. CONCLUSION

This paper studies the problem of delay (performance) driven layer assignment under multi-tier interconnect structure, which arises during 3D (multilayer) global routing. The problem requires to transform a 2D global routing result into a 3D one and minimize the delays and via count, under the wire congestion constraints. This paper presents an effective algorithm for the problem, and the resistances and capacitances of metal wires and vias are considered in the corresponding RC model. The algorithm can be especially applicable to circuits in which the interconnecting layers have drastically different electrical characteristics. Encouraging experiment results have been provided to support the proposed algorithm.

7. REFERENCES

[1] J. Cong, "An interconnect-centric design flow for nanometer technologies," *Proceedings of IEEE,* vol. 89, pp. 505-528, April 2001.

[2] J. A. Davis and J. D. Meindl, *Interconnect Technology and Design for Gigascale Integration*, Kluwer Academic Publishers, 2003.

[3] Z. Li, C. J. Alpert, S.Hu, T. Muhmud, S. T. Quay, and P. G. Vilarubia, "Fast interconnect synthesis with layer assignment," *Proceedings of International Symposium on Physical Design,* pp. 71-77, 2008.

[4] C. J. Alpert, C. Chu, and P. G. Villarrubia, "The coming age of physical design," *Proceedings of International Conference on Computer-Aided Design,* pp. 246-249, 2007.

[5] M. D. Moffitt, "Global routing revisited," *Proceedings of International Conference on Computer-Aided Design,* pp. 805-808, 2009.
http://web.eecs.umich.edu/~mmoffitt/iccad2009.alternate.html

[6] ISPD 2007 Global Routing Contest. [Online]. Available: http://archive.sigda.org/ispd2007/contest.html

[7] ISPD 2008 Global Routing Contest. [Online]. Available: http://archive.sigda.org/ispd2008/contests/ispd08rc.html

[8] M. Cho, K. Lu, K. Yuan, and D. Z. Pan, "BoxRouter 2.0: Architecture and implementation of a hybrid and robust global router," *Proceedings of International Conference on Computer-Aided Design,* pp. 503-508, 2007.

[9] J. A. Roy and I. L. Markov, "High-performance routing at the nanometer scale," *IEEE Transactions on Computer-Aided Design of Integrated Circuits and Systems,* 27(6), pp. 1066-1077, 2008.

[10] T.-H. Lee and T.-C. Wang, "Congestion-constrained layer assignment for via minimization in global routing," *IEEE Transactions on Computer-Aided Design of Integrated Circuits and Systems,* 27(9), pp. 1643- 1656, 2008.

[11] K.-R. Dai, W.-H. Liu, and Y.-L. Li, "Efficient simulated evolution based rerouting and congestion-relaxed layer assignment on 3-D global routing," *Proceedings of Asia and South Pacific Design Automation Conference,* pp. 570-575, 2009.

[12] Y. Xu, Y. Zhang and C. Chu, "FastRoute 4.0: global router with efficient via minimization," *Proceedings of Asia and South Pacific Design Automation Conference,* pp. 576-581, 2009.

---

[1] The wire overflow values listed in Table 2 are two times the wire overflow defined in Section 2.

Table 2. Comparison between NVM [15], Greedy and DLA Algorithm on NTHU-Route 2.0

circuit	NVM [15]					Greedy[d]			DLA (S1[d])			DLA (S1[d] +S2)[e]		
	VC^a (e4)	WO total[f]	WO max[g]	TD^b (e5)	MD^c	VC (e4)	TD (e5)	MD	VC (e4)	TD (e5)	MD	VC (e4)	TD (e5)	MD
adaptec1	85.43	0	0	3.86	2767	117.26	3.45	1131	85.34	3.34	873	85.39	3.33	270
adaptec2	94.97	0	0	3.95	764	130.84	3.35	500	94.91	3.35	411	95.04	3.34	84
adaptec3	170.63	0	0	11.4	1689	238.02	9.70	1016	171.33	8.47	283	171.42	8.46	156
adaptec4	156.35	0	0	10.56	3352	212.24	8.77	3386	156.63	7.84	417	156.66	7.83	268
adaptec5	243.61	0	0	18.92	2792	337.09	15.98	1182	244.31	14.92	439	244.37	14.91	271
bigblue1	93.37	0	0	6.25	517	128.89	5.38	369	93.29	5.27	227	95.37	5.16	67
bigblue2	187.75	104	4	3.62	1984	257.81	3.32	1014	187.43	3.38	476	187.48	3.38	209
bigblue3	259.47	38	6	17.22	1054	379.71	15.40	998	260.30	12.08	235	261.54	12.05	108
bigblue4	518.67	394	4	32.51	10224	758.27	28.61	2299	519.65	24.76	859	519.69	24.76	737
newblue1	105.18	200	2	2.05	123	141.14	1.91	160	105.01	1.91	69	105.36	1.89	34
newblue2	126.31	0	0	7.88	764	171.40	6.95	1198	126.38	6.68	242	126.40	6.68	184
newblue4	229.82	262	4	9.08	1256	316.18	7.82	828	228.95	7.81	287	229.00	7.81	173
newblue5	407.69	20	2	22.17	991	549.79	18.97	870	407.41	18.36	274	408.20	18.31	114
newblue6	342.82	0	0	20.91	997	466.64	18.09	997	343.32	16.90	311	343.46	16.89	110
newblue7	847.27	86	2	56.13	6116	1246.30	41.87	1285	846.38	40.04	907	847.77	39.96	484
ratio	1			1	1	1.409	0.837	0.487	1.000	0.773	0.178	1.002	0.772	0.092

[a] VC: via count, [b] TD: total delay of the set of nets, [c] MD: maximum delay among the set of nets, [d] λ=0.15, [e] S1/S2: Stage 1/2

[f] total WO: total wire overflow, [g] max WO: maximum wire overflow

Table 3. Comparison of CPU Time (seconds)

circuit	NVM[15]	Greedy	DLA(S1)	DLA(S1+S2)
adaptec1	121	137	195	199
adaptec2	96	93	153	157
adaptec3	325	312	471	481
adaptec4	234	205	321	333
adaptec5	338	452	519	529
bigblue1	131	273	213	221
bigblue2	196	162	309	316
bigblue3	312	286	515	525
bigblue4	556	488	980	993
newblue1	74	69	121	124
newblue2	123	101	171	176
newblue4	244	244	376	384
newblue5	486	482	718	733
newblue6	354	386	534	544
newblue7	965	1212	1677	1730
ratio	1	1.076	1.60	1.634

[13] C.-H. Hsu, H.-Y. Chen, and Y.-W. Chang, "Multi-layer global routing considering via and wire capacities," *Proceedings of International Conference on Computer-Aided Design*, pp. 350-355, 2008.

[14] T.-H.. Lee and T.-C. Wang, "Robust layer assignment for via optimization in multi-layer global routing," *Proceedings of International Symposium on Physical Design*, pp. 159-166, 2009.

[15] W.-H. Liu and Y.-L. Li, "Negotiation-based layer assignment for via count and via overflow minimization," *Proceedings of Asia and South Pacific Design Automation Conference*, pp. 539-544, 2011.

[16] T.-H. Lee and T.-C. Wang, "Simultaneous antenna avoidance and via optimization in layer assignment of multi-layer global routing," *Proceedings of International Conference on Computer-Aided Design*, pp. 312-318, 2010.

[17] W.-H. Liu, Y.-L. Li, "Optimizing the antenna area and separators in layer assignment of multi-layer global routing," *Proceedings of International Symposium on Physical Design*, pp. 137-144, 2012.

[18] S. Hu, Z. Li, and C. J. Alpert, "A polynomial time approximation scheme for timing constrained minimum cost layer assignment," *Proceedings of International Conference on Computer-Aided Design*, pp. 112-115, 2008.

[19] Y.-J. Chang, T.-H. Lee, and T.-C. Wang, "GLADE: a modern global router considering layer directives," *Proceedings of International Conference on Computer-Aided Design*, pp. 319-323, 2010.

[20] T.-H. Lee, Y.-J. Chang, and T.-C. Wang, "An enhanced global router with consideration of general layer directives," *Proceedings of International Symposium on Physical Design*, pp. 53-60, 2011.

[21] M. J. Ciesielski, "Layer assignment for VLSI interconnect delay minimization," *IEEE Transactions on Computer-Aided Design of Integrated Circuits and Systems*, 8(6), pp. 702-707, 1989.

[22] C.C. Chang and J. Cong, "An efficient approach to multilayer layer assignment with applications to via minimization," *IEEE Transactions on Computer-Aided Design of Integrated Circuits and Systems*, 18(5), pp. 608-620, 1999.

[23] P. Saxena and C. L. Liu, "Optimization of the maximum delay of global interconnects during layer assignment," *IEEE Transactions on Computer-Aided Design of Integrated Circuits and Systems*, 20(4), pp. 503-515, 2001.

[24] D. Wu, J. Hu, R. Mahapatra, "Coupling aware timing optimization and antenna avoidance in layer assignment," *Proc. International Symposium on Physical Design*, 2005.

[25] W.C. Elmore. "The transient response of damped linear networks with particular regard of wideband amplifiers," *J. Appl. Phys.*, 19(1):55-63, Jan. 1948.

[26] L. McMurchie and C. Ebeling, "Pathfinder: a negotiation-based performance- driven router for FPGAs," *Proc. ACM Symposium on FPGAs*, pp. 111-117, 1995.

[27] Y.-J. Chang, Y.-T. Lee, and T.-C. Wang, "NTHU-Route 2.0: a fast and stable global router," *Proceedings of International Conference on Computer-Aided Design*, pp. 338-343, 2008.

[28] "Nangate 45nm open cell library, http://www.nangate.com," 2008.

SRP: Simultaneous Routing and Placement for Congestion Refinement

Xu He, Wing-Kai Chow and Evangeline F.Y. Young
Department of Computer Science and Engineering
The Chinese University of Hong Kong, Shatin, N.T., Hong Kong
email: {xhe, wkchow, fyyoung}@cse.cuhk.edu.hk

ABSTRACT

In this paper, an effective simultaneous routing and placement refinement tool called SRP is proposed for routability improvement. SRP is independent of any placer and global router. Based on a given placement layout and global routing result, SRP relocates problematic cells by considering routing and placement simultaneously. Not only overflow from local nets, but overflow from global and semi-global nets can be solved by SRP. A cell will be relocated and its associated nets will be rerouted if its connections go across any congested region, even if the cell is not in the congested region. Therefore, our method can reduce the overflow effectively. Given the layouts generated by the top four routability-driven placers in the DAC Contest 2012, our method can still reduce the total overflow by 32.6% in average while the routed wirelength and HPWL are not increased obviously.

Categories and Subject Descriptors

B.7.2 [**Hardware**]: Integrated Circuits— *esi n Aids*

Keywords

Placement and Routing, Routability

1. INTRODUCTION

Today, the routability problem has become one of the most important issues in placement. The existence of macro blockages and the large problem size with up to millions of standard cells and nets make the routability problem more and more challenging. In traditional VLSI design flow, the processes of placement and routing are done separately. A placer determines the positions of cells and generates a row-based layout. The major objective is to minimize wirelength which is often estimated by the half-perimeter wirelength (HPWL) model. After placement, a router determines the routing path to connect all the pins of the cells in the same signal net. Since the routing capacity of metal layers is limited, the router not only need to route all signal nets with small wirelength, but

also has to satisfy routing resource constraints. Therefore, although minimizing HPWL in placement can reduce the average routing wirelength, the wires may be distributed unevenly. Over optimization in wirelength during placement may lead to unroutability of some nets.

Routability-driven placement problem has been studied in many previous works. Various techniques could be performed during 1) global placement, 2) detailed placement, and 3) refinement in post placement process. During global placement, several techniques model the congestion information in the objective function [19][10][9][4][18] to address the routability issue. Besides, some works try to distribute the routing demand evenly by allocating whitespace in the congested regions explicitly or implicitly [5][11][20][12][2][17][6]. In order to obtain accurate congestion map, IPR [15] and SimPLR [11] integrate global router FastRoute [14] and BFG-R [8] respectively during their placement. In detailed placement, the most common approach is to change the objective of cell swapping to reduce congestion [5][11][7][15][21]. In post-placement process, there are some previous works considering congestion refinement. The most recent ones, CROP [21] and CRISP [17], address routability in post-placement process directly. After obtaining the congestion map, CROP [21] adjusts the boundary of each G-Cell and shifts the modules according to the new G-Cell boundary. CRISP [17] applies cell inflation and spreading to allocate cells more sparsely in the congested regions.

Global router overlays a regular grid (G-Cell) on the chip, and constructs a global routing on the grid. In a global routing grid graph, each G-Cell is represented by a node, and there are global routing edges to connect adjacent G-Cells. If the usage of routing tracks on an edge e is above e's capacity, this edge e is considered as over-congested. Given a globally routed layout, the congestion of a region can be caused by three types of nets: (1) local nets connecting cells within the region, (2) semi-global nets coming in/out of the region, and (3) global nets passing through the region without connecting any cell of the region. Many previous works [21][17] can directly solve the overflow from local nets by relocating cells in congested G-Cells. However, as we know , the overflow caused by global or semi-global nets is usually more than that caused by local nets. Therefore, we should consider congestion caused by all types of nets in our routability refinement method. Since many placers have already used whitespace allocation method during placement, we choose to do cell relocation directly during our refinement process. Besides, other refinement tools can also integrate with SRP to further improve the congestion.

In this work, a fast and effective simultaneous routing and placement refinement tool called SRP is proposed for routability improvement. SRP is independent of the placer and global router being used. Based on a given placement layout and a global routing result, three major steps are used in our tool to do congestion refinement. **First**, we identify and rip-up problematic cells by considering congestion caused by all types of nets mentioned above. Different from many previous works that only focus on reducing the overflow caused by local nets, our method may relocate a cell even when its routing path obtained by the global router has just run across congested regions. **Second**, we search for new location for a cell by multi-source propagation. Unlike other works that only search the surrounding region around the cell's original location, we will find a new location for the problematic cell by searching around the G-Cells passed through by the routing paths of the associated nets of the cell. **Third**, we will connect the problematic cell from its new location to its associated nets by a multi-subnet maze routing algorithm.

Our major contributions are summarized as follows:

- We present SRP, a placement routability refinement tool that is independent of any placer and router. Other refinement tools can be integrated with our relocating and routing algorithm for further improvement.

- We relocate problematic cells by considering routing and placement resources simultaneously. A cell will be moved if the routing paths connecting to its new location can relieve the congestion.

- Not only overflow from local nets, but overflow from global and semi-global nets can be relieved by SRP. We relocate and reroute a cell if its connections run across congested regions, even if the cell is not lying in the congested region. Therefore, our method can reduce overflow effectively.

The remainder of this paper is organized as follows. In Section 2, we give an overview of our post-placement process. Section 3 describes the details of our simultaneous relocating and rerouting algorithm. Section 4 gives further discussion on SRP. Experimental results and a conclusion will be given in Section 5 and Section 6 respectively.

2. OVERVIEW OF SRP

Based on a given placement layout and its global routing result, we try to relocate cells and reroute nets to reduce overflow. Fig. 1 shows the overall flow of SRP. The input is a placement layout and its global routing result. Before relocating and rerouting, we need to obtain the routing capacity and the routing demand according to the global routing result. If the routing demand of a region is larger than its routing capacity, this region is a congested region. During cell relocation and net rerouting, every problematic cell that has some connections going across some congested regions will be identified and ripped up first. Since not only the cells in the congested region, but also the cells that have connections going across congested regions will be identified and relocated, the congestion cause by local nets, semi-global nets and global nets are all handled. Second, the problematic cell will be relocated and its associated nets rerouted. If the total overflow is improved, the cell will stay at its new location, and the congestion map will be updated according to the rerouted paths; otherwise, the cell will go back to its original position. In each iteration, all the problematic cells will be processed one after another.

The whole post-placement process will be finished if there is no more obvious reduction in overflow. To trade off between congestion improvement and runtime, the number of iterations is 3 in our implementation.

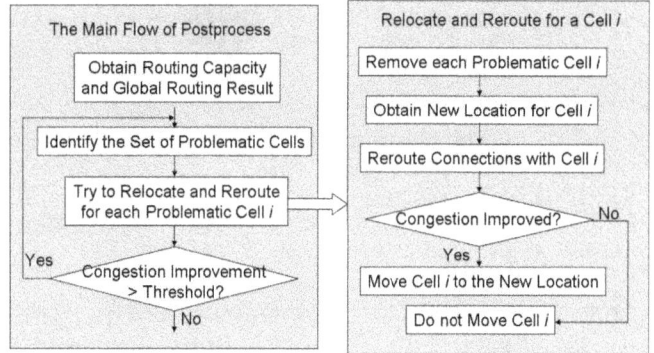

Figure 1: The whole flow of postprocess.

3. SIMULTANEOUS CELL RELOCATION & NET REROUTING

The chip is divided uniformly into $m \times n$ G-Cells. Each pair of adjacent G-Cells have a edge connecting them. If the pins of a net is in different G-Cells, there is a routing path going through edges to connect these pins. Each edge e has a capacity, which is the total number of available tracks for wires to go through. If the total number of used tracks (routing demand) is larger than the capacity of the edge, there will be overflow on this edge.

3.1 Identify Problematic Cell

After obtaining the routing result, the cells with connections going across congested regions should be relocated to reduce the overflow of those regions. Different from previous works which only relocate the cells in the congested regions, we also identify the cells whose connections have passed through congested regions. By doing this, not only the overflow caused by local nets but also the overflow caused by semi-global nets and global nets can be reduced.

Given the routing result of a net, we will first identify the congested G-Cells passed through by the routing path of this net. If a congested G-Cell has at least one pin of this net, the cell of this pin will be marked as a problematic cell. Otherwise, we will follow the routing path from this congested G-Cell until reaching other G-Cells that contain some pins of this net, and then the cells of these pins will be marked as problematic cells. After identifying all the problematic cells, we will relocate them to reduce overflow.

Take net A in Fig. 2 as an example. Net A has three cells: cell 1, cell 2 and cell 3. The routing path (in grey color) of net A is shown in Fig. 2 (a). This path passed through a congested routing edge (in red color) connecting $tile(1, 2)$ and $tile(2, 2)$. Since $tile(1, 2)$ has a pin of cell 1, cell 1 will be marked as a problematic cell. Since $tile(2, 2)$ doesn't contain any pin of net A, we will follow the routing path until we reach $tile(4, 3)$ that contains a pin of cell 2. Then we stop propagating, and mark cell 2 as a problematic cell.

3.2 Remove Problematic Cell

Before problematic cell relocation and net rerouting, we first need to remove the problematic cell and rip up the routing

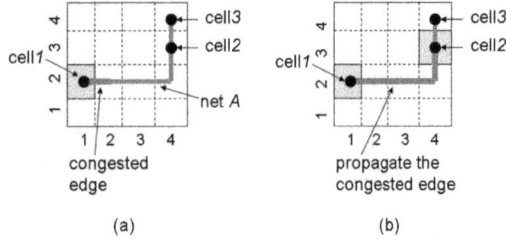

(a)

(b)

Figure 2: (a)The routing path of net A passes through a congested routing edge connecting $tile(1,2)$ and $tile(2,2)$ (b) We propagate from $tile(1,2)$ and $tile(2,2)$ until reaching any pins of net A. Cell 1 and cell 2 are marked as problematic cells.

paths connecting this cell. There are three cases need to be considered (shown in Fig. 3) when we delete the routing path connecting with a problematic cell:

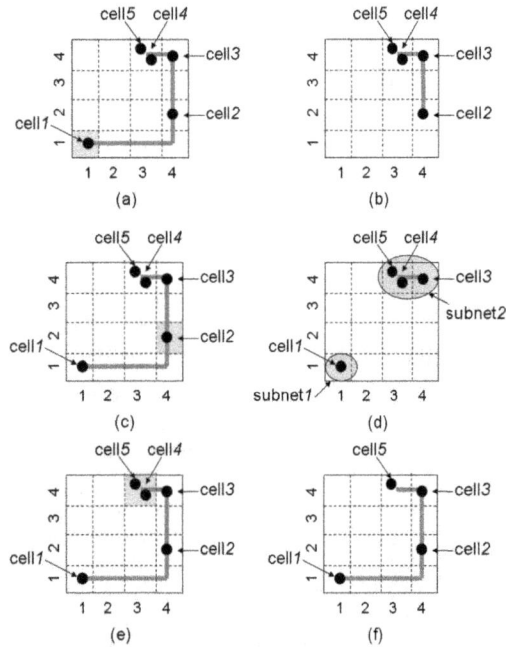

(a)

(b)

(c)

(d)

(e)

(f)

Figure 3: (a) $Tile(1,1)$ contains a problematic cell 1 (b) The remaining path and cells after removing cell 1 (c) $Tile(4,2)$ contains a problematic cell 2, and the degree of this tile is two (d) After removing cell 2, the remaining path has two subnets (e) $Tile(3,4)$ has two cells, and cell 4 is the current problematic cell (f) The remaining path and cells after removing cell 4.

Case 1: In Fig. 3 (a), cell 1 is a problematic cell. Since $tile(1,1)$ contains cell 1 which only has degree one, we will remove the cell directly, and delete the routing path connecting to $tile(1,1)$. After deleting the path from $tile(1,1)$ to $tile(2,1)$, the degree of $tile(2,1)$ becomes one. Since there is no pin in $tile(2,1)$, we can delete the path from $tile(2,1)$ to $tile(3,1)$ as well. This deleting process will continue until reaching a G-Cell that contains a pin or a steiner point of this net. The remaining routing path of this net after removing cell 1 is shown in Fig. 3 (b).

Case 2: In Fig. 3 (c), cell 2 is a problematic cell. $Tile(4,2)$

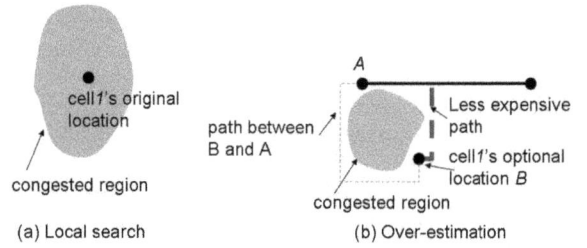

(a) Local search

(b) Over-estimation

Figure 4: Potential problems of some previous approaches.

contains cell 2 which has degree more than one ($1 \leq degree \leq 4$). In order to reduce the overflow caused by the paths connected to cell 2, we choose to delete the path from $tile(4,2)$ to $tile(4,4)$ and the path from $tile(4,2)$ to $tile(1,1)$. After deleting these paths connected to cell 2, the net becomes two subnets (shown in Fig. 3 (d)). We will discuss how to connect cell 2 and these subnets together after searching a new position for cell 2 in section 3.3 and section 3.4.

Case 3: In Fig. 3 (e), cell 4 is a problematic cell. However, $tile(3,4)$ contains cell 4 and cell 5 which are both on this net. After removing cell 4, we cannot delete the routing path connecting to this G-Cell, since cell 5 is still in this G-Cell. Fig. 3 (f) shows the remaining path and cells of this net after removing cell 4.

3.3 Searching New Location

During global placement and detailed placement, most previous works [15][12][16][20] find the new locations of the problematic cells by searching the surrounding G-Cells of the cells' original locations. Besides, before evaluating the cost of each optional G-Cell, the multi-pin nets are first decomposed into two-pin subnets. These two scenarios are illustrated in Fig. 4 to show the potential problems of their strategies.

Algorithm 1 Multi-Source Propagation

1: **for** net $i \in$ associateNet($curCell$) **do**
2:   **for** $tile_a \in$ path(net i) **do**
3:     $cost_i(tile_a) = 0$
4:     enqueue($tile_a, Q_i$)
5:   **end for**
6: **end for**
7: **while** $numOptionTile < optionThreshold$ **do**
8:   **for** net $i \in$ associateNet($curCell$) **do**
9:     $tile_a$=dequeueMin(Q_i)
10:     **if** $!isVisitTile(tile_a, i)$ **then**
11:       $isVisitTile(tile_a, i)$=true
12:       $numNetReach(tile_a)+ = 1$
13:     **end if**
14:     **if** $numNetReach(tile_a) ==$ |associateNet($curCell$)| **then**
15:       $numOptionTile+ = 1$
16:     **end if**
17:     **for** $tile_b$ adjacent with $tile_a$ & $!isVisitTile(tile_b, i)$ **do**
18:       $cost_i(tile_b) = cost_i(tile_a) + cost_i(tile_a, tile_b)$
19:       enqueue($tile_b, Q_i$)
20:     **end for**
21:   **end for**
22: **end while**

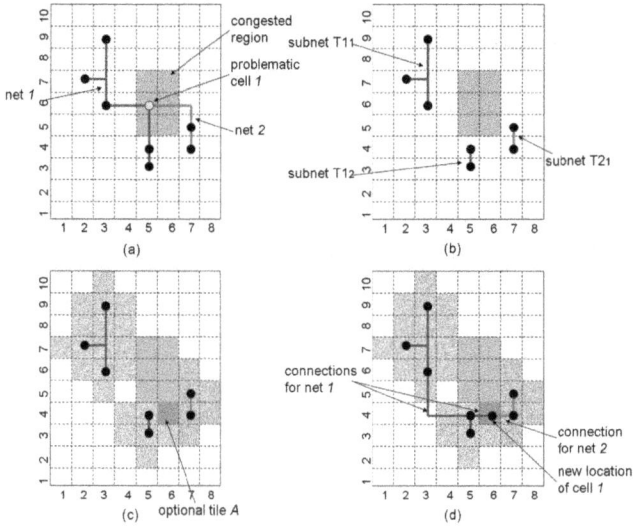

Figure 5: (a) The problematic cell 1 has two associated nets: net 1 and net 2. (b) After removing cell 1 and its connections from net 1 and net 2, the remaining path is partitioned into two subnets: $T1_1$ and $T1_2$; while net 2 becomes subnet $T2_1$. (c) Obtain an optional location for cell 1 by using multi-source propagation method from both net 1 and net 2. (d) Find the shortest path to connect the new location and the subnets of together.

Local search: Consider the scenario in Fig. 4 (a). The problematic cell 1's location and its surrounding regions are all congested. If we only consider the surrounding regions as optional new locations, either the cell will not be moved (because of no overflow improvement) or the cell has to be moved several times until it reaches a good position out of the congested region.

Over-estimation: Consider the scenario in Fig. 4 (b). After decomposing a net into two-pin subnets, problematic cell 1 is connected with the pin at location A directly. When we evaluate the cost of an optional location B, we will try to connect B with A. However, besides point A, B can actually be connected to any other G-Cell passed by the net (thickened blue line), and the cost may become smaller. Therefore, the cost of locating cell 1 in B may be over-estimated.

As we can see in these two scenarios, the remaining routing path of the net is not utilized in an effective way when searching for a new location for the problematic cell. Not only the positions around the cell's original location, but also the positions surrounding any G-Cells passed through by the remaining routing path of the net should be considered. Besides, when connecting the problematic cell to its associated net, we only need to find a path between its new location to any G-Cells passed through by the remaining routing path of the net.

Aware of this problem, we propose a multi-source propagation method to obtain the new location. After removing a problematic cell from its associated nets, the G-Cells passed through by the remaining routing paths of its nets will be treated as sources. Then, we apply the multi-source propagation method on each associated net i of the problematic cell. The sources on each net i will propagate simultaneously until a G-Cell is reached by all the associated nets. As shown in Fig. 5 (a), the problematic cell 1 is associated with two nets: net 1 and net 2. After ripping up cell 1 and deleting the paths

at these two nets connecting with it, net 1 is partitioned into two subnets $T1_1$ and $T1_2$; while net 2 becomes subnet $T2_1$ (shown in Fig. 5 (b)). All the G-Cells passed through by $T1_1$ and $T1_2$ are considered as sources of net 1. Similarly, the G-Cells passed through by $T2_1$ are the sources of net 2. We will propagate from these two sets of sources until a G-Cell A is reached from these two nets. Then G-Cell A will be an optional position for cell 1.

Algorithm 1 shows our multi-source propagation method. Given a problematic cell $curCell$, for each associated net i of $curCell$ ($i \in associateNet(curCell)$), a priority queue Q_i is maintained. Initially, the sources of net i are enqueued into Q_i, with the cost equal to zero (line 1-6). From each Q_i, we dequeue a tile $tile_a$ with the minimum cost (line 9-13). Once a tile $tile_a$ has been visited by all the associated nets, $tile_a$ can be an optional location for the problematic cell $curCell$ (line 14-16). After dequeuing $tile_a$ from Q_i, we will propagate from $tile_a$ to its adjacent tile $tile_b$ (line 17-20). The cost of visiting $tile_b$ is the cost of $tile_a$ added by the cost $tile_a$ to $tile_b$. Both the overflow and the wirelength of the edge connecting from $tile_a$ to $tile_b$ are considered when calculating the cost from $tile_a$ to $tile_b$. The cost function is shown in equation (1). The loop in line 8-21 is similar to the Dijkstra's algorithm. The loop in line 7-22 will be repeated until the number of optional locations is equal to a threshold $optionThreshold \geq 1$, (e.g., $optionThreshold=2$ in our experiments).

$$
\begin{aligned}
cost_i(tile_b) &= cost_i(tile_a) + cost_i(tile_a, tile_b) \\
cost_i(tile_a, tile_b) &= of(tile_a, tile_b) + wl(tile_a, tile_b)
\end{aligned}
\tag{1}
$$

$$
cost(tile_k) = \sum_{net_i \in N} cost_i(tile_k)
\tag{2}
$$

We use equation (2) to evaluate each optional location $tile_k$. The tile with the minimum cost will be chosen as a new location for the cell. The method to generate new paths between the cell's new location and its associated subnets will be illustrated in section 3.4.

3.4 Connections for New Location

After finding a new location for the problematic cell, we can trace back according to the propagation process (section 3.3) to find the shortest path from this new location to the cell's associated nets. However, there are two reasons not to use this method.

First, not all the subnets of a net can be connected together. An associated net may be partitioned into more than one subnets after ripping up the problematic cell. Since tracing back can only find a path from the new location to one G-Cell passed through by one subnet. Therefore, only one of the subnets can be connected. As shown in Fig. 6 (a), the new location of cell 1 can only be connected with subnet 1 when tracing back, while subnet 2 cannot be connected.

Second, the path may not be the shortest path when tracing back. If the problematic cell has more than one associated net, the path obtained by tracing back may not be the shortest one since the costs on the edges may change. As shown in Fig. 6 (b), the problematic cell 4 has three associated nets: net 1, net 2, and net 3. After removing cell 4, the remaining cells of these three nets are cell 1, cell 2, and cell 3 respectively. Since both net 1 and net 2 will pass through $tile(1,3)$ and $tile(2,3)$ after tracing back, the overflow along these two paths will be increased comparing the values used during the propagation

Algorithm 2 Multi-Subnets Maze Routing

1: **while** $numSubnet > 1$ **do**
2:    **for** $tile_a \in$ path(subnet 1) **do**
3:      $cost(tile_a) = 0$
4:      enqueue($tile_a$, Q);
5:    **end for**
6:    **while** $|Q| > 0$ **do**
7:      $tile_a$=dequeueMin(Q)
8:      **if** $!isVisitTile(tile_a, i)$ **then**
9:        $isVisitTile(tile_a, i)$=true
10:        **if** isSink($tile_a$) **then**
11:          i=subnetID($tile_a$)
12:          break
13:        **else**
14:          **for** $tile_b$ adjacent $tile_a$ & $!isVisitTile(tile_b, i)$ **do**
15:            $cost(tile_b) = cost(tile_a) + cost(tile_a, tile_b)$
16:            enqueue($tile_b$, Q)
17:          **end for**
18:        **end if**
19:      **end if**
20:    **end while**
21:    path(subnet 1)=path(subnet 1) $\cup$ path(subnet i) $\cup$ path($tile_a$, subnet 1)
22:    updateRoutingDemand(path($tile_a$, subnet 1))
23:    $numSubnet - = 1$
24: **end while**

process. Therefore, the paths obtained by tracing back the propagation process may not be the shortest ones for net 1 and net 2 any more.

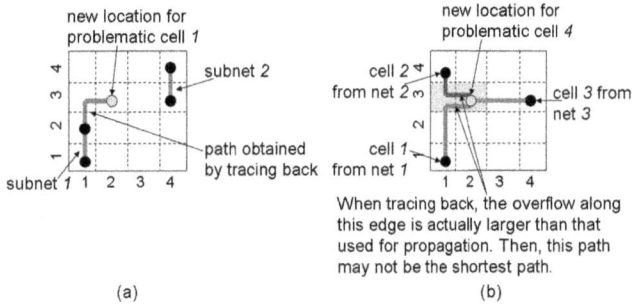

(a)   (b)

Figure 6: (a) The path obtained by tracing back can only connect the problematic cell 1 to one of its subnets (subnet 1), and subnet 2 is not connected. (b) Problematic cell 4 has three associated nets. Since both net 1 and net 2 will pass through $tile(1,3)$ and $tile(2,3)$ after tracing back, the overflow on these two paths will be larger comparing with the value in the propagation process. Therefore, the paths obtained by tracing back the propagation may not be the shortest paths for net 1 and net 2 respectively.

Therefore, we will connect the new location to each associated subnet one by one, and update the overflow of the routing edges after obtaining the shortest path for each net.

In order to find the shortest path to connect all the subnets of a net, we use multi-subnets maze routing. If the new location of the problematic cell is not passed through by any subnets, this new location will be seen as a subnet as well. The algorithm of multi-subnet maze routing is shown in Al-

gorithm 2. We first take all the G-Cells on one subnet (e.g., subnet 1) as the sources (line 2-5), and all the G-Cells on other subnets as sinks. We propagate from the sources by using the cost function in equation (1) (line 6-17). Once we reach a sink $tile_a$ (line 8-10), we will get the subnet ID i in $tile_a$. Then subnets 1 and subnet i will be connected as a new subnet by the $path(tile_a, subnet1)$. The overflow of all the routing edges passed through by $path(tile_a, subnet1)$ will be updated. All the G-Cells on this new subnet will become sources for the next propagation process (line 18-19). This propagation process will repeat until all the subnets are connected.

Fig. 5 (d) shows the paths connecting between cell 1's new location and the other subnets.

4. FURTHER DISCUSSION

Most placers have already used whitespace allocation method as a major technique to reduce overflow during global placement or detailed placement. If we use similar technique in the refinement process, the improvement in congestion may not be large. Therefore, we choose to relocate and reroute problematic cell directly.

The number of cells which can be moved to reduce congestion successfully is much smaller than the number of problematic cells. The major reason is that even a new location and a new routing path are found, the overflow caused by the new routing path is hard to be smaller than that of the original one. In our implementation, we change the relocation order of the problematic cells. We sort the problematic cells by the total overflow value of their routing paths in a non-increasing order. The cells with larger overflow values will be relocated and rerouted earlier. Besides, in order to reduce runtime, we ignore the problematic cells whose probability of reducing overflow is too small. For example, these include cells whose routing path overflow value is less than a threshold $thr1$, and the cells whose minimum distance between its original location and the overflow G-Cell its routing paths passing through is larger than a threshold $thr2$ (e.g., $thr1 = 2$, $thr2 = 100$ in our experiments).

We also tried to move multiple cells together. However, we find that the congestion map is inaccurate when many cell locations and routing paths are updated simultaneously. It is very difficult to determine how to move many cells and reroute many nets simultaneously. Actually, in our implementation, the overflow improvement of moving multiple cells together is even smaller than that of moving cells one by one.

5. EXPERIMENTAL RESULT

Our implementation is written in C++ and compiled with g++ 4.1.2. All the benchmarks are run on an Intel Core 2 Duo Linux workstation with 2.8GHz and 4 GB RAM, using one CPU core. We use the global router NCTUgr [13] to evaluate our placement result.

We use the top four placer results in DAC 2012 Contest to test SRP. The input placement results are obtained by the top four placers NTUplacer [7], Ripple [5], SimPLR [11] and mpl12 [3] in the routability-driven placement DAC Contest 2012 [1]. The input global routing results are given by NC-TUgr [13] which is machine independent.

After using SRP, we use NCTUgr [13] to route the output layouts. We show the total normalized overflow of all metal layers in Table 1. The total overflow after using SRP is improved by 12.35%, 8.39%, 75.01%, 34.57% for the placers NTUplacer [7], Ripple [5], SimPLR [11] and mpl12 [3] respec-

Table 1: Overflow Comparison and Runtime of SRP

Benchmark	NTUplacer Input OF	NTUplacer Post OF	NTUplacer Post CPU(s)	Ripple Input OF	Ripple Post OF	Ripple Post CPU(s)	SimPLR Input OF	SimPLR Post OF	SimPLR Post CPU(s)	mpl12 Input OF	mpl12 Post OF	mpl12 Post CPU(s)
sb2	4527	4320	454	51038	47451	1444	114562	90674	1935	248259	185867	1651
sb3	11744	10182	346	39068	36295	974	16191	13689	515	34903	28891	485
sb6	4723	4215	319	7011	6087	467	2753	2506	280	4189	3241	300
sb7	3013	2650	168	13604	13239	381	299288	123573	671	19031	14199	215
sb9	4893	4648	127	13592	12479	342	8389	7420	207	18812	15380	281
sb11	2386	2176	107	2884	2686	200	21617	15953	171	13224	11341	231
sb12	77	59	81	1975	1718	365	2878	2524	298	105472	67374	939
sb14	410	400	45	1750	1592	66	5479	4074	155	2666	2373	77
sb16	7410	6359	105	3831	2850	144	43027	32677	348	5498	4308	145
sb19	1606	1297	47	4352	3939	148	3908	2949	91	21382	18850	100
Avg.	+12.35%	1.00	179.9	+8.39%	1.00	453.1	+75.01%	1.00	467.1	+34.57%	1.00	442.4

Table 2: Routed Wirelength and HPWL Comparison

Bench mark	NTUplacer RoutedWL Input (×E6)	NTUplacer RoutedWL Post (×E6)	NTUplacer HPWL Input (×E8)	NTUplacer HPWL Post (×E8)	Ripple RoutedWL Input (×E6)	Ripple RoutedWL Post (×E6)	Ripple HPWL Input (×E8)	Ripple HPWL Post (×E8)	SimPLR RoutedWL Input (×E6)	SimPLR RoutedWL Post (×E6)	SimPLR HPWL Input (×E8)	SimPLR HPWL Post (×E8)	mpl12 RoutedWL Input (×E6)	mpl12 RoutedWL Post (×E6)	mpl12 HPWL Input (×E8)	mpl12 HPWL Post (×E8)
sb2	17.89	17.89	6.11	6.12	19.33	19.34	6.51	6.52	18.77	18.77	6.29	6.31	21.20	21.22	7.11	7.13
sb3	10.90	10.90	3.27	3.28	11.66	11.66	3.34	3.35	12.57	12.56	3.63	3.63	12.32	12.32	3.64	3.65
sb6	10.45	10.45	3.30	3.31	10.77	10.77	3.35	3.35	11.18	11.18	3.50	3.50	12.22	12.22	3.87	3.87
sb7	12.74	12.74	3.91	3.91	14.09	14.09	4.23	4.23	14.34	14.29	4.29	4.29	14.38	14.38	4.47	4.48
sb9	7.66	7.66	2.37	2.37	8.44	8.44	2.58	2.58	8.26	8.26	2.54	2.54	9.01	9.01	2.78	2.78
sb11	10.32	10.32	3.42	3.42	10.71	10.70	3.56	3.56	10.55	10.55	3.47	3.47	12.70	12.69	4.23	4.23
sb12	10.98	10.98	3.12	3.12	11.78	11.77	3.33	3.33	12.19	12.19	3.55	3.55	10.85	10.86	3.24	3.26
sb14	6.93	6.93	2.25	2.25	7.09	7.09	2.27	2.27	7.47	7.47	2.38	2.39	8.08	8.08	2.58	2.58
sb16	7.90	7.90	2.62	2.62	7.78	7.78	2.59	2.59	8.13	8.15	2.67	2.68	8.68	8.68	2.89	2.89
sb19	4.68	4.68	1.50	1.50	4.91	4.91	1.59	1.59	4.94	4.94	1.58	1.58	5.36	5.36	1.74	1.74
Avg.	-.002%	1.00	-.04%	1.00	+.002%	1.00	-.05%	1.00	+.03%	1.00	-.10%	1.00	-.05%	1.00	-.15%	1.00

tively. The runtime of SRP is shown in table 1. If we use the runtimes of these four placers in the DAC 2012 Contest [1] as reference, the runtime of SRP takes only 8.42% of the total runtime of the placement process on average.

As shown in table 2, the routed wirelength by NCTUgr after using SRP is not affected obviously than that of the input layout. Besides, the increased HPWL after using SRP is always very small compared with the input placement.

6. CONCLUSION

In this paper, a fast and effective simultaneous routing and placement refinement tool called SRP is proposed for routability improvement. SRP is independent of any placer and global router. Based on a given placement layout and global routing result, we relocate problematic cells by considering routing and placement simultaneously. Not only overflow from local nets, but overflow from global and semi-global nets can be removed by SRP. Results show that our method can reduce overflow effectively without worsening HPWL obviously.

7. REFERENCES

[1] DAC 2012 Routability-Driven Contest. http://archive.sigda.org/dac2012/contest/dac2012_contest.html.
[2] U. Brenner and A. Rohe. An effective congestion-driven placement framework. TCAD, 22(4):387–394, 2003.
[3] T. Chan, J. Cong, J. Shinnerl, K. Sze, and M. Xie. mpl6: Enhanced multilevel mixed-size placement. In ISPD, pages 212–214. ACM, 2006.
[4] Y. Chuang, G. Nam, C. Alpert, Y. Chang, J. Roy, and N. Viswanathan. Design-hierarchy aware mixed-size placement for routability optimization. In ICCAD, pages 663–668. IEEE, 2010.
[5] X. He, T. Huang, L. Xiao, H. Tian, G. Cui, and E. Young. Ripple: An effective routability-driven placer by iterative cell movement. In ICCAD, pages 74–79. IEEE, 2011.
[6] W. Hou, H. Yu, X. Hong, Y. Cai, W. Wu, J. Gu, and W. Kao. A new congestion-driven placement algorithm based on cell inflation. In ASP-DAC, pages 605–608. IEEE, 2001.
[7] M. Hsu, S. Chou, T. Lin, and Y. Chang. Routability-driven analytical placement for mixed-size circuit designs. In ICCAD, pages 80–84. IEEE, 2011.
[8] J. Hu, J. Roy, and I. Markov. Completing high-quality global routes. In ISPD, pages 35–41. ACM, 2010.
[9] Z. Jiang, B. Su, and Y. Chang. Routability-driven analytical placement by net overlapping removal for large-scale mixed-size designs. In DAC, pages 167–172. ACM, 2008.
[10] A. Kahng and Q. Wang. Implementation and extensibility of an analytic placer. TCAD, 24(5):734–747, 2005.
[11] M. Kim, J. Hu, D. Lee, and I. Markov. A simplr method for routability-driven placement. In ICCAD, pages 67–73, 2011.
[12] C. Li, M. Xie, C. Koh, J. Cong, and P. Madden. Routability-driven placement and white space allocation. TCAD, 26(5):858–871, 2007.
[13] W. Liu, W. Kao, Y. Li, and K. Chao. Multi-threaded collision-aware global routing with bounded-length maze routing. In DAC, pages 200–205. ACM, 2010.
[14] M. Pan and C. Chu. Fastroute: A step to integrate global routing into placement. In ICCAD, pages 464–471. ACM, 2006.
[15] M. Pan and C. Chu. Ipr: an integrated placement and routing algorithm. In DAC, pages 59–62. ACM, 2007.
[16] J. Roy and I. Markov. Seeing the forest and the trees: Steiner wirelength optimization in placement. TCAD, 26(4):632–644, 2007.
[17] J. Roy, N. Viswanathan, G. Nam, C. Alpert, and I. Markov. CRISP: congestion reduction by iterated spreading during placement. In ICCAD, pages 357–362. ACM, 2009.
[18] P. Spindler and F. Johannes. Fast and accurate routing demand estimation for efficient routability-driven placement. In DATE, pages 1–6. IEEE, 2007.
[19] K. Tsota, C. Koh, and V. Balakrishnan. Guiding global placement with wire density. In ICCAD, pages 212–217, 2008.
[20] X. Yang, B. Choi, and M. Sarrafzadeh. Routability-driven white space allocation for fixed-die standard-cell placement. TCAD, 22(4):410–419, 2003.
[21] Y. Zhang and C. Chu. Crop: Fast and effective congestion refinement of placement. In ICCAD, pages 344–350. IEEE, 2009.

Case Study for Placement Solutions in ISPD11 and DAC12 Routability-Driven Placement Contests

Wen-Hao Liu[12], Cheng-Kok Koh[2], and Yih-Lang Li[1]

[1]Department of Computer Science, National Chiao-Tung University, Hsin-Chu, Taiwan

[2]School of Electrical and Computer Engineering, Purdue University, West Lafayette, IN, USA

dnoldnol@gmail.com, chengkok@purdue.edu, ylli@cs.nctu.edu.tw

ABSTRACT

Routability is a critical issue in VLSI design flow. To address this issue, the routability-driven placement contests [1, 2] held at ISPD11 and DAC12 promote the development of routability-driven placers such as those in [4-6]. ISPD11 and DAC12 contests adopt metrics that are based on global routing solutions to evaluate the routability of placement solutions. However, such global-routing-based metrics typically ignore local congestion, and they cannot evaluate the actual routability effectively. In this work, we develop a translator that allows us to feed the placement solutions of mPL [3], NTUplace [4], Ripple [5], and SimPLR [6] into a commercial router for detailed routing. We then analyze the detailed routing result of each placement solution to better understand the issues that may cause routing violations. Moreover, we examine the suitability of using the ISPD11 and DAC12 metrics in predicting routability. Our findings indicated that the metrics might not reliably predict actual routability, in terms of the number of (detailed) routing violations.

Categories and Subject Descriptors

B.7.2 [**Integrated Circuits**]: Design Aids - Placement and Routing.

General Terms

Algorithms, Design.

Keywords

Routability, Placement, Global Routing, Detailed Routing

1. INTRODUCTION

In advanced technology nodes, the main contributing factors to the increasingly more challenging routing problem include the high number of metal layers, wide metal thickness range, and complex design rules. To address the routability issue, it is getting more common for a placement tool to incorporate features that may improve the routability of a design. Indeed, the routability-driven placement contests [1, 2] held at ISPD11 and DAC12 have attracted many researchers to develop routability-driven placers; the most notable ones are, in alphabetical order, mPL [3], NTUplace [4], Ripple [5], and SimPLR [6].

In ISPD11 and DAC12 contests, the routability of a placement solution is evaluated by the metrics defined based on a global routing result of the placement solution. The aforementioned

This work was partially supported by the National Science Council of Taiwan by Grant Nos. NSC 101-2220-E-009-043 and 101-2917-I-009-004, and National Science Foundation of USA by Grant CCF-1065318.

placers in [3-6] use these global-routing-based metrics to evaluate placement solutions and guide the optimization process. However, many studies [7-11] indicate that a big mismatch between global routing and detailed routing exists. In other words, a good global routing result does not imply that a feasible detailed routing result exists. Even if a placer uses global-routing-based metrics to guide the placement process in the hope that the routability of the design is improved, the effective routability of the design in the context of detailed routing may not have improved. Indeed, while it is acceptable to use some routability estimation metrics to guide the placement optimization process, it is crucial that detailed routing, rather than estimation metrics, is used to evaluate the actual routability. This is akin to using a simple delay model, such as the Elmore delay model, to optimize a routing topology. The quality of the optimized solution, however, has to be evaluated by, for example, SPICE simulations.

Motivated by this principle, this work evaluates the placement solutions of mPL, NTUplace, Ripple, and SimPLR by their detailed routing results. We develop a translator that allows us to feed the placement solutions of mPL, NTUplace, Ripple, and SimPLR into a commercial router for detailed routing. The development of the translator is a non-trivial task because of the format issue and the lack of design rules in the benchmark suites in [1, 2]. Moreover, to evaluate the efficacy of the ISPD11 and DAC12 metrics in improving the effective routability of a placement solution, this work investigates whether the routing violations reported by detailed routing and the routability estimated by the ISPD11 and DAC12 metrics are correlated. We also attempt to analyze the factors in these placement solutions that make or break the routability of these designs. Moreover, as different placement solutions would be generated when the optimization process is guided by different metrics, we compare the detailed routing results of the placement solutions that are optimized by a similar placer but based on the ISPD11 and DAC12 metrics, respectively. It turns out that for some circuits, the placement solutions optimized based on the DAC12 metric have more routing violations in their detailed routing results than those placement solutions optimized based on the ISPD11 metric.

2. ROUTABILITY ESTIMATION METRICS USED IN ISPD11 AND DAC12 CONTESTS

ISPD11 and DAC12 contests use two different global-routing-based metrics to evaluate the routability of placement solutions. Before providing the details of the ISPD11 and DAC12 metrics, we first briefly introduce the global routing problem. In multi-layer global routing, a given placement is typically partitioned into a 3D array of uniform global cells modeled in the form of a 3D grid graph. Each grid node in the grid graph is called a global cell. There are two types of edges: a grid edge corresponds to a boundary between two abutting global cells in the same layer, and a via edge connects two abutting global cells in two adjacent layers. The capacity $c(e)$ of grid edge e is the number of routing tracks that can legally cross the abutting boundary, and the demand $d(e)$ of e is the number of global routing paths passing through e. The overflow of e is defined as max(0, $d(e) - c(e)$). Non-zero overflow in e implies that too many global routing paths are passing through e, resulting in global congestion. Thus, global congestion can be measured in terms of $d(e)$ and $c(e)$. It is typical that the global routing problem ignores the capacity limitation of via edges and ignores the routes within global cells. Such simplifications may make the routing of nets within each global cell difficult because of local congestion. Local congestion may cause a good global routing solution to be non-detailed-routable, i.e., a mismatch between global routing and detailed routing exists.

2.1. ISPD11 Metric

ISPD11 contest uses global router *coalesCgrip* [12] to route each placement solution within a time budget (15 min in ISPD11 contest). The total overflow, by summing up all overflows in grid edges as reported by the global router, is used to evaluate the routability of each placement solution, i.e., lower total overflow implies better routability. However, considering only total overflow may encourage placers to use routings with very long detour to reduce a small amount of overflows. Such snaking routings may add to the difficulty in detailed routing. In addition, as the peak overflow is not considered, the routing solution may have very congested hot-spots.

2.2. DAC12 Metric

In DAC12 contest, each placement solution P is global routed by BFG-R [13] or NCTU-GR 2.0 [14]. For each global routing solution G_P, a global congestion metric $PWC(G_P)$ is computed by first calculating the congestion cost of each grid edge e in G_P, and then adding up the average congestion costs of the top 0.5%, 1%, 2%, and 5% congested grid edges. Consequently, $PWC(G_P)$ can take the peak overflow into account. If P contains a very congested hot-spot, $PWC(G_P)$ will be high for P, giving P a high penalty because of high congestion. The score of a placement solution, $S(P)$, in the DAC12 contest is then computed as follows:

$$S(P) = HPWL(P) \times (1 + 0.03 \times PWC(G_P)), \quad (1)$$

where $HPWL(P)$ denotes the total half-perimeter wirelength of P. A lower $S(P)$ implies better routability. The DAC12 metric considers total wirelength, total overflow, and peak overflow together to overcome the perceived limitations of the ISPD11 metric, but no experiments have compared these two metrics.

3. FRAMEWORK FOR PERFORMING DETAILED ROUTING

The benchmark suites used in the ISPD11 and DAC12 placement contests can be found in [1] and [2], respectively. However, these benchmark suites omit many physical design rules. Therefore, one cannot easily perform detailed routing on the placement solutions of [1, 2]. This work develops a translator based on the 28nm technology node to translate the placement solutions of [1, 2] to LEF/DEF files and then feed the LEF/DEF files to commercial router Wroute [14] (version 3.0.61) to obtain detailed routing results. Because the details of the design rules of the 28nm technology node are not readily available, we consult with several senior engineers from various companies (vendors, design companies, and manufacturers) to set reasonable routing parameters in the translator, in which the minimum wire width and via size are set to 42 *nm* and 56×56 *nm²*, respectively. The minimum spacing rule is set to 42 *nm*. Based on the default setting in the benchmark suites [1, 2], metal layers 1-4, 5-7, and 8-9 have 1X, 2X, and 4X the minimum wire width, via size, and spacing, respectively.

In the benchmark suites of [1, 2], some pins are not accessible because these pins are blocked by big macros. Thus, the translator promotes these pins to the top layer of the macros to make these pins accessible. In addition, some pads are located at the same positions, which would cause routing violations. Therefore, the translator only reserves the pad that appears first in the benchmark files and removes the others that are located at the same position. As only few pads are removed, the effect of such removals should be negligible. In practice, many small obstacles and cell routings occupy the bottom layer and they consume most of the routing resources there, leaving very little routing resource available for the detailed routing of other components in the circuit. As [1, 2] does not provide such information in the bottom layer, the translator conservatively assumes that the bottom layer is not available for the detailed routing. However, vias are allowed to connect to pins at the bottom layer. The detailed routing results should improve if the obstacle information is available to allow detailed routing in the bottom layer.

After obtaining the translated LEF/DEF files, Wroute is invoked to perform (global and detailed) routing in two iterations. The first iteration is the default routing mode and the second iteration is the post routing mode. In the default mode, Wroute reports the number of unroutable nets (NUN) after the global routing stage and the unroutable nets are not routed in the detailed routing stage of the first iteration. NUN can be regarded as an indicator to estimate the routability of the given placement. However, NUN is based on global routing information, and it cannot predict the local congestion precisely.

In the post routing mode, Wroute detailed routes *every* net and further minimizes the number of routing violations, the total wirelength, and the number of vias of the routing result generated in the previous iteration. The runtime limitation of each iteration is 24 hours. We have tried to run Wroute for more iterations to further optimize the detailed routing results. However, we typically observe only slight improvement over the solutions obtained using only two iterations. We conclude that for the purpose of evaluating the routability of placement solutions, running two iterations of Wroute strikes a good balance between routing quality and runtime.

4. OBSERVATIONS AND ANALYSES

The top four placers in both ISPD11 and DAC12 placement contests are, in alphabetical order, mPL, NTUplace, Ripple, and SimPLR. This work performs detailed routing on their placement solutions in the contests and then analyzes the detailed routing results. Because ISPD11 and DAC12 contests adopt different evaluation metrics, the contestants develop different versions of placers to optimize the placement solutions based on the different metrics. To distinguish the versions for ISPD11 and DAC12 contests, the versions of mPL, NTUplace, Ripple, and SimPLR for ISPD11 contest are named as mPL_i, $NTUplace_i$, $Ripple_i$, and $SimPLR_i$, respectively. Similarly, the versions for DAC12 contest are named as mPL_d, $NTUplace_d$, $Ripple_d$, and $SimPLR_d$. Sections 4.1 and 4.2 show the detailed routing results of the placement solutions in ISPD11 and DAC12 contests, respectively. Section 4.3 compares the placement solutions of $Ripple_i$ and $Ripple_d$ to show the difference between the solutions optimized by a similar placer but based on different routability metrics.

4.1. ISPD11 Placement Solutions

ISPD11 contest performs global router *coalesCgrip* on each placement solution within 15 min to obtain the total overflow. The first major column in Table 1 shows the total overflow of each

TABLE 1 COMPARING THE PLACEMENT SOLUTIONS IN ISPD11 CONTEST BASED ON THE ROUTABILITY ESTIMATION METRICS

	ISPD11 metric				DAC12 metric (10^7)				NUN			
	$Ripple_i$	mPL_i	$SimPLR_i$	$NTUplace_i$	$Ripple_i$	mPL_i	$SimPLR_i$	$NTUplace_i$	$Ripple_i$	mPL_i	$SimPLR_i$	$NTUplace_i$
s1	816	89176	78	170314	27.97	32.76	29.98	41.58	2070	2817	1269	2244
s2	1128906	1849664	2138796	1453774	85.34	108.20	127.50	100.07	1086	3228	1353	1273
s4	118850	159584	443324	256632	26.86	31.01	49.87	38.89	632	2983	989	2568
s5	143580	499582	223944	765852	39.12	53.46	42.70	75.03	1761	3594	2130	2897
s10	1010058	1159416	1311688	616424	78.70	80.46	78.90	62.40	657	689	535	505
s12	542786	2272764	514614	3147446	48.50	136.41	47.19	148.56	501	2147	681	3866
s15	143580	171184	345284	767310	41.13	43.04	54.20	69.17	985	1850	2086	3719
s18	514886	52498	72426	470266	44.09	17.54	20.65	30.97	2515	1834	1701	2571
AR	1.750	2.500	2.625	3.125	1.625	2.625	2.625	3.125	1.625	3.375	2.000	3.000
ARD	0.8125				0.8125				0.6875			

TABLE 2 ROUTING VIOLATIONS IN DETAILED ROUTING RESULTS

	$Ripple_i$	mPL_i	$SimPLR_i$	$NTUplace_i$
s1	99	1675	59	5985
s2	820	40967	79293	241252
s4	242	2448	27740	496819
s5	805	182008	314573	9235
s10	837	31801	4643	29370
s12	179	7181846	657749	14152052
s15	118	117	6033	133975
s18	64478	165549	9818	309269
AR	1.375	2.625	2.375	3.625

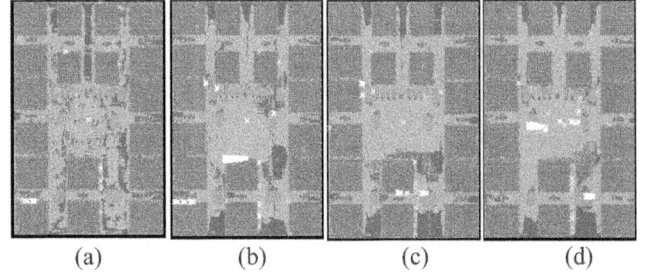

(a)   (b)   (c)   (d)

Fig. 1. The placement solutions of s10 obtained by (a) $Ripple_i$; (b) mPL_i; (c) $SimPLR_i$; (d) $NTUplace_i$.

placement solution in ISPD11 contest. To compare the routability of each placement solution, we follow the convention used in the ISPD11 contest and rank the placement solutions of each benchmark from 1 (least total overflow or best) to 4 (most total overflow or worst). After that, we average the rankings of the solutions in each column to get the average ranking (AR) for each placer. In ISPD11 contest, a placer with smaller average ranking implies that this placer produces better routability placement solutions. However, the focus of this work is not to rank the placers. Instead, the goal is to evaluate the fidelity of routability metrics. To do that, we examine whether the ranking order based on the routability metrics and that based on the number of routing violations reported by detailed routing are similar.

Moreover, we also use the DAC12 metric to evaluate these placement solutions, and then analyze the consistency between each metric and the routing violations reported by detailed routing. The second major column in Table 1 shows the routability score $S(P)$ obtained by Eq. (1) for each placement solution P, in which G_P is obtained by NCTU-GR 2.0 with the default setting.

Furthermore, the third major column shows the NUN reported by the global routing stage of Wroute (in the first iteration of the two-iteration detailed routing process). Although the average ranking of each placer changes when the three different evaluation metrics, i.e., ISPD11 metric, DAC12 metric $S(P)$, and NUN, are adopted, these metrics show a similar trend. For example, although the ranking order of the solutions on s12 and s18 based on the different metrics are different, the gap between the leading placement solutions and others on s12 and s18 is obvious. This implies that the routability estimated by ISPD11 and DAC12 metrics is similar to that estimated by the commercial router. This may indicate that metrics based on global routing solutions are likely to be similar.

However, a different story is revealed in Table 2. Table 2 shows the numbers of routing violations in the detailed routing results obtained by Wroute, in which a placement solution with fewer routing violations gets a better ranking. By comparing Tables 1 and Table 2, we can find a mismatch of ranking orders between Table 1 and Table 2. For example, $NTUplace_i$ obtains the

placement solution of s10 with the best total overflow, routability score $S(P)$, and NUN in Table 1, but has the highest number of routing violations, i.e., the worst performance, in Table 2. On benchmark s12, $Ripple_i$ and $SimPLR_i$ obtain solutions with similar total overflow, routability score $S(P)$, and NUN, but $Ripple_i$ has much fewer routing violations than $SimPLR_i$. On benchmark s18, mPL_i has the better total overflow, routability score $S(P)$, and NUN than $Ripple_i$, but $Ripple_i$ has fewer routing violations.

To analyze the accuracy of each metric in predicting routability, the average ranking discrepancy (ARD) in Table 1 is calculated to show the difference between each metric and the routing violations. If a placement solution has rank r_e based on an estimation metric and rank r_v based on the number of routing violations, the ranking discrepancy of this solution is $|r_e - r_v|$. In Table 1, ARDs are the average of the ranking discrepancies of all placement solutions in each major column. A metric with lower ARD means that this metric is more accurate in estimating the effective routability.

The reason for using ARD to evaluate the accuracy of a metric is that if a metric determines that the routability of placement solution P_1 is better than solution P_2, the detailed routing result of P_1 should have fewer routing violations than that of P_2. Table 1 reveals that the ISPD11 and DAC12 metrics have similar ARDs, and NUN has smaller ARD than the ISPD11 and DAC12 metrics. It is no surprise to see that NUN has smaller ARD because NUN and routing violations both are reported by Wroute. NUN is estimated based on the global routing result obtained by Wroute, and then the detailed routing result obtained by Wroute is based on its global routing result. Although ISPD11 and DAC12 metrics have similar ARDs, using the DAC12 metric to estimate routability is faster than the ISPD11 metric. Recall that the DAC12 metric is calculated based on an initial global routing result obtained by a global routing with only few routing iterations, whereas the calculation of the ISPD11 metric requires one to perform several iterations of global routing to reduce overflows.

TABLE 3 WIRELENGTH, VIAS AND RUNTIME COMPARISION OF THE PLACEMENT SOLUTIONS IN ISPD11 CONTEST

	$Ripple_i$			mPL_i			$SimPLR_i$			$NTUplace_i$		
	WL(10^7)	Via(10^6)	Runtime	WL(10^7)	Via(10^6)	Runtime	WL(10^7)	Via(10^6)	Runtime	WL(10^7)	Via(10^6)	Runtime
s1	33.70	10.23	09:58:37	35.04	10.81	11:29:32	34.05	10.24	08:17:36	36.71	11.19	16:45:59
s2	78.83	12.75	14:40:57	76.53	14.32	41:15:07	78.43	13.46	24:53:58	80.35	14.10	49:02:39
s4	27.50	6.87	06:07:22	27.35	7.46	11:15:59	31.69	7.28	13:20:42	25.92	7.32	48:30:12
s5	41.97	9.23	11:23:19	44.33	10.29	46:30:12	42.84	9.51	29:51:25	49.70	10.32	18:06:48
s10	67.88	13.72	15:26:41	68.82	14.86	22:25:37	67.59	13.71	17:30:56	65.66	14.83	28:59:10
s12	55.25	18.24	11:23:43	53.47	21.73	51:28:56	50.98	18.66	31:18:03	47.53	22.96	52:05:18
s15	44.71	13.70	09:12:07	40.46	13.78	09:16:59	43.75	13.89	14:32:30	40.13	13.91	29:07:13
s18	32.03	7.99	20:41:57	22.60	7.39	21:09:52	27.54	7.81	12:09:08	21.37	7.57	26:11:38
$Ratio_{ind}$	1	1	1	0.957	1.073	2.236	0.992	1.020	1.672	0.953	1.085	3.176
$Ratio_{sum}$	1	1	1	0.965	1.085	2.172	0.987	1.020	1.536	0.962	1.102	2.718

TABLE 4 COMPARING THE PLACEMENT SOLUTIONS IN DAC12 CONTEST BASED ON THE DAC12 METRIC, NUN AND VIOLATIONS

	DAC12 metric (10^7)				NUN				Violations			
	$Ripple_d$	mPL_d	$SimPLR_d$	$NTUplace_d$	$Ripple_d$	mPL_d	$SimPLR_d$	$NTUplace_d$	$Ripple_d$	mPL_d	$SimPLR_d$	$NTUplace_d$
s2	78.21	115.5	87.67	64.3	1743	2068	553	2169	81227	113991	876	725813
s3	44.28	46.08	40.75	37.6	566	649	487	1450	243	231	194	988
s6	36.93	41.12	36.94	36.68	267	892	443	921	232	231	361	637
s7	47	51.86	131.04	40.52	703	1525	518	419	300	170	5402	168
s9	30	33.8	28.93	26.7	125	573	786	1112	8136	68	22	89
s11	36.27	44.66	38.51	34.73	115	531	979	313	433	11009	1840	697
s12	37.38	53.03	37.69	31.68	167	994	715	120	155	343637	241	94
s14	23.89	27.45	25.68	22.96	1220	1717	1459	2352	19086	271799	224239	18446
s16	27.23	30.7	35.81	28.27	129	127	434	78	38	36	135	24
s19	16.95	22.78	16.63	15.33	518	811	510	414	110	276	72777	15114
AR	2.2	3.8	2.9	1.1	1.9	3.1	2.4	2.6	2.3	2.6	2.7	2.4
ARD	1				0.8				-			

TABLE 5 TOTAL LOCAL MANHATTAN DISTANCE WIRELENGTH

	$Ripple_d$	mPL_d	$SimPLR_d$	$NTUplace_d$
s2	9714291	7932550	9898958	9948203
s3	10383310	8429011	9864931	10052341
s6	11860333	9338853	11539937	11710891
s7	17110222	14415218	16826775	17527856
s9	10268725	8496641	10311523	10318484
s11	10053468	7992076	10265926	10269487
s12	17881504	16059810	8769507	17761563
s14	6538557	5100532	6325446	6686377
s16	8770145	7053053	8769507	8882537
s19	5663757	4729536	5658416	5896041
AR	3	1.1	2.2	3.7

Table 3 compares the total wirelength, via count, and Wroute's runtime (hh:mm:ss) of each detailed routing result, in which the results of $Ripple_i$ are treated as the baseline for comparison. $Ratio_{ind}$ is the average of the ratio of individual entries in the same column, while $Ratio_{sum}$ denotes the ratio of the sum of each column. Because $Ripple_i$ spreads cells more evenly, the routing wirelength is longer than other placers. However, the detailed routing results of $Ripple_i$ and $SimPLR_i$ have fewer vias; the number of vias is also an important factor to influence routability, but global routing model usually ignores via size and via spacing, contributing in part to the mismatch between global routing and detailed routing. Note that both $Ripple_i$ and $SimPLR_i$ are based on a lower-upper-bound framework [16], and mPL_i and $NTUplace_i$ are based on a multilevel framework [17]. It is interesting to see that the placers based on the same frameworks have similar results.

4.2. DAC12 Placement Solutions

Table 4 compares the placement solutions of mPL_d, $NTUplace_d$, $Ripple_d$, and $SimPLR_d$ based on the DAC12 metric, NUN, and the number of routing violations, respectively. As the global routing results of these placement solutions are all overflow free, there is no effective way to rank the solutions based on the ISPD11 metric. In DAC12 contest, global routers BFG-R and NCTU-GR 2.0 are used to evaluate placement solutions. Thus, we use BFG-R and NCTU-GR 2.0 to get two global routing results G_1 and G_2 for each placement solution, and then compute congestion penalty $PWC(G_P)$ by averaging $PWC(G_1)$ and $PWC(G_2)$. Finally, the routability score for each placement solution is obtained via Eq. (1), which is shown in the first major column in Table 4; the ARD is 1. However, if we use only the routing results of either BFG-R or NCTU-GR 2.0 to compute the routability score via Eq. (1), the ARD become 1.1 and 1.05, respectively. This implies that using multiple global routers to evaluate routability is more accurate than using only one.

By comparing Table 4 and Table 1, we found that the ARDs of DAC12 metric and NUN in Table 4 are higher than those in Table 1. A higher ARD means that the accuracy in predicting routability decreases. It is worthwhile to discuss why the predictability decreases. We have performed NCTU-GR 2.0 to route every placement solution in Table 1 and Table 4 until either an overflow-free routing result is obtained or overflows cannot be reduced anymore. While NCTU-GR 2.0 cannot obtain overflow-free routing results for more than half of the placement solutions in Table 1, it can obtain an overflow-free result for every placement solution in Table 4. In other words, any global congestion issues in the placement solutions in Table 4 can be easily resolved by global routers. Consequently, local congestion becomes the primary issue that determines which placement solution has better routability. However, DAC12 metric and NUN are global-routing-based metrics; there is no easy way to use them to predict local congestion consistently. Thus, DAC12 metric and NUN have higher ARD in Table 4 than those in Table 1. To a certain extent, one may argue that the placers may have accentuated the local congestion issue when the primary goal of

	Ripple$_d$			mPL$_d$			SimPLR$_d$			NTUplace$_d$		
	WL(10^7)	Via(10^6)	Runtime	WL(10^7)	Via(10^6)	Runtime	WL(10^7)	Via(10^6)	Runtime	WL(10^7)	Via(10^6)	Runtime
s2	72.77	12.67	32:26:26	80.08	13.40	40:44:26	69.67	12.47	14:37:14	67.80	12.28	48:33:08
s3	43.55	11.58	10:47:19	44.93	11.56	10:32:16	45.59	11.71	10:26:24	39.99	11.05	13:06:50
s6	40.96	11.77	10:35:16	45.42	12.22	10:13:03	41.54	11.80	10:26:33	39.17	11.49	11:02:01
s7	53.83	17.77	13:28:38	55.55	18.44	14:13:09	55.44	18.15	15:40:00	48.31	16.93	13:17:31
s9	32.46	10.03	09:21:58	34.22	10.32	08:34:57	31.09	9.94	08:53:54	29.19	9.62	08:49:18
s11	40.22	10.48	11:14:55	47.48	11.31	22:16:53	39.27	10.49	15:41:12	38.63	10.19	11:52:41
s12	47.13	17.91	12:05:03	47.58	19.18	48:49:04	46.98	17.80	12:04:46	42.73	17.01	11:26:41
s14	27.79	7.59	11:16:24	31.10	7.96	36:07:43	28.65	7.71	24:24:14	26.81	7.46	21:57:51
s16	29.17	7.92	06:13:12	31.90	8.00	07:11:05	30.31	7.97	06:42:16	29.19	7.72	06:59:38
s19	19.35	6.21	05:37:44	20.74	6.43	05:57:55	18.72	6.12	17:55:16	18.11	5.99	16:42:50
Ratio$_{ind}$	1	1	1	1.080	1.040	1.661	1.002	1.001	1.334	0.937	0.966	1.373
Ratio$_{sum}$	1	1	1	1.078	1.043	1.662	1.000	1.002	1.112	0.933	0.963	1.331

TABLE 7 ROUTING VIOLATIONS IN DETAILED ROUTING RESULTS

	Ripple$_i$					
	NUM	Violation	WL(10^7)	Via(10^6)	Runtime	TLMD(10^6)
s1	2070	99	33.70	10.23	09:58:37	8.81
s2	1086	820	78.83	12.75	14:40:57	7.77
s4	632	242	27.50	6.87	06:07:22	5.83
s5	1761	805	41.97	9.23	11:23:19	6.53
s10	657	837	67.88	13.72	15:26:41	10.23
s12	501	179	55.25	18.24	11:23:43	15.98
s15	985	118	44.71	13.70	09:12:07	11.73
s18	2515	64478	32.03	7.99	20:41:57	5.50

	Ripple$_d$					
	NUM	Violation	WL(10^7)	Via(10^6)	Runtime	TLMD(10^6)
s1	830	45	33.13	10.25	09:29:02	8.85
s2	1066	4989	71.91	12.61	18:05:02	8.39
s4	378	238	26.05	6.76	06:28:59	6.15
s5	1248	506	39.55	8.98	10:59:48	6.60
s10	416	398	63.74	13.51	15:37:18	10.38
s12	992	78519	50.09	17.93	19:50:53	17.55
s15	270	195	37.28	12.92	08:02:07	12.93
s18	1311	31435	24.57	7.40	19:06:16	6.14
Ratio$_{ind}$	0.762	56.177	0.904	0.973	1.095	1.060
Ratio$_{sum}$	0.638	1.721	0.907	0.974	1.088	1.064

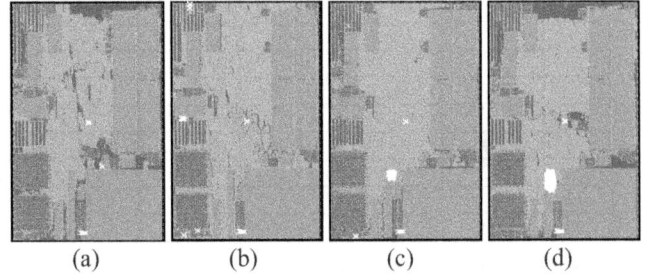

(a) (b) (c) (d)

Fig. 2. The placement solutions of s19 obtained by (a) Ripple$_d$; (b) mPL$_d$; (c) SimPLR$_d$; (d) NTUplace$_d$.

the placers is to reduce global congestion. The placers may have moved cells or macro blocks such that many more nets are now local nets residing within a global cell, instead of global nets spanning two or more global cells.

In Table 4, NTUplace$_d$ has a much lower AR than other placers based on the DAC12 metric, which implies that NTUplace$_d$ reduces both HPWL and global congestion well to obtain better routability scores $S(P)$ (Eq. (1)). However, based on the number of routing violations, NTUplace$_d$ has a similar AR compared to other placers. We suspect that local congestion is the main cause of such behavior. To estimate the local congestion for each placement solution, we decompose every net into two-pin nets based on the topology of a rectilinear minimum spanning tree and then add up the Manhattan distance between the terminals of each local two-pin net to get the total local Manhattan distance (TLMD). If both terminals of a two-pin net are within a global cell, the two-pin net is regarded as a local two-pin net. Table 5 shows the TMLD of each placement solution, as well as the AR for each placer based on TMLD; a solution with shorter TMLD gets a better ranking. Table 4 and 5 reveal that the orders of AR based on the DAC12 metric and TLMD are reversed. This implies that a placement solution that has a better routability score evaluated by the DAC12 metric has longer TLMD. But, a longer TLMD may

lead to local congestion and higher number of routing violations, which would cancel the benefit of performing well on the DAC12 metric. Thus, the four placers have similar ARs based on the number of violations.

Table 6 compares the detailed routing results of the placement solutions obtained by mPL$_d$, NTUplace$_d$, Ripple$_d$, and SimPLR$_d$. As Ripple$_d$'s results have the least number of routing violations, they are treated as the baseline. NTUplace$_d$ continues to perform well in terms of wirelength. Its solutions have the shortest wirelength and the lowest via count compared to other placers. However, Wroute spends the shortest runtime to route Ripple$_d$'s placement solutions. Fig. 2 shows the placement solutions of s19 obtained by each placer.

4.3. Comparing Ripple$_i$ and Ripple$_d$

Generally speaking, one would expect a newer version of a routability-driven placer to have placement solutions that are more routable than the solutions of the older version, if the placer optimizes routability properly. In this section, we compare the routing results of the placement solutions obtained by the ISPD11 version and DAC12 version of Ripple. Because ISPD11 and DAC12 contests are based on different benchmark circuits, we cannot directly compare the placement solutions in ISPD11 contest and in DAC12 contest. Fortunately, the authors of Ripple$_d$ graciously made available to us their placement solutions of ISPD11 benchmarks, allowing us to compare the detailed routing results of Ripple$_i$ and Ripple$_d$ for ISPD11 benchmark circuits.

Table 7 compares the detailed routing results of Ripple$_i$ and Ripple$_d$ obtained by Wroute, the routing results of Ripple$_i$ are treated as the baseline. It shows that Ripple$_i$ has shorter wirelength, fewer vias, and lower NUN than Ripple$_i$. However, the placement solutions of Ripple$_d$ have more routing violations for benchmark circuits s2 and s12. We suspect that the DAC12 metric steers Ripple$_d$ in a direction that actually increases the

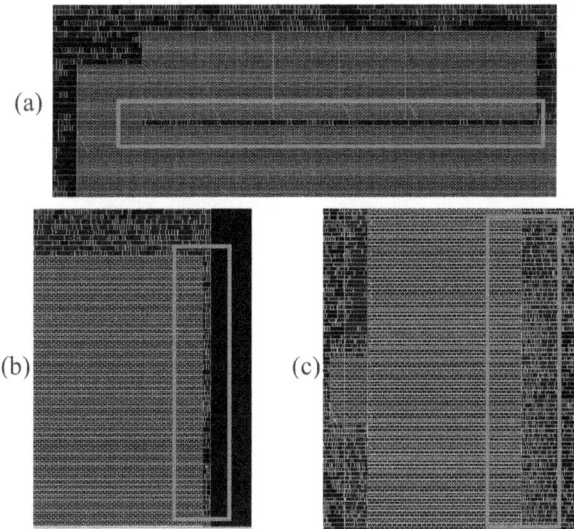

Fig. 3. The local views of the most congested region in the placement solution of (a) mPL_i for s10 with 21072 routing violations; (b) $Ripple_i$ for s18 with 29551 routing violations; (c) $NTUplace_d$ for s19 with 11957 routing violations.

routing violations in s2 and s12. Because the DAC12 metric takes only HPWL and global congestion into account but ignores local congestion, $Ripple_d$, which uses the DAC12 metric to evaluate placement solutions, may have pushed more nets into a global cell to reduce HPWL and global congestion. However, pushing many nets into a global cell would worsen local congestion, resulting in higher number of routing violations. In fact, Table 7 shows that $Ripple_d$ has longer TLMD than $Ripple_i$, and longer TLMD increases the probability of higher local congestion.

5. WHAT CAUSES ROUTING VIOLATIONS

As placers are now getting better in resolving global congestion issue, it is more critical than ever that they consider local congestion in the optimization process. Otherwise, the placers may further accentuate the local congestion problem by turning global nets into local nets. To consider local congestion in the early stage, the works in [7, 8] developed local congestion estimation metrics based on the pin density; however, they did not consider the effect of blockages. Figs. 1 and 2 show that most routing violations occur near blockages. This implies that incorporating the effect of blockages into local congestion estimation metrics may improve the accuracy of the metrics in predicting local congestion.

We list some reasons why most routing violations near blockages based on our observations. (1) The global placement stage in placers would place cells on blockages. After legalization, these cells are pushed to the fringe of blockages, resulting in high congestion around the fringe of these blockages. (2) Routing wires cannot access pins that are too close to blockages because of spacing rules. (3) Although routing wires can stride over blockages via the high routing layers, the routing wires hardly use the high layers in the congested regions. (4) Routing wire bypassing blockages would increase via count, which may make designs harder to route. But, the placement and global routing stages usually ignore the routing overhead incurred by vias.

To take a deeper look why blockages cause routing violations, Fig. 3 provides the local views of the most congested regions in some placement solutions. In Fig. 3(a), many cells are placed in a

narrow channel between blockages. In Fig. 3(b), many cells are placed into a narrow channel between a blockage and the design's boundary. In Fig. 3(c), many cells are adjacent to a big blockage. Note that we do not show the routing violations in Fig. 3 because doing so would make it impossible to view the cells and blockages, as the number of routing violations are just too high. Such patterns in Fig. 3, which cause routing violations, are commonly found in the placement solutions in ISPD11 and DAC12 contests. Even if a placement solution can be easily globally routed, routing violations still exist in the placement solution if the patterns shown in Fig. 3 exist. We believe that routing violations can be reduced if placers avoid placing too many cells into narrow channels and preserve a small space between cells and big blockages.

6. SUMMARY

This work develops a translator to feed the placement solutions of mPL, NTUplace, Ripple, and SimPLR into a commercial router, in order to get the detailed routing results for the evaluation of their routability. To examine the accuracy of the ISPD11 and DAC12 metrics in predicting routability, this work uses the average ranking discrepancy (ARD) to show the difference between routing violations in the detailed routing results and the routability estimated by the metrics. Moreover, this work compares the placement solutions of $Ripple_i$ and $Ripple_d$ to show how the ISPD11 and DAC12 metrics affect routability of solutions optimized by a similar placer but with different metrics. Finally, this work takes a closer look at where the routing violations occur, which suggests that the effect of blockages for local congestion should be addressed.

REFERENCES

[1] N. Viswanathan et al, "The ISPD-2011 Routability-driven Placement Contest and Benchmark Suite", in Proc. ISPD, 2011.

[2] N. Viswanathan et al, "The DAC 2012 Routability-driven Placement contest and benchmark suite", in Proc. DAC, 2012.

[3] T. F. Chan et al, "mPL6: Enhanced Multilevel Mixed-Size Placement", in Proc. ISPD, 2006.

[4] M.-K. Hsu et al, "Routability-driven analytical placement for mixed-size circuit designs", in Proc. ICCAD, 2011.

[5] X. He et al, "Ripple: an effective routability-driven placer by iterative cell movement", in Proc. ICCAD, 2011.

[6] M.-C. Kim et al, "A SimPLR method for routability-driven placement", in Proc. ICCAD, 2011.

[7] T. Taghavi et al, "New Placement Prediction and Mitigation Techniques for Local Routing Congestion", in Proc. ICCAD, 2010.

[8] Y. Wei et al, "GLARE: Global and Local Wiring Aware Routability Evaluation", in Proc. DAC, 2012.

[9] C. J. Alpert et al, "What Makes a Design Difficult to Route", in Proc. ISPD, 2010.

[10] Z. Li et al, "Guiding a Physial Design Colsure System to Produce Easier-to-Route Designs with More Predictable Timing", in Proc. DAC, 2012.

[11] Y. Zhang and C. Chu, "GDRouter: interleaved global routing and detailed routing for ultimate routability," in Proc. of DAC, 2012.

[12] H. Shojaei et al, "Congestion analysis for global routing via integer programming", in Proc. ICCAD, 2011.

[13] J. Hu et al, "Completing High-quality Global Routes,"In Proc. ISPD, 2010.

[14] W.-H. Liu et al, "Multi-Threaded Collision-Aware Global Routing with Bounded-Length Maze Routing" in Proc. DAC, 2010.

[15] http://www.cadence.com/products/di/soc_encounter/

[16] M.-C. Kim et al, "SimPL: An Effective Placement Algorithm," in Proc. ICCAD, 2010.

[17] T. Chan et al, "Multilevel Optimization for Large-Scale Circuit Placement," in Proc. of ICCAD, 2000.

A Compiler for Scalable Placement and Routing of Brain-like Architectures

Narayan Srinivasa
HRL Laboratories LLC
Malibu, CA, USA
310-317-5870
nsrinivasa@hrl.com

ABSTRACT

The challenging aspect of building neuromorphic circuits in mature CMOS technology to match brain-like architectures is two-fold: *scalability* and *connectivity*. Scalability means that the circuits have to be expandable to match biological brains in terms of synaptic and neuronal densities. The challenge here is to implement 10^6 neurons and 10^{10} synapses with an average fanout of 10^4, in a square cm of CMOS [1]. Connectivity means that the circuit has to offer the capability to have both short and long range (by physical distance) connections between neurons. A large part of this challenge is how to implement a connectivity of 10^4 synapses per neuron [2]. Unfortunately, even the exponential transistor density growth being experienced today is not sufficient to realize such massive connectivity and synaptic densities in a traditional CMOS process. Recent approaches to address these challenges have been to integrate CMOS with nanotechnology [3, 4] in order to achieve the required synaptic densities. These solutions use crossbar architectures predominantly but the connectivity challenge still remains a daunting task for such solutions [2, 5]. To meet these challenges, a novel synaptic time-multiplexing (STM) concept was developed along with a neural fabric design [6]. This combination has the advantage of offering greater flexibility and long range connectivity. It also provides a method to overcome the limitations of conventional CMOS technology to match the synaptic density and connectivity requirements found in mammalian brains while maintaining non-linear synapses and learning.

In order to program neuromorphic hardware [7] for any desired brain architecture, the topology would first have to be converted into a connectivity matrix or a graph representation. This matrix along with the statistics on the number of neurons and synapses is provided as input to a neuromorphic compiler. The neuromorphic compiler compiles the neural network structure description into: 1) an assignment of the network's neurons and synapses to hardware neurons and virtual (multiplexed) synapses, and 2) a STM compatible routing schedule with switch states for the neural fabric at each STM timeslot.

For each neuron, the exact location on the chip on the neural fabric should be determined. This is the *placement problem*. The quality of neuron placement can affect the ability of the routing algorithm to efficiently find the needed synaptic pathways to cover all the synapses within a STM duty cycle. For each synaptic pathway, a set of required grid lines from an output axon of the presynaptic neuron to an input dendrite of the postsynaptic neuron

should be determined, and the switches on the way must be set to the ON state. This is the *routing problem*.

The problem of routing and placement is closely related to problems in other programmable hardware such as the FPGA (Field Programmable Gate Array). There are some interesting differences between the neuromorphic solution proposed in this work and those designed for other programmable hardware such as the FPGA. In such applications, most current algorithms [8-10] for placing and routing expect a single timeslot and therefore do not have to address the immense routing demands placed by the problem described here. Unlike FPGA circuits, the neuromorphic hardware is expected to use every neuron device during routing. However, a study of FPGA architecture [11] show that on reconfigurable hardware 100% device utilization results in almost a 200% routing area increase due to congestion problems. A single neuromorphic chip is expected to house 10^6 neurons while the largest FPGA in 2011, the Xilinx Virtex-6 LX760 contains less than 800 thousand logic cells [12] and only utilizes 60% of them at any given time. The connectivity on our neuromorphic chip is two orders of magnitude higher than on current FPGA's. This required us to develop a completely new neuromorphic compiler which can scale to support large scale network architectures in hardware.

To achieve this we have developed three key concepts. First, we developed an efficient and automatic method for generating brain-like architectures that have small-world network like topological properties [13] to evaluate our algorithms. Second, a placement algorithm was developed that could deal with the extremely large connectivity and finally a parallel router was developed to enable routing of billions of synapses. The details of these concepts will be described during the presentation.

Categories and Subject Descriptors

C.3.3 [**Computer Systems Organization**]: Special-Purpose and Application-Based Systems.

General Terms

Algorithms, Design.

Keywords

Routing, Placement, Brain-like Architectures, Compiler.

1. ACKNOWLEDGMENTS

This work was performed under DARPA SyNAPSE grant HRL011-09-C-0001. This work is approved for public release and distribution is unlimited. The views, opinions, or findings contained in this abstract are those of the author and should not be interpreted as representing the official views or policies, either expressed or implied, of the DARPA or Department of Defense.

2. REFERENCES

[1] Srinivasa, N., and Cruz-Albrecht, J. 2012. Neuromorphic adaptive plastic scalable electronics: Analog Learning Systems. *IEEE Pulse*, vol. 3, no. 1, pp. 51–56.

[3] Bailey, J. and Hammerstrom, D. 1988. Why VLSI Implementation of Associative VLCNs Require Connection Multiplexing. *IEEE International Conference of Neural Networks*. pp. 173-180.

[4] Strukov, D. B. and Likharev, K. K. 2005. Prospects for terabit-scale nanoelectronic memories. *Nanotechnology*, 16(1): p. 137-148.

[5] Jo, S., Chang, T., Ebong, I., Bhavitavya, B., Mazumder, P., and Lu, W. 2010. Nanoscale Memristor Device as Synapse in Neuromorphic Systems. *Nano Letters*, vol. 10, pp. 1297-1301.

[6] Gao, C. and Hammerstrom, D. 2007. CMOL-based cortical models. Emerging Brain-Inspired Nano-Architectures. *IEEE Transactions on Systems and Circuits -1*, vol. 54, no. 11, pp. 2502 – 2515.

[7] Minkovich, K., Srinivasa, N., Cho, Y. K., Nogin, A. and Cruz-Albrecht, J. 2012. Programming Time-Multiplexed Reconfigurable Hardware using a Scalable Neuromorphic Compiler. *IEEE Transactions on Neural Networks and Learning Systems*, vol. 23, no. 6, pp. 889-901.

[8] Cruz-Albrecht, J., Yung, M., and Srinivasa, N. 2012. Energy Efficient Neuron, Synapse and STDP Integrated Circuits.

[9] Luu, J., Kuon, I., Jamieson, P., Campbell, T., Ye, A., Fang, W., and Rose, J. 2009. VPR 5.0: FPGA cad and architecture exploration tools with single driver routing, heterogeneity and process scaling. *Proc. ACM/SIGDA Int. Symp. Field Program Gate Arrays*, pp. 1–10.

[10] Viswanathan, N., Pan, M., and Chu, C. 2007. FastPlace 3.0: A fast multilevel quadratic placement algorithm with placement congestion control. *Proc. Asia South Pacific Design Autom. Conf.*, pp. 135–140.

[11] Padmini, G., Li, X., and Pileggi, L. 2006. Architecture-aware FPGA placement using metric embedding, *Proc. 43rd Annu. Design Autom. Conf.*, pp. 1–6.

[12] DeHon, A. 1999. Balancing interconnect and computation in a reconfigurable computing array. *Proc. ACM/SIGDA 7th Int. Symp. Field Program Gate Arrays*, pp. 1–10.

[13] *Virtex6 Datasheet* [Online]. Available: http://www.xilinx.com/publications/prod_mktg/Virtex6_Product_Table.pdf

[14] Achard, S., Salvador, R., Whitcher, B., Suckling, J., and Bullmore, E. 2006. A resilient, low-frequency, small-world human brain functional network with highly connected association cortical hubs. *J. Neurosci.*, vol. 26, pp. 63–72.

IEEE Trans. on Biomedical Circuits and Systems, vol. 6, no. 3, pp. 246-256.

Physical Design for Debug: Insurance Policy for IC's

John Giacobbe
Intel Corporation
1900 Prairie City Rd
Folsom, CA 95630
(1) 916-377-4971
john.giacobbe@intel.com

ABSTRACT
Physical Design for Debug (PDFD) is a method to enable root cause and validation of design errata and process defects during silicon debug, failure analysis, and fault isolation. This keeps the backend portion of the product life cycle (Post Si to product qualification) on schedule. Specifically, Optical probers and Focused Ion Beam (FIB) systems rely on pre-placed PDFD features for navigation, logic and timing modifications, and signal probing. Typical PDFD features include bonus logic and sequentials, fiducial markers, mechanical probe access points, FIB access points designed into standard cells like clock buffers, and cut and connect options to enable spare circuits/features. Why should a Design or Automation engineering care about PDFD and post Si debug? Because this insurance policy consumes considerable resources required to build and validate library cells, develop insertion tools and flows, and simulate timing, performance, and power impact.

Categories and Subject Descriptors
B.6.2 [**LOGIC DESIGN**]: Reliability and Testing - *Redundant design, Testability.*

General Terms
Algorithms, Measurement, Documentation, Performance, Design, Reliability, Experimentation, Standardization, Verification.

Keywords
Design for Debug, Focused Ion Beam, Optical probe, Si debug, navigation, circuit edit.

A Top-Down Synthesis Methodology for Flow-Based Microfluidic Biochips Considering Valve-Switching Minimization

Kai-Han Tseng, Sheng-Chi You, Jhe-Yu Liou, Tsung-Yi Ho
Department of Computer Science and Information Engineering
National Cheng Kung University, Tainan, Taiwan

ABSTRACT

Designs of flow-based microfluidic biochips have emerged as a popular alternative for laboratory experiments because they replace conventional biochemical paradigms on a chip. As the applications are becoming more complicated, a flow-based microfluidic biochip requires more valves to manipulate the sample flow for the large-scale and concurrent experiments. However, current synthesis methodologies still use full-custom and bottom-up procedures to synthesize a biochip. These manual steps are time consuming and would lead to dispensable valve-switching. According to recent studies, frequent switching of the valves may reduce the reliability. To minimize the valve-switching activities, we propose a top-down synthesis methodology for flow-based microfluidic biochip. We develop a *set-based minimum cost maximum flow* (SMCMF) resource binding algorithm and an *incremental cluster expansion* (ICE) placement algorithm in architecture-level and physical-level synthesis, respectively. The experimental results show that our methodology not only makes significant reduction of valve-switching amount but also diminishes the application completion time for both real-life applications and a set of synthetic benchmarks.

Categories and Subject Descriptors

B.7.2 [**Integrated Circuits**]: Design Aids - Layout, Place and Route

Keywords

Biochip, microfluidics, synthesis, valve minimization

1. INTRODUCTION

Recently, microfluidics-based biochips are receiving much attention and becoming more practical for the biochemical experiments. Unlike conventional biochemical analyzers, which consist of fluid-handling robots, microfluidics-based biochips integrate different biochemical functionalities on a chip [5], offering a number of benefits such as high portability, high throughput, high sensitivity, less human intervention and low sample volume consumption [2]. These composite microsystems are also known as lab-on-a-chip, replacing cumbersome equipments to miniaturized and integrated systems. Several applications are already demonstrated on microfluidics-based platforms (e.g., point-of-care diagnosis, environmental monitoring, etc) which pave a generic and consistent way for miniaturization, integration, automation and parallelization of biochemical processes [4].

Different from droplet-based microfluidic biochips, flow-based microfluidic biochips adopt valves to manipulate the fluids continuously on the pre-defined micro-channels [5]. Since the valve is the most basic unit of flow-based microfluidic biochips, by combining several valves, more complex units such as mixer, reactor, separator, detector, filter can be built [3]. Therefore, the mainstream for microfluidics-based system now is flow-based microfluidic biochip.

However, as the requirements and the design complexity rapidly increase, the manufacture and the biochemical analysis of flow-based microfluidic biochip become more complicated. According to recent study [8], the biochips can now use more than 25,000 valves and about a million features to run 9,216 parallel polymerase chain reactions. Moreover, the number of mechanical valves per square inch for flow-based microfluidic biochips has grown exponentially and four times faster than the reflection of Moore's Law [1].

Although the scale of flow-based microfluidic biochips are enlarging and the amount of the valves fabricated on a chip also growing significantly, computer-aided design (CAD) tools are still in their infancy today. Designers still use bottom-up and full-custom design approaches to manually adjust the components and the interconnection to satisfy the steps of desired biochemical applications [7]. As a result, the development of explicit design rules and strategies allowing modular top-down synthesis methodologies are needed, in order to provide the same level of CAD support for the biochip designers as the one that are currently done for the semiconductor industry [5].

In addition, to the best of our knowledge, the current synthesis methodologies for flow-based microfluidic biochips ignore the issue of valve-switching. Because the valves play the most fundamental role on flow-based microfluidic biochip, a biochemical application running on an inappropriately biochip would result in dispensable switching of the valves. It should be averted to have such redundant activities because a valve operates reliably only for a few thousands of actuations [7]. Therefore, a top-down synthesis methodology for flow-based microfluidic biochips considering valve-switching minimization is necessary for maintaining the reliability.

1.1 Our Contribution

In this paper, we propose the first top-down synthesis methodology considering valve-switching minimization for flow-based microfluidic biochips. We separate the synthesis procedures into three steps including architecture-level synthesis, physical-level synthesis and explicit scheduling. Our contributions can be summarized as follows:

- We model the valve-switching procedures for functional components such as mixer and storage, and identify how to reduce the valve-switching amount by binding the continuous operations to the same component.

- Unlike traditional methodology, our top-down synthesis methodology minimizes the valve-switching amount and the application completion time in a global view.

- In architecture-level synthesis, we develop a set-based minimum cost maximum flow (SMCMF) resource binding algorithm to minimize the valve-switching activities and application completion time and maximize the component parallelization.

- In physical-level synthesis, we develop an incremental cluster expansion (ICE) placement algorithm to separate the components into different clusters to simplify the biochip architecture and applying Dijkstra shortest path algorithm for routing. It can reduce intersections and the total length of flow-channels, i.e., total valve-switching amount and application completion time.

The experimental results show that our top-down synthesis methodology not only makes significant reduction of valve-switching activities but also diminishes the application completion time for both real-life applications and a set of synthetic benchmarks.

The remainder of this paper is organized as follows: Section 2 introduces the system model of the flow-based microfluidic biochips and the manipulations of the valves. Section 3 formulates the biochip synthesis problem. Section 4 introduces our top-down synthesis methodology. Section 5 - 6 show our experimental results and conclusions.

2. PRELIMINARY

2.1 Biochemical Application Model

The biochemical application is modelled as a sequential graph which is directed, acyclic [6]. There are many vertices in the graph, the one having no predecessors is the source, another one having no successors is the sink and the others represent the operations $O_i \in O$ containing the operation types and the execution time. For the directed edges between each operation O_i and O_j indicate that the the output of O_i is the input of O_j. An operation can't be executed until all the inputs are arrived. Fig. 1(a) shows an example of an application containing 13 mixing operations.

2.2 Biochip Architecture Model

The flow-based microfluidic biochips are manufactured using multilayer soft lithography. Basically, the biochip can contain multiple layers which can be logically divided into two layer including the flow layer and the control layer. [5].

We construct the biochip architecture with a set of vertex, edge, and flow path, and the schematic view of it is

shown in Fig. 1(d). A vertex can be either an intersection(e.g., P_1) or a component with input/output ports (e.g., $Mixer_1$ accompanied with In_1 and Out_1). An edge represents a flow-channel from a vertex to another vertex and the transportation between two vertices can be seen as a set of permissible routes on the flow-channels. The transportation time among each edge is defined as T, while T can be evaluated after the biochip architecture is determined. For example, in Fig. 1(d), one of the flow paths between $Mixer_2$ and $Storage_1$ can be combined by the edges of $Mixer_2 \rightarrow P_1$ and $P_1 \rightarrow Storage_1$.

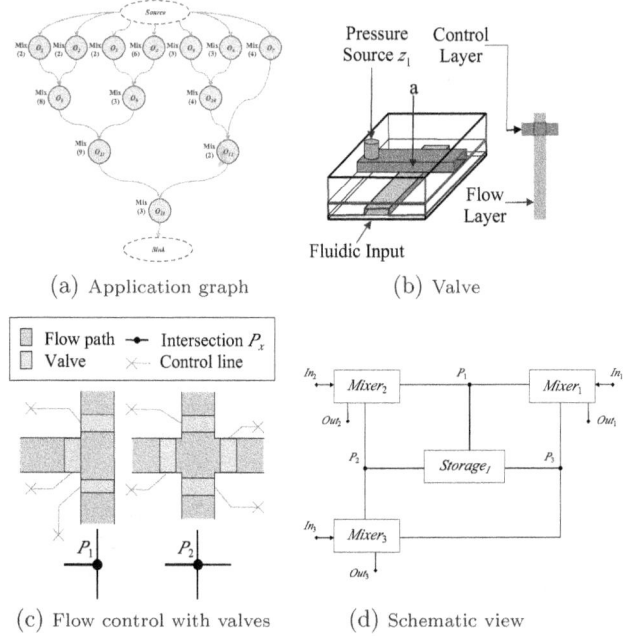

(a) Application graph   (b) Valve

(c) Flow control with valves   (d) Schematic view

Figure 1: Elements of microfluidic biochip

2.2.1 Micro-valve Model

The most basic unit of the flow-based biochip is a valve, which is used to control the fluid in the flow layer by switching the valves [5]. In Fig. 1(b), the control layer (red) is connected to an external air pressure source z_1. The flow layer (blue) is connected to a fluid reservoir through a pump. Without the activation of pressure source, the fluid can flow through freely. Otherwise, high pressure through the control layer pinches the underlying flow layer (i.e., a in Fig. 1(b)) and blocks the fluid flow. Current technology can fabricate small size of valves (100 x 100 μm^2) [6]. By combining these valves, more complex units, such as switches, multiplexers, mixers, can be built. Fig. 1(c) shows two cases that flow-channels intersect with each other. 3 and 4 valves are required for the intersections P_1 and P_2 respectively [5].

2.2.2 Component Model

There are many different components of flow-based microfluidic biochip. In this paper, we will take the storage and mixer as examples to compute the valve-switching activities because both of them are the most common used and important components in the biochemical experiments.

A storage acts as a buffer, which is used to temporarily save the samples. It is very common used because the given number of components is not enough, we should save the

TABLE I: Valve-switching for a mixing operation

Phase	V_1	V_2	V_3	V_4	V_5	V_6	V_7	V_8	V_9
In_1	0	0	1	0	0	0	0	0	1
In_2	0	1	0	0	0	0	1	0	0
Mix	1	0	0	x	x	x	0	1	0
Out_1	0	0	1	0	0	0	0	0	1
Out_2	0	1	0	0	0	0	1	0	0

output of an operation in the buffer until the component is idle. A storage contains many cells which are addressed using a microfluidic multiplexer, which is a powerful tool that addresses large number of valves with a small number of connections from the chip to the outside world [5]. A storage having N cells can be controlled by $2 * log_2 N$ control channels. Once a transportation into/from a storage occurred, $2 * log_2 N$ valve-switching are needed.

Fig. 2 shows a pneumatic mixer [5]. It contains 9 valves, each valve has a control line connecting with the control layer. Because it is very common to mix some samples together in biochemical experiments, the mixer becomes the most important component for flow-based microfluidic biochip. The input and the output controllers can be formed by the valves v_1, v_2, v_3 and v_7, v_8, v_9. The valves v_4, v_5, v_6 act as an on-chip pump to perform mixing activities.

2.3 Valve-switching Minimization

According to recent studies, a valve can only be actuated in thousands of times [7]. Because the flow-based microfluidic biochips are manipulated by the valves, more valve-switching would reduce the reliability.

In order to minimize the valve-switching amount, in this paper, we take the mixer as an example to explain how to reduce the valve-switching activities by grouping the continuous mixing operations together. In Fig. 2, there are three samples A, B, and C, where C is the mixing result of A and B. As shown in Table I, a mixing operation typically contain 5 phases and the states of the valves in each phase are represented by 9 boolean numbers. Here, the number "0" represents the valve is opened while the number "1" represents the valve is closed. The special states "x" on 3rd phase for the valves v_4, v_5, v_6 represent the mixing activities of the micro-pump. The first two phases called In_1 and In_2 store sample A and B into the mixer by closing (v_3, v_9) and (v_2, v_7), respectively (Fig. 2(b) and (c)). The 3rd phase mixes two stored samples together into sample C by closing (v_1, v_8) and switching the valves in the order of (v_4, v_5, v_6) repeatedly which acts as an on-chip micro-pump (Fig. 2(d)). The last two phases, Out_1 and Out_2, are used to output the mixing sample C. Out_1 outputs half of C as the waste by closing (v_3, v_9), and Out_2 outputs the other half of C as the mixing result by closing (v_2, v_7), which would be transported to other component to become the input for the further processing (Fig. 2(e) and (f))[6].

Therefore, 20 times of valve-switching $(0 \rightarrow 1$ or $1 \rightarrow 0)$ are needed for a mixing operation.

2.3.1 Group the Operations Continuously

Assume that there are two mixing operations and the output of the first operation can become one of the inputs for the second operation. Considering Fig. 2(f) again, if the output of the mixing result could be stayed in the mixer, then it can become one of the input for the second operation without transportation. This situation can

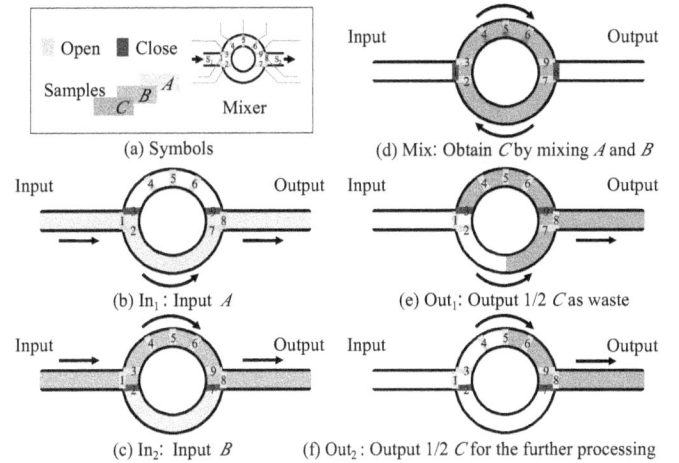

(a) Symbols
(d) Mix: Obtain C by mixing A and B
(b) In_1: Input A
(e) Out_1: Output 1/2 C as waste
(c) In_2: Input B
(f) Out_2: Output 1/2 C for the further processing

Figure 2: Illustration for a mixer operation

be illustrated in more detail in Fig. 3. The output of $Operation_1$ can become one of the inputs for $Operation_2$. So, it will reduce 12 times of valve-switching. Similarly, the output of $Operation_2$ can become the one of the inputs for $Operation_3$. So, it will totally reduce 24 times of valve-switching if we bind these 3 continuous operations to the same component. Besides, Binding continuous mixing operations to the same component not only reduces the valve-switching of the mixer but also removes the unnecessary transportation cost. The transportation cost includes the sample flowing time and valve-switching which is used to control the intersections on the flow path. Moreover, if the sample is transported to the storage, the valve-switching amount would be even more since we should consider the valve-switching on the storage which takes $2 * log_2 N$ times of valve-switching. Therefore, binding the operations continuously to the same components can have a significant reduction on valve-switching.

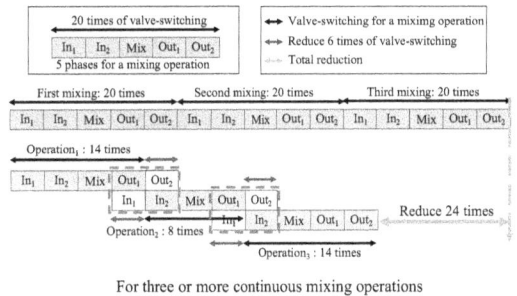

For three or more continuous mixing operations

Figure 3: Continuous valve-switching on a mixer

3. PROBLEM FORMULATION

Based on previous discussions, the problem addressed in this paper can be formulated as follows:

Input: *A biochemical application modelled as a sequencing graph and the component library including different types of components.*

Constraints:

1. Component constraints: *The allocated components cannot exceed the maximum allowable numbers.*
2. Design rules: *The size and distance limit for flow-channels and components.*

Objective: *Synthesizing a flow-based microfluidic biochip architecture. Under the above constraints, the total valve-switching amount and the application completion time for the biochemical application are minimized*

4. SYNTHESIS METHODOLOGY

To make an efficient synthesis procedures, our top-down synthesis methodology can be hierarchically divided into three stages including architecture-level synthesis, physical-level synthesis and explicit scheduling.

4.1 Architecture-Level Synthesis

The inputs for architecture-level synthesis are the application graph, component library and the component constraints. The outputs are the resource binding and scheduling results. In this paper, we develop a set-based minimum cost maximum flow (SMCMF) algorithm in architecture-level synthesis. The term set-based here means that we will firstly group the continuous operations together and then use minimum cost maximum flow algorithm to maximize the component parallelization and minimize the application completion time.

In this subsection, we will use the application graph shown in Fig. 1(a) under the component constraint of 3 mixers as an example to present our SMCMF algorithm. The SMCMF algorithm can be hierarchically separated into three steps. In the first step, we apply a depth-first search (DFS) technique to compute the estimated execution time of each operation and then group the continuous operations into continuous sets $S_i \in S$. In the second step, we will construct a flow network $G = (V, E)$ for S where V represents a set of vertices and E represents a set of directed edges in G. Then, we will apply minimum cost maximum flow algorithm on G to obtain the concatenated-sets $CS_i \in CS$. In the last step, we will first sort CS by the priorities and split low-priority CS into single operations, and then insert them into the high-priority CS.

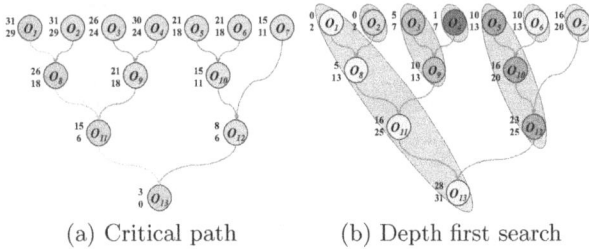

(a) Critical path   (b) Depth first search

Figure 4: Separate operation strategy

4.1.1 Separating Continuous Operation by DFS

Our SMCMF algorithm can be started by applying a depth-first search (DFS) on the application graph from the last operation to find the critical path of the estimated execution time. In Fig. 1(a), the last operation is O_{13}, so we will start the DFS on it from 0 second. Because the execution time for O_{13} is 3 seconds, after searching the O_{13}, we can get the execution time slice for it from 0-3 seconds. Then, the DFS will search to O_{11} in this case. Because in the real cases, we should consider the transportation time between two operations and the time for the input and output. For example, in Fig. 1(a), the output for O_{11} will be transported to become

one of the inputs for O_{13} which takes some transportation time, and we should also waiting for the other input for O_{13}, too. So, in order to make our estimated execution time more practical, we should consider the transportation between each operation. However, the biochip architecture is not yet determined, we can't obtain the exact T now. So, we will use a constant time to estimate the transportation time in this stage. In this case, we set the transportation time T to 3 as an example to elaborate our algorithm.

After the DFS, we can find the time slice for each operation and find a critical path $(O_1, O_8, O_{11}, O_{13})$ having a length of 31 in this case, which is shown in Fig. 4(a). Because we use DFS to obtain the time slice from the last operation and search back to the source vertex, the time slices are now inverse. So we will use the length of the critical path to inverse the time slice for each operation. After that, we can obtain the estimated execution time for each operation as shown in in Fig. 4(b). For each operation O_i, it will obtain an estimated starting time st and an estimated ending time ed, which is represented as $O_i.st$ and $O_i.ed$ respectively. These estimated execution time will be used later to build the flow network G is the next stage.

As depicted in Section 2.3, binding continuous operations to the same component can minimize the valve-switching activities. Therefore, after doing DFS, we can also group the continuous operations together to obtain the continuous set S as shown in 4(b). So, we can obtain 7 continuous set including $S_1\{O_1, O_8, O_{11}, O_{13}\}$, $S_2\{O_2\}$, $S_3\{O_3, O_9\}$, $S_4\{O_4\}$, $S_5\{O_5, O_{10}, O_{12}\}$, $S_6\{O_6\}$, $S_7\{O_7\}$. Each $S_i \in S$ will inherit the estimated starting time $S_i.st$ from the first operation and the estimated ending time $S_i.ed$ from the last operation.

4.1.2 Minimum Cost Maximum Flow Formulation

Though we already grouped operations together, the given number of mixers is still insufficient. Nevertheless, the execution time for some continuous sets are disjointed, so we can concatenate those sets together. Therefore, we build a flow network for the continuous sets and apply minimum cost maximum flow to maximize on it to maximize the component parallelization and minimize the application completion time.

The rules for construct the flow network $G = (V, E)$ can be listed into 5 steps as follows:

Step #1: Create the vertex
1. Create a source and destination vertex Src and Dst.
2. For each $S_i \in S$, create a corresponding vertex V^i and inherit the estimated starting time and ending time from S_i to get $V^i.st$ and $V^i.ed$.

Step #2: Split V into two vertices
1. For each vertex V^i, split V^i into an in-vertex V^i_{in} and an out-vertex V^i_{out}. Both of V^i_{in} and V^i_{out} inherit the estimated starting time and ending time from V^i represented by $(V^i_{in}.st, V^i_{in}.ed)$ and $(V^i_{out}.st, V^i_{out}.ed)$.
2. Create a directed edge for each vertex pair $V^i_{in} \rightarrow V^i_{out}$ with one unit capacity and *zero* cost per unit flow.

Step #3: Split V into two vertices
1. Split the destination vertex Dst into an in-vertex Dst_{in} and an out-vertex Dst_{out}.
2. Create a directed edge $Dst_{in} \rightarrow Dst_{out}$ with the capacity equal to the given component numbers and *zero* cost per unit flow.

Step #4: Create the edges

1. For each in-vertex V_{in}^i, create a directed edge $Src \rightarrow V_{in}^i$ with *one* unit capacity and *zero* cost per unit flow.
2. For each out-vertex V_{out}^i, create a directed edge $V_{out}^i \rightarrow Dst.in$ with *one* unit capacity and *zero* cost per unit flow.
3. For an out-vertex V_{out}^i and an in-vertex V_{in}^j if $V_{out}^i.ed + T \leq V_{in}^j.st$, create a directed edge $V_{out}^i \rightarrow V_{in}^j$ with *one* unit capacity and $f(x)$ cost per unit flow, where T is the transportation time, x is defined by $| V_{out}^i - V_{in}^j |$ and f is the cost, which is proportional to x.

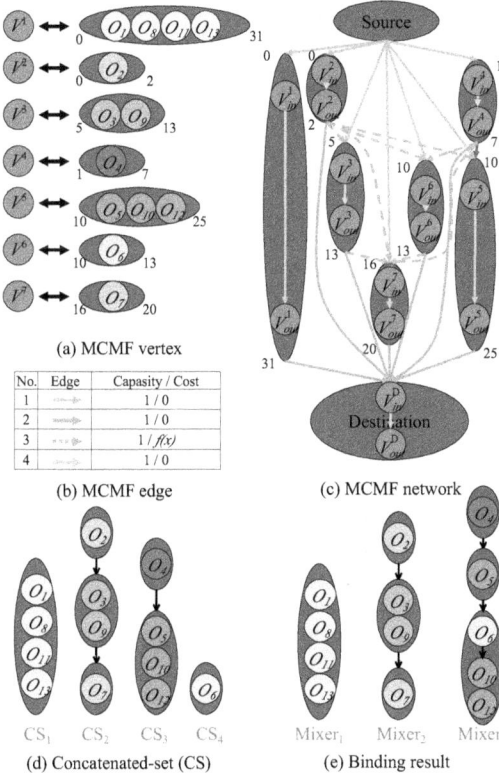

(a) MCMF vertex

No.	Edge	Capacity / Cost
1	⟶	1 / 0
2	⟶	1 / 0
3	⟶	1 / $f(x)$
4	⟶	1 / 0

(b) MCMF edge

(c) MCMF network

(d) Concatenated-set (CS)

(e) Binding result

Figure 5: Set-based minimum cost maximum flow

Fig. 5(a) shows the corresponding operations for each vertex in G and Fig. 5(b) shows the cost and capacity for each kind of edge.

In the first three steps, we will create the vertices and split them into two vertices except the source vertex. Each vertex will inherit the execution time computed by DFS in the first stage. The reason for split each V^i into two vertices is that the flow paths in our SMCMF algorithm represents the components. Those continuous sets passed through by a flow path means that they should be bound to this component. In order to avoid a set passing by two flow paths, we separate the V^i into two vertices having a unit capacity to make sure there is no conflict in the graph. For the destination vertex, we should also split it into two vertices having the unit capacity the same as given component numbers, which is also refer to the maximum parallelization that we can do in the same time.

For the last step, we will build the edges between source and each V^i and also build the edges between each V^i to the destination. Then, we will build the edges between two

V^i if their time difference is more than the transportation time.

In the end, we construct the flow network shown in Fig. 5(c) and apply minimum cost maximum flow algorithm on the graph to obtain some 3 flow paths on the graph and find that V^5 is isolated. For those continuous set concatenated by the flow path or isolated in the flow network, both them we call concatenated-sets CS. So, we can finally find 4 concatenated sets as shown in Fig. 5(d).

4.1.3 Component Allocation

Nevertheless, the given number of mixers is still insufficient since there are only 3 mixers but 4 concatenated-sets. So, the goal in this step is to allocate the finite components to the high-priority CS and then split the low-priority CS into single operations to insert the other CS. Here, we will first compare the continuous operation numbers and adopt the total operation numbers as the tight breaker. In Fig. 5(d), CS_1 has the highest priority while CS_4 is the lowest. After assigning the components to the high-priority CS, we can find that CS_4 has the low-priority, and the operation O_6 in CS_4 should be inserted into the other CS. Because we want to minimize the valve-switching activities, we can insert O_6 before it's next operation O_{10} to maintain the continuous operation numbers, as shown in Fig. 5(e). Finally, we obtain the resource binding and scheduling results.

4.2 Physical-Level Synthesis

The inputs for physical-level synthesis are resource binding and scheduling results from architecture-level synthesis and design rules which include the size and distance limit for the flow-channels and the components. The objective for physical level synthesis is to obtain a biochip architecture such that the valve-switching amount and the application completion time for the given application are minimized. The procedures for physical level synthesis can be divided into two steps including placement and routing. Since more intersections relate to more valve-switching on the valves and the longer flow-channels means more transportation time for the application, we will try to minimize the intersection numbers and the total length for the flow-channels in physical-level synthesis.

In order to simplify the biochip architecture, we propose an incremental cluster expansion (ICE) algorithm for the placement step. The idea of ICE algorithm is similar to internet structure which participates the end users into different regions, and connect the regions by the routers. By doing this way, the connection complexity can be extremely reduced. This idea can be easily porting to physical-level synthesis. We will firstly separate the components into different clusters depending on the given number of storages, and then build the interconnections to connect the storages together. For those transportations in a cluster, it will be transported by either a dedicated flow-channels or the storage in the cluster, and for those transportations between different clusters, it will be transported by the interconnections between storages.

4.2.1 Relational Graph

In order to obtain the transportations between different clusters, we will first build a relational graph to identify the transportations for the components. The relational graph can be obtained by using the information from resource binding result and the application graph. For example, in Fig. 5(e), we know that O_1 is bound to $Mixer_1$ and O_2 is bound to $Mixer_2$. So, there is a transportation from $Mixer_1$ to $Mixer_2$ since that O_2 is the next operation of O_1 in Fig. 1(a). By applying this rule, we can obtain a relational graph which shows the transportation times for the components.

4.2.2 Incremental Cluster Expansion Algorithm

The ICE algorithm can be separated into two phases including the clustering phase and the placement phase. we will find the cluster in clustering phase and then place the components on the biochip in the placement phase. The details for each phase can be listed as follows:

Clustering phase:

1. Each component is regarded as an unique cluster $C_i \in C$ initially.

2. In each iteration, each cluster will try to combine with other cluster having the biggest transportations, which is represent as $MaxT(C_i)$ in the relational graph. However, if $MaxT(C_i)$ is already combined to other cluster, it will not be allowed to be combined with until the next iteration.

3. The clustering will be terminated until the cluster numbers are the same as the storage numbers.

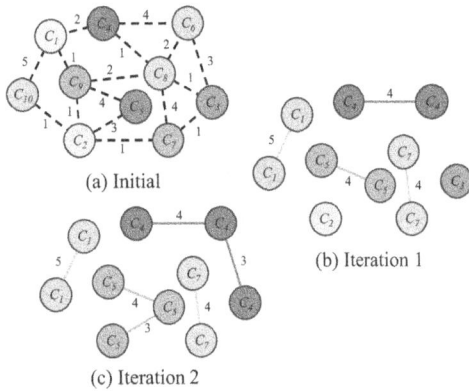

(a) Initial

(b) Iteration 1

(c) Iteration 2

Figure 6: Clustering

Since the relational graph for the given application is too simple, we will use another relational graph which contain 10 Mixers to illustrate our ICE algorithm in the clustering phase, and the given storages is equal to 4 in this case. Fig. 6 shows the steps for seeding the cluster in the relational graph. The edges in the graph represent the corresponding edges for the components in the relational graph. Here, the solid edges indicate the selected edges while the dotted edges represent the non-selected edges. Fig. 6(a) shows the initial state of each cluster C_i with the non-selected edges in the relational graph. All of the clusters are regarded as independent clusters. In the first iteration, we find $MaxT(C_9)$ is C_5, so we can combine this two clusters together. However, though we find $MaxT(C_2)$ is also C_5, we can't combine them together since C_5 is already combined to C_9. So, we can find 6 clusters in this iteration. By applying the same rules, we can find 4 clusters in the end.

Placement phase:

1. Create a platform that is used to temporarily place the components, and the size of the platform is bigger than the given chip size to make sure each cluster has the enough space to expand.

2. Place the storages on the platform circularly having a diameter of half of the platform to make sure the space between each cluster is sufficient.

3. For each cluster, it will circularly expand to find the space which meets the design rules for placing all the components belong to the cluster on the platform to minimize the space of local placement.

4. The clusters will compress by the center of the platform to minimize the global placement.

5. Find a bounding box on the platform that can include all the components.

6. Place the components on the biochip with the corresponding place in the bounding box.

The details of the placement phase are shown above. The first two steps place the center of the clusters on the platform and also make sure there is enough space for each cluster to expand. The third step place the component circularly surrounding with the storages to minimize the placement region in each cluster, and the forth steps is applied to make a global compression of all the clusters. By doing steps 3 and 4, we can obtain a smaller placement region. In the final two steps, we will place the components on the chip to get the placement result and then go to the routing step to create the flow-channels as the interconnections.

4.2.3 Routing Constraints

We apply the Dijkstra shortest path algorithm for building the flow-channel between components. Here, we define the costs in different situations and apply Dijkstra algorithm to find out a lowest cost flow-channel. In the worst case, the Dijkstra's algorithm might need to search for the whole chip size. By applying our cluster-based method, we only need to search for the chip area inside the cluster, which reduces the execution time of the algorithm.

As described before, our cluster-based method is similar to the internet structure. So, we only build the flow-channels for those transportations in the same clusters and build the flow-channels between storages. After the Dijkstra algorithm, we can determine the biochip architecture. Finally, the physical-level synthesis is finished.

4.3 EXPLICIT SCHEDULING

After the resource binding and scheduling results and architecture are obtained, the next step is to obtain the explicit scheduling for the application graph. Since the biochip architecture is now explicit, we can use the real lengths of the flow-channels to calculate the exact T for each transportation. The problem of explicit scheduling is related to resource constrained scheduling with non-uniform weights, which is NP-complete [9]. We modify a list scheduling-based approach to solve the problem in a computationally efficient manner [6]. Since that we already have the sequences for the scheduling, the thing that we should do here is to decide the transportations for the operations. However, a flow-channel can only allow a transportation at the same time. Once the collision occurred, one of the transportation have to wait until the other is finished. After this step, we can obtain the valve-switching amount and the application completion time for the given application graph.

128

TABLE II: PCR real-life application with SMCMF algorithm

Arc. No.	Allocated Units	Total Valves	Valve-Switching On Flow / Op.	Total Wait / Time
1	(3,3,3,0,0,1)	54	40 / 104	5.4 / 20.1
2	(4,4,4,0,0,1)	72	14 / 125	3.1 / 17.0

TABLE III: PCR real-life application with baseline method

Arc. No.	Allocated Units	Total Valves	Valve-Switching On Flow / Op.	Total Wait / Time
1	(3,3,3,0,0,1)	54	40 / 116	15.6 / 20.9
2	(4,4,4,0,0,1)	72	18 / 137	9.0 / 18.3

5. EXPERIMENTAL RESULTS

We evaluate both real-life cases and a set of synthetic benchmarks by using our SMCMF algorithm for binding, then construct the architecture by ICE algorithm and finally schedule these applications to determine the valve-switching amount and application completion time. These algorithms are implemented in C++, running on a PC with Core2 Quad processors at 2.66GHz and 3.25GB of RAM.

Table II and III show our experimental results on polymerase chain reaction (PCR) real-life case, which contains 7 mixing operations. We implement both SMCMF algorithm and baseline method to construct the binding information for different architectures. The baseline method is modified by a state-of-the-art scheduling approaches for flow-based microfluidic biochip in [6], which calculates the urgency criteria by applying a topological sort and bind the most urgent operation to the component at each moment.

Column 1 presents the number of the architectures and column 2 shows the list of allocated components, in the format of (Input ports, Output ports, Mixers, Heaters, Filters, Storage). Columns 3 shows the total number of valves on the chip. Columns 4 presents the valve-switching amount on flow-channels and components. Columns 5 shows the total idle time and the application completion time. These two tables show that our SMCMF algorithm obtains less valve-switching amount and shorter total idle time/application completion time than the baseline method.

The following four line charts are experimental results on synthetic benchmarks under different considerations. Fig. 7 shows the distribution of different number of mixers and one storage for valve-switching amount and application completion time. We use a synthetic application containing 1023 mixing operations as a benchmark and two binding information made by SMCMF algorithm and baseline method. It shows 43% and 54% improvements for valve-switching amount and application completion time on average. When an architecture contains more mixers, the completion time and valve-switching amount will increase because only one storage in these architectures. However, the increasing rate of SMCMF algorithm is still smaller and more stable than the baseline method.

Fig. 8 shows the distribution of different number of storages and 100 mixers for valve-switching amount and application completion time. We also use the benchmark of 1023 mixing operations and two binding information from SMCMF and baseline method. When an architecture contains more storages, the completion time and valve-switching amount will both decrease because more synchronism could be executed on the architecture. Moreover, when comparing SMCMF algorithm with the baseline method, it shows 40% and 64% improvements on valve-switching amount and application completion time on average.

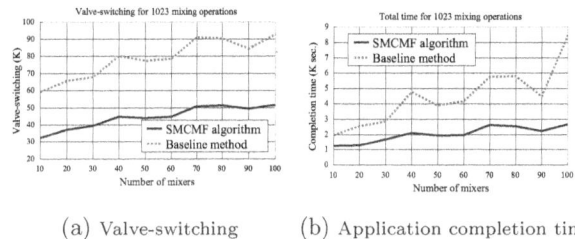

(a) Valve-switching (b) Application completion time

Figure 7: The results on different number of mixers

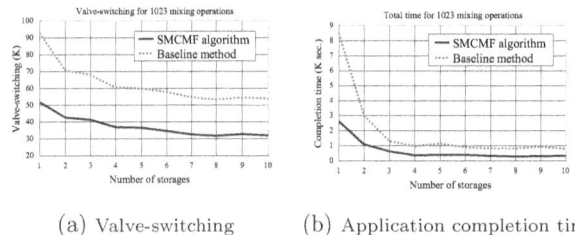

(a) Valve-switching (b) Application completion time

Figure 8: The results on different number of storages

6. CONCLUSIONS

In this paper, we proposed a top-down synthesis methodology for flow-based microfluidic biochips. Given a application graph, we used SMCMF and ICE algorithms to get the resource binding and scheduling results and the biochip architecture. By using a list-scheduling based approach to evaluate the results, we can find that our our top-down synthesis methodology not only minimized the valve-switching amount but also diminished the application completion for both real-life case PCR and a set of synthetic benchmarks. The experimental results also showed that our cluster-based ICE algorithm simplified the biochip architecture significantly to met the objective. In addition, to the best of our knowledge, this was the first synthesis methodology considering the issue of valve-switching and application completion time simultaneously.

7. REFERENCES

[1] J. W. Hong and S. R. Quake, "Integrated nanoliter systems," *Nature Biotechnology*, 21:1179–1183, 2003.

[2] T.-W. Huang, T.-Y. Ho, and K. Chakrabarty, "Reliability-oriented broadcast electrode-addressing for pin-constrained digital microfluidic biochips," *IEEE/ACM ICCAD*, pp. 448–455, 2011.

[3] Y. C. Lim, A. Z. Kouzani, and W. Duan. Lab-on-a-chip: a component view. *Journal of microsystems technology*, 16(12), December 2010.

[4] D. Mark, S. Haeberle, G. Roth, F. von Stetten, and R. Zengerle. Microfluidic lab-on-a-chip platforms: requirements, characteristics and applications. *Chem. Soc. Rev.*, 39:1153–1182, 2010.

[5] J. Melin and S. Quake, "Microfluidic large-scale integration: The evolution of design rules for biological automation," *Annual Reviews in Biophysics and Biomolecular Structure*, 36:213–231, 2007.

[6] W. H. Minhass, P. Pop, and J. Madsen, "System-level modeling and synthesis of flow-based microfluidic biochips," in *IEEE/ACM CASES*, 2011.

[7] W. H. Minhass, P. Pop, J. Madsen, M. Hemmingsen and M. Dufva, "System-Level Modeling and Simulation of the Cell Culture Microfluidic Biochip ProCell," *IEEE DTIP*, pp 91–98, 2010.

[8] J. M. Perkel, "Microfluidics - bringing new things to life science," *Science*, November 2008.

[9] D. Ullman, "NP-complete scheduling problems," *Journal of Computing System Science*, no. 10, pp. 384–393, 1975.

Designing VeSFET-based ICs
with CMOS-oriented EDA Infrastructure

Xiang Qiu, Malgorzata Marek-Sadowska
ECE Department
University of California, Santa Barbara
Santa Barbara, CA, 93106
{xqiu,mms}@ece.ucsb.edu

Wojciech Maly
ECE Department
Carnegie Mellon University
Pittsburgh, PA, 15213
maly@ece.cmu.edu

ABSTRACT
Medium volume VeSFET-based ASICs can fill the gap between high cost microprocessors and low performance FPGAs. Circuits can be customized onto pre-manufactured VeSFET canvases by properly designed interconnects. In this paper, we propose *chain canvases,* a family of VeSFET canvases for which CMOS-oriented EDA tools can be easily adapted. Footprint area, wire length, via usage, performance and power are compared between *chain canvas-* and *basic canvas-*mapped benchmarks. Experimental results show that chain canvas-mapped circuits outperform those mapped to basic canvases.

Categories and Subject Descriptors
B.7.2 [**Integrated Circuits**]: Design Aids–*Layout*

General Terms
Design

Keywords
VeSFET; regular layout; canvas; EDA infrastructure;

1. Introduction
During the last three decades, an exponential growth in circuit complexity and speed were both possible through the continuous downscaling of device and wire geometries. Now, conventional lithography-based CMOS technology faces serious challenges at both manufacturing and design levels. The steep cost increases at each new technology node and the existing design and manufacturing models favor monopolization of both fabrication resources and design innovation. This leads to market domination by two kinds of devices: FPGAs and microprocessors. Their costs make them unfriendly to medium volume products.

In [1, 2], the new VeSFET-based design and manufacturing paradigm was proposed. Its objective was to increase the density of manufactured transistors at a hugely reduced cost. In this paper, we assess the VeSFET paradigm in the context of the effort needed to build the design automation infrastructure necessary for creating useful circuits. We ask: Is it possible to increase the useful transistor density without a big investment in design infrastructure?

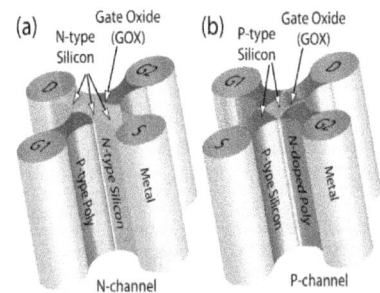

Figure 1. Vertical Slit Field Effect Transistor

We demonstrate that indeed, it is possible to achieve high density of useful transistors without a big investment into EDA tools. We accomplish this by properly selecting the topology of contacted and isolated transistors that are pre-manufactured on the chip. The proposed transistor array topology referred to as *chain canvas* can be used for arbitrary designs and customized only by interconnects.

In Section 2, we provide a brief summary of the VeSFET-based IC paradigm. In Section 3, we introduce *chain canvases (CCs).* In Section 4, we discuss standard cell design for *CC*. In Section 5, we study CMOS-like ASIC flow for *CC*. Section 6 concludes the paper.

2. VeSFET-based IC Paradigm
2.1 Vertical Slit FET (VeSFET)
The nucleus of the proposed paradigm is **Ve**rtical **S**lit **F**ield **E**ffect **T**ransistor (VeSFET). The VeSFET, shown in Fig. 1 [1, 2], is a twin-gate, junction-less transistor that can be fabricated using an SOI-like process. Its four terminals (drain, source and two gates) are implemented as metal pillars. A lightly doped vertical slit connects the drain and source. When voltages on two symmetrical MOS gates are greater than the threshold voltage, bulk current is carried by majority carriers flowing through the slit. VeSFET can be used in Tied-Gate (TG) mode with two gates shorted to form a traditional three-terminal transistor, or Independent-Gate (IG) mode with the gates individually connected such that either one can be used for threshold voltage control in a wide range of voltages. Finally, note that the manufactured *n*-type devices [3] show excellent electrical characteristics: I_{ON} = 20μA, I_{ON}/I_{OFF} ratio = 10^9, SS (subthreshold slope) = 65mV/decade, DIBL (drain-induced barrier lowering) = 13mV/V. VeSFET also has much smaller gate capacitance than a MOSFET and a very small leakage current.

The VeSFET transistor height h is equivalent of channel width of a MOSFET. VeSFET's electrical properties are determined by h, the gate oxide thickness t_{ox}, and the radius of a pillar r that also defines the minimum feature size. Note, that all transistors manufactured on one chip must have the same height. Therefore, transistor sizing is achieved by duplicating minimum size transistors and shorting electrically equivalent terminals with metal wires. The geometry of a VeSFET transistor is built on a basis of the minimum-feature-radius circle, which defines the dimensions of the terminals and the spacing between each pair of neighboring terminals. Such geometry enables highly regular mask patterns and has the potential to enable scaling to 22 nm and below [1, 2].

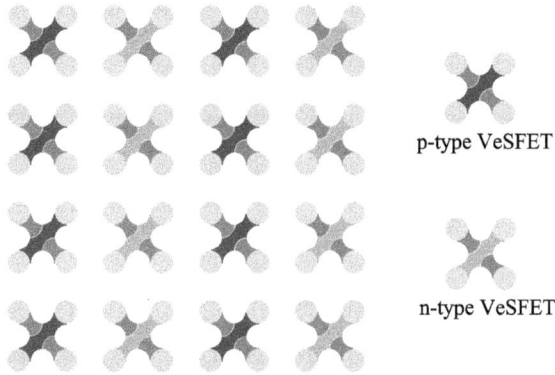

Figure 2. The topology of Basic Canvas

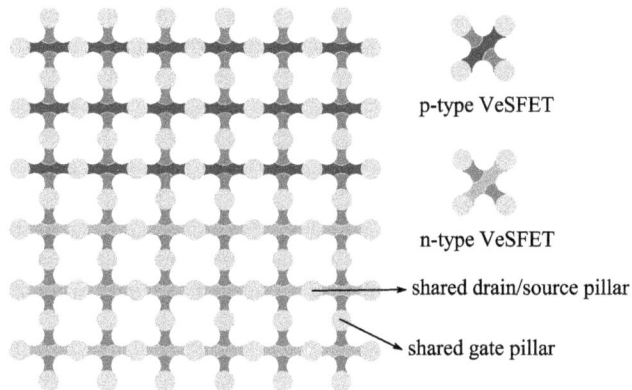

Figure 3. The *CC-3-3* canvas.

2.2 VeSFET Canvases

VeSFETs can be arranged as an array of geometrically identical devices and configured into useful circuits by customizing interconnects in an OPC-free metallization process [4]. Metal layers can be horizontal, vertical, 135-, and 45-degrees.

The transistors in an array may share terminals. The individual and terminal-shared transistors correspond to isolated and contacted transistors in conventional technolo-gies. *P*- and *n*-type transistors can be positioned anywhere within the array. Various topologies of transistor arrays are possible. We will refer to such different topology transistor arrays, with possibly some pre-manufactured interconnects, as *canvases* [1, 2]. In Fig. 2 we show *basic canvases* (*BCs*) in which all transistors are isolated. Member canvases of this family are defined by the specific positions of *p*- and *n*- transistors. Although wires on each metal layer must be strictly parallel and no wire bends on the same layer are allowed, results in [5] suggest that properly designed *BCs* can achieve very high transistor utilization.

Canvases with shared pillars may integrate more transistors than canvases in the *BC* family. We will demonstrate that careful selection of isolated and shared pillar transistor positions within the array can greatly simplify the design automation process.

3. Chain Canvas-based IC Paradigm

In this Section, we introduce a class of canvases—the *chain canvases* (*CCs*) in which transistors in rows share drain /source (*D/S*) terminals and in columns—gate (*G*) terminals. In canvases of *CC* family, each pillar is shared by two transistors. Thus the transistor density of *CCs* is two times greater than in *BCs*. We denote as *CC-x-y* a specific chain canvas with alternating *x* rows of *p*- and *y* rows of *n*- transistors. In Fig. 3, we show a specific canvas in the *CC* family, the *CC-3-3* in which *p*- and *n*- transistors are arranged in rows of three devices tall. *CCs* are naturally compatible to the well-established CMOS-oriented EDA infrastructure. We demonstrate that standard CMOS EDA tools can be used to map circuits in ASIC style to *CC* canvases. The performance and footprint of *CC*-mapped circuits are compared to their implementations in *BCs* [5].

CCs contain contacted transistors which sometimes may need to be disconnected. This can be accomplished by electrically disabling some transistors or by physically removing them. For example, in Fig. 4(a), if transistors *T1* and *T2* which are not to be connected are mapped as shown, the embedded transistor *Tp* will introduce undesired connectivity. *Electrical isolation* can be implemented in several ways. For example, we can either apply a negative gate bias to one of the twin gates to cut off the transistor (no matter what the voltage on the other gate is) as shown in Fig. 4(b), or short the drain and source terminals, as shown in Fig. 4(c). Electrical isolation usually disables more than one transistor at a specific location. It does not change the transistor array, but may introduce some area and performance overhead.

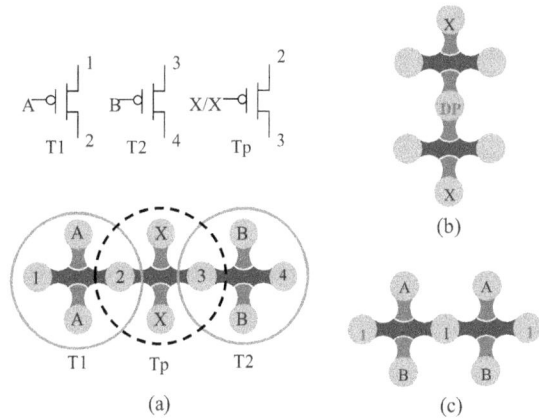

Figure 4. (a) Placing *T1* and *T2* adjacently is unfeasible because *Tp* introduces an undesirable connection (b) The two *p*- transistors are disabled by applying cutting-off voltage *DP* on their common gate; (c) The two *p*-transistors are disabled by assigning the same drain/source voltages that on node 1.

If the unwanted transistors can be physically removed from a canvas, the area overhead can be eliminated. We call this a *physical isolation*. Various configurations of physically removed transistors along with positions of *p*- and *n*- devices define different canvases. Essentially, electrical and physical isolation can be viewed as manufacturing cost versus area and performance trade-offs: with electrical isolation, the same pre-manufactured canvas can be used for various designs; but with physical isolation, different designs need different canvases.

4. Standard Cell Design
4.1 *CC*-based Standard Cell Generation

In CMOS technology, transistors of the same type and common *D/S* connections are usually placed adjacently with abutted diffusion regions to increase transistor density and save local wires. Layouts of static CMOS gates can be automatically generated by finding dual Euler paths in their pull-up and pull-down networks [6]. This strategy guarantees optimal footprint area by abutting *D/S* terminals and aligning corresponding gate terminals of PMOS and NMOS transistors.

VeSFET gates mapped to *CCs* and to CMOS may have similar layout patterns. Fig. 5(a) shows the stick diagram of an NAND2 gate in CMOS, and 5(b) shows its layout on *CC*. The VeSFET placement corresponds to the stick diagram. *D/S* pillar sharing is equivalent to diffusion abutment and *G* pillar sharing of *n*- and *p*-type VeSFETs is similar to the alignment of poly gates. In CMOS, if dual Euler paths cannot be found, more than one diffusion strip is needed and the footprint area increases. Similar situation happens in *CCs*. If two transistors placed adjacently in a row do not have a common *D/S*, electrical or physical isolation must be applied, as shown in Fig. 6, and the disabled or removed transistors increase the footprint.

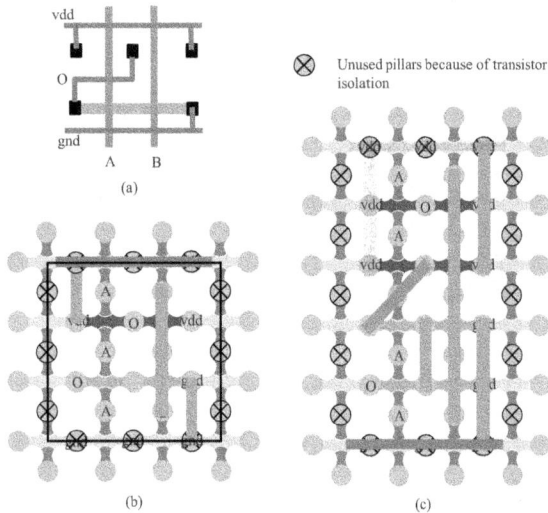

Figure 5. (a) Stick diagram of NAND2 layout in CMOS; (b) NAND2 with unit size transistors on *CC*; and (c) NAND2 with size 2X transistors on *CC*.

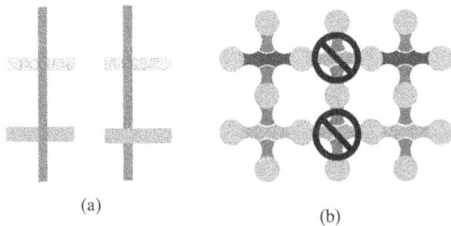

Figure 6. If dual Euler paths do not exist: (a) diffusion break is introduced in CMOS; (b) isolated transistors on *CC*.

Transistor sizing in CMOS can be achieved by adjusting the width of diffusion regions and the transistor placement does not change. On *CCs*, transistor sizing is achieved by using multiple transistors and maintaining the same transistor order. In Fig. 5(c), NAND2 with *2X* size transistors is shown. Slightly different from CMOS

layouts, metal wires are needed to short electrically equivalent but physically separated pillars. As shown in the figure, such wires can be routed straightforwardly without increasing routing complexity.

Therefore, just like CMOS, static logic gates on *CCs* can also be automatically generated with optimal footprint areas determined by the dual Euler path method.

4.2 Cell Level Comparison of *BC* and *CC*
4.2.1 Logic Cells

Lin *et. al.* described transistor level placement and routing methodologies for building standard cells on *BCs* with one-sided interconnect [5]. The footprint of those cells is more than 50% reduced compared to low power standard CMOS cells [7]. At the same time, the power of VeSFET cells is much lower than that of their CMOS counterparts with a comparable performance [7]. Experiments were performed for 65nm CMOS technology and VeSFETs with $r = 50nm$, $h = 200nm$ and $t_{ox} = 4nm$ [7]. We use the same VeSFET technology parameters in all experiments reported in this paper.

In this Section, we compare footprint area, wire length, via usage, performance and power of six basic logic gates: INV, BUF, NAND2, NOR2, AOI21, and OAI21 with *1X*, *2X* and *4X* transistor sizes mapped on *CC* and *BC*.

The gate area comparisons on both canvases are listed in Table 1. In the table, we use the number of pillars occupied by a cell to represent its footprint area. Note that for cells mapped to *CCs*, peripheral pillars are used for transistor isolation, as shown in Fig. 5. Those pillars can be shared by neighboring cells. Thus, we only count half of them towards the cell area. For example, a complete NAND2X1 cell layout shown in the box in Fig. 5(b) occupies 24 pillars. Among them, 12 pillars are actually used by the cell whereas the other 12 along the boundaries may be wasted or used for electrical transistor isolation. The 12 unused pillars can be shared by other cells, as shown in Fig. 8. (The row-based standard cell placement will be described in detail in the next Section.) Therefore, we count the footprint of this cell as 12+12/2=18. The average footprint areas are normalized to *BC*-mapped cells. Results in Table 1 indicate that small gates with unit size transistors, like INVX1, NAND2X1, are slightly greater on *CC* than on *BC*. The reason is that pillar sharing is limited for small gates, but some pillars are wasted for transistor isolation. As gates have more inputs or the transistor size becomes greater, *CC* layouts are denser than those on *BC* because more pillars can be shared and the transistor isolation costs are amortized. In general, if a static gate has dual Euler paths of length *M*, and the transistor size is *N*, the footprint area of this gate on *CC* is *2(M+1)(2N+1)*, while on *BC* it is *8MN*. *CC* cells have smaller footprint than *BC* cells if *2MN>2N+M+1*.

Routing is difficult for *BCs* because transistors are densely packed and all connections have to be made by metal wires [5]. For *CC*-mapped cells, routing may be easier because all gate inputs are aligned and pillar sharing eliminates many wires. The total length of local signal wires and total number of intra-cell vias for both canvases are compared in tables 2 and 3. Note that wires and vias of VDD/GND are not counted here. The data shows that for cells with unit size transistors, *CC*-mapped layouts have 32% shorter wires and 36% fewer vias than those mapped to *BC*. Furthermore, *CC* saves more wires and vias as transistor size increases. As a result, the parasitic capacitance of local interconnects may be much smaller for *CC*-mapped cells. Regarding that parasitic capacitances play critical role in performance and power of VeSFET-based circuits [8], *CC*-mapped cells may have better performance and power than cells built on *BC*.

Table 1. Area of Cells mapped on *BC* and *CC*

CELL	BC			CC		
	1X	*2X*	*4X*	*1X*	*2X*	*4X*
INV	8	16	32	12	18	30
BUF	16	32	64	18	30	54
NAND2	16	32	64	18	30	54
NOR2	16	32	64	18	30	54
AOI21	24	48	96	24	40	72
OAI21	24	48	96	24	40	72
AVG	1	1	1	1.15	0.93	0.83

Table 2. Total Wire Lengths of Cells mapped on *BC* and *CC*

CELL	BC			CC		
	1X	*2X*	*4X*	*1X*	*2X*	*4X*
INV	5.6	14.0	33.0	4.2	9.8	21.0
BUF	12.2	29.0	67.0	10.8	22.0	44.4
NAND2	13.2	33.2	74.4	7.6	17.4	37.0
NOR2	13.2	33.2	74.4	7.6	17.4	37.0
AOI21	20.2	49.4	115.6	13.2	27.6	59.4
OAI21	20.2	49.4	115.6	13.2	27.6	59.4
AVG	1	1	1	0.68	0.60	0.55

Table 3. #Vias of Cells mapped on *BC* and *CC*

CELL	BC			CC		
	1X	*2X*	*4X*	*1X*	*2X*	*4X*
INV	8	17	46	5	9	17
BUF	18	36	95	15	23	39
NAND2	21	54	107	12	20	34
NOR2	21	54	107	12	20	34
AOI21	30	75	166	19	31	53
OAI21	30	75	166	19	31	53
AVG	1	1	1	0.64	0.46	0.34

Table 4. Performance of Cells Mapped on *BC* and *CC*

CELL	BC (ps)			CC (ps)		
	1X	*2X*	*4X*	*1X*	*2X*	*4X*
INV	328	405	414	305	295	284
NAND2	617	694	647	442	436	425
NOR2	740	819	767	539	520	508
AOI21	872	1015	984	702	692	673
OAI21	862	1004	953	689	663	635
AVG	1	1	1	0.80	0.67	0.67

Table 5. Power-Delay-Product of Cells on *BC* and *CC*

CELL	BC (fJ)			CC (fJ)		
	1X	*2X*	*4X*	*1X*	*2X*	*4X*
INV	2.46	6.06	12.34	2.28	4.43	8.55
NAND2	3.46	7.79	14.58	2.47	4.93	9.64
NOR2	3.63	8.01	15.08	2.61	5.09	9.99
AOI21	3.89	8.94	17.31	3.06	6.07	11.82
OAI21	3.77	8.74	16.61	3.00	5.73	11.09
AVG	1	1	1	0.79	0.67	0.67

To compare the performance and power, we simulate a 5-stage ring oscillator composed of each cell (except BUF). We extract the parasitic capacitance using *Raphael* tool [9]. Table 4 reports the periods of ring oscillators and Table 5 shows the Power-Delay-Products (PDPs). The experimental results demonstrate that the average speed and PDP of unit size cells on *CC* are 20% better than on *BC*. The difference increases to 33% when transistor size is doubled, because *CC* saves more wires and vias with transistors of size *2X*. However, there is no further improvement when transistor size increases to *4X*, possibly because the extra wire and via savings do not further reduce parasitic capacitances.

In summary, although the footprint areas of small gates mapped to *CC* are slightly larger than on *BC*, *CC*-mapped cells show great performance and power improvements due to reduced wire lengths and number of vias.

4.2.2 Flip-Flops

Latches and Flip-Flops (*FFs*) are necessary components of sequential circuits. Here, we map a minimum size C^2MOS *FF* [10] to *BC* and *CC*, and compare their footprint areas, parasitic capacitances and performance.

The layouts of *BC*- and *CC*-mapped designs are shown in Fig. 7. Note that this *FF* circuit is a cascade of two clocked inverters. There are no dual Euler paths for both stages. Thus, for *CC*-mapped layout, our cell generation algorithm simply aligns the transistors with same gate inputs to maximize gate pillar sharing. Although *CCs* have twice as great transistor density as *BCs*, the footprint area of *CC*-mapped *FF* is 50% greater than that of *BC*-mapped, as listed in Table 6. It is so because not many pillars can be shared in this circuit and quite a few pillars are wasted for transistor isolation. The larger footprint of *CC*-mapped *FF* results in 13% greater wire length than for *BC*-mapped *FF*. However, the advantages of *CC*-mapped FF are that it needs one fewer metal layer and 20% fewer vias.

The capacitances extracted by *Raphael* [9] show that *BC*- and *CC*-mapped *FFs* have comparable capacitances for *data-in* and *-out* wires. However, for the *clock* wire, *CC*-mapped *FF* has only a half capacitance of *BC*-mapped one. This means significant power saving for *CC*-based *FFs* because clock is the most active signal on the whole chip. Also, we see comparable *setup time* and *clock-to-q* time for these two layouts. Note that in Table 6, the *clock-to-q* times are obtained with 4 minimum size inverters driven by an *FF*.

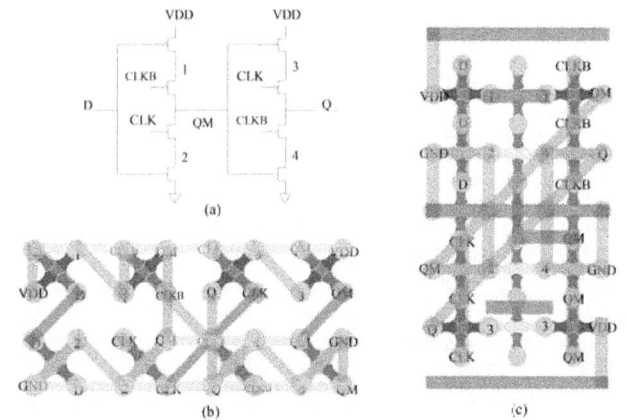

Figure 7. (a) C^2MOS Flip-Flop design; (b) Layout mapped to *BC*; and (c) Layout mapped to *CC*.

Table 6. *BC-* and *CC*-mapped Flip-Flop comparison

	Footprint Area	# Metal Layer	Wire Length	#Vias
BC	32	4	30.6	47
CC	48	3	34.6	38
	Parasitic Capacitance (fF)			
	CLK	**CLKB**	**D**	**Q**
BC	0.27	0.29	0.22	0.22
CC	0.16	0.16	0.17	0.26
	Performance (ps)			
	Clock-to-q (0)	**Clock-to-q (1)**	**Setup time**	**Hold time**
BC	82	96	158	0
CC	85	100	170	0

5. Row Based Standard Cell Design Flow

For ASIC designs based on TG-VeSFETs, the logic synthesis and functional verification steps are the same as in CMOS flow. Although the physical design of VeSFET and CMOS-based ICs are different, in this Section, we show that CMOS physical design tools can be easily re-used for VeSFET based ASICs.

5.1 Row Based Standard Cell Placement

Row based standard cell placement is widely used in CMOS ASIC design flow. VeSFET ICs mapped on *CCs* can also adopt this design style and use the well-developed existing placers. Fig. 8 shows an example how VeSFET standard cells can be placed on the canvas. The cells not only have the same height but also the same rows of *p/n*-type transistors. For example, the cells shown in Fig. 8 have one row of *p-* and *n*-type transistors each. Between two neighboring cells in the same row, a column of *G* pillars cannot be utilized because of needed transistor isolation. Similarly, a row of *D/S* pillars at the boundary of neighboring cell rows will be wasted. All unused pillars are shared by two cells in densely packed layout.

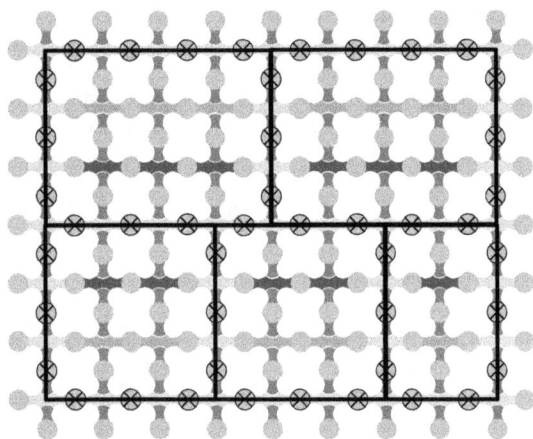

Figure 8. Row based standard cell placement on a *CC*

If we flip every other cell row, as shown in Fig. 8, the rows of pillars at the boundary naturally form VDD/GND straps for two cell rows, as it is normally also done for CMOS ASICs. Furthermore, all embedded transistors at the boundary will be electrically disabled. Therefore, different designs based on cell library with unit size transistors can be fitted into the same pre-manufactured canvas *CC-3-3* shown in Fig. 3. Circuits based on different cell libraries may choose other *CC-x-y* canvases with a different number of *p-* and *n-* transistors in rows.

In CMOS, layouts can be further compacted by breaking the cell boundary and merging neighboring cells if they have common *D/S* connections. This technique can also be applied in *CCs* as common *D/S* pillars of neighboring cells can be shared. Large cells occupying multiple rows are also allowed, but the transistor placement must match the *p-/n-* row configuration of the canvas.

In summary, we can easily produce industrial standard LEF files for VeSFET based cell library and apply commercial tools to obtain row-based standard cell placements for *CCs*.

5.2 Inter-cell routing

To reduce manufacturing cost, interconnects of VeSFET based ICs are strictly parallel on each metal layer. Vias are aligned to the transistor pillars [1, 2]. Commercial routers can be adapted for inter-cell routing of *BC* based circuits [11, 12].

Compared to *BCs*, canvases in *CC* family are rotated by 45°. Thus the horizontal/vertical wires on *CCs* are actually diagonal wires on *BCs*. Unlike *BC,* which has one uniform routing grid, the horizontal/vertical routing tracks on *CC* form two separate grids because the vias are aligned with the metal pillars of transistors, as shown in Fig. 9. One grid only covers gate pillars (*G*-grid), and the other only covers *D/S* pillars (*D/S*-grid). To connect a *G* pillar to a *D/S* pillar, a diagonal wire (horizontal or vertical in *BCs*) must be used. We call them *jumper wires*. Most inter-cell nets contain a single *D/S* pillar (the output of the driving gate) and one or several *G* pillars (the inputs of the driven gates). Ideally, we can route a portion of a net on *G*-grid and the rest on *D/S*-grid using jumper wires to connect them. To adapt commercial routers with the least effort, we assume that each net is routed completely on either *G*-grid or *D/S*-grid, and the jumper wires are implemented within the cell layout. For example, if a net is assigned to *G*-grid, we place the diagonal wire for the driving gate output such that it can be accessed on the *G*-grid, as shown in Fig. 5(c). If a net is assigned to *D/S*-grid, jumper wires will be added locally to the driven gates input so they are accessible by the *D/S*-grid. Thus, before we can use a commercial router, we need to partition inter-cell nets into two groups, each to be routed on a different grid.

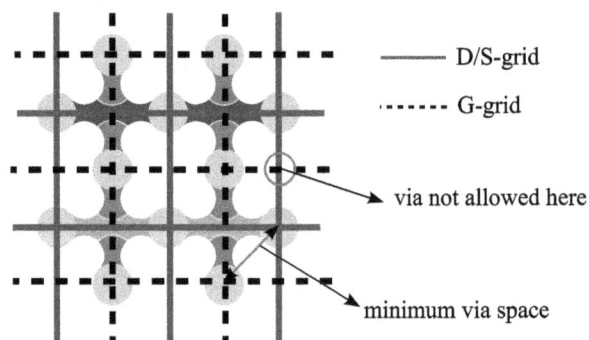

Fig. 9. Disjoint routing grids

To minimize the number of metal layers, we attempt to fully utilize both grids by balancing the routing congestion. In this paper, we adopt the net partitioning algorithm in [12]: we divide the placed circuit into small bins and probabilistically estimate track usage for each net [13]. For each bin, we define the partition cost as the difference of track usage on *G-* and *D/S*-grids. Pin access in densely packed layouts may be a problem [12, 14]. Thus, we try to balance the number of access pins on both grids. We include the pin accessing cost for each bin, and the total bin cost is obtained by adding up the costs of all the bins. A greedy algorithm is used for partitioning [12].

The metal pitch of *CC*-inter-cell wires is only 70% of that for *BC* wires [1, 2]. Thus, the coupling capacitance of *CC*-inter-cell wires is 40% greater for the same wire length. On the other hand, we have 40% more routing tracks on each layer. To achieve better performance and power, we may only route inter-cell nets on *G*-grid or *D/S*-grid, in which case the space between neighboring wires is about *2X* compared to *BCs*, but the number of routing tracks on each layer is 30% less. In the next Section, we show the trade-off between wire lengths and metal layer usage versus power and performance.

5.3 Circuits Mapped to *BCs* and *CCs*

In this Section, we study LGSynth91 benchmark circuits implemented on *CC-3-3* by applying the described standard cell based ASIC flow. A similar flow is applied on *BC* and the results are compared. We use *ABC* tool [15] to map the benchmark circuits to the six-minimum-size-cell library discussed in Section 3. The technology mapping results are reported in Table 7. Commercial placer is used to obtain row-based placements with 10% white space in each circuit. The footprint areas for circuits implemented on both canvases are listed in the last two columns of Table 7. The table shows that at circuit level, *CC-3-3*-mapped circuit areas are about 13% greater than *BC*-mapped because the library cells on *CC-3-3* are greater.

Commercial router is used for global and detailed routing, with strictly parallel wires enforced on each metal layer. To illustrate the trade-off between routability and performance on *CC-3-3*, we route the benchmarks on *G*-grid only and on both grids with partitioned nets. The routing results are listed in Table 8. Here we only count wires and vias for inter-cell nets (for *CC-3-3* jumper wires are included). Also, the power/ground and clock networks are pre-routed as special nets and we only consider the wires and vias of signal nets. In the table, *CC-G* columns refer to the results for *G*-grid only routing on *CC-3-3*, and *CC-G/D* to routing on both grids.

We report the minimum routing layers for all benchmarks (*#ML_min* in the table). As expected, *CC-G/D* needs the least number of metal layers because on each layer, more routing tracks are available. For larger designs like *pdc, s38417* and *s38584*,

metal layer saving is significant. On the other hand, *CC-G* needs more metal layers because only half of the routing tracks are available as compared to *CC-G/D*, and 30% fewer than for *BC*.

As pointed out in [11], to achieve optimal wire length, more than minimum number of metal layers might be needed. Thus, to determine the number of metal layers for best wire length and via usage, we gradually increase the number of layers until no significant improvement can be achieved. The number of metal layers to achieve the minimum wire length is reported in Table 8 as *#ML_opt*, and a similar trend as for *#ML_min* can be found.

For most benchmarks, *CC-3-3* based circuits have longer wires than on *BC* because the footprint area is greater. Via saving of *CC-G/D* is significant: *CC-G/D* uses 21% fewer vias than *CC-G* and 19% fewer than *BC* because of fewer metal layers and detour paths. Interestingly, although *CC-G* needs more metal layers and produces longer wires, it's via count is comparable to *BC*. This may be attributed to better pin accessibility for *CC-3-3* cells as compared to cells on *BC*.

We extract the capacitance of inter-cell nets after detailed routing using StarRC [16]. In Table 8, we list the total wire capacitances for designs implemented on both canvases to estimate the possible power consumption caused by circuit level interconnects. As expected, the total capacitance of *CC-G* is 17% less than *CC-G/D* and 22% less than *BC*, although it has the longest wires and the most number of vias. Surprisingly, for most benchmarks except *des, frisc* and *pdc*, *CC-G/D* mapped circuits show smaller capacitances than those mapped to *BC*. It is so because for circuits with abundant routing resources, wires are sparsely packed and inter-cell interconnect coupling is smaller. *CC-3-3* has thinner wires, thus the wire capacitance to ground plane is smaller. For tightly packed wires as in *des, frisc,* or *pdc*, coupling may dominate the total capacitance. Thus *CC-G/D* implementations have greater capacitance for these three benchmarks.

Static timing analysis is applied to find the longest path delay for each benchmark circuit. The timing libraries for both canvases are obtained by HSPICE simulation. For each cell, the timing library includes pin to pin propagation delays of rising and falling transitions, as well as the rising and falling slews of the output signal with different load capacitances and input slews.

Table 7. Synthesis and Placement Results of LGSynth91 Benchmarks

Benchmarks	# INV	# BUF	# NAND2	# NOR2	# AOI21	# OAI21	# FFs	# cells	# IOs	# nets	Area(μm^2)	
											BC	CC
bigkey	361	8	468	572	678	231	224	2542	460	2805	2162	2428
C6288	61	139	2292	2	29	0	0	2523	64	2555	1780	2009
C7552	582	159	909	337	177	514	0	2678	315	2885	1945	2164
clma	658	0	1566	1311	1501	1300	33	6369	466	6766	5318	5588
des	1096	315	1910	1869	248	558	0	5996	501	6252	4166	4723
dsip	347	192	243	455	678	231	224	2370	426	2598	2043	2292
frisc	627	0	1366	823	866	1000	886	5568	137	5588	5027	5845
i10	723	138	1035	700	148	345	0	3089	481	3346	2116	2408
pdc	1077	421	2648	1386	236	266	0	6034	56	6050	4083	4667
s38417	1188	322	2134	1736	1059	1203	1462	9104	136	9133	7885	9331
s38584	1412	222	2267	1813	1221	1135	1260	9330	344	9398	7867	9197
spla	394	0	951	980	374	335	0	3034	62	3050	2266	2456
AVG											**1**	**1.13**

Table 8. Layout and Performance of VeSFET Canvases

Bench marks	#ML_min / #ML_opt			Total Wire Length(mm)			# Vias			Total capacitance (pF)			Longest Path Delay (ns)		
	BC	CC_G	CC_G/D	BC	CC_G	CC_G/D	BC	CC_G	CC_G/D	BC	CC_G	CC_G/D	BC	CC_G	CC_G/D
bigkey	5/6	6/6	4/4	14.9	16.2	15.1	21.7k	20.8k	17.2k	1.51	1.23	1.39	11.3	9.0	9.3
C6288	4/4	4/4	4/4	8.5	9.6	9.0	17.9k	17.0k	15.0k	1.26	1.06	1.15	13.8	11.0	11.1
C7552	4/4	5/6	4/4	13.9	14.9	14.0	20.8k	20.0k	16.4k	1.90	1.36	1.66	8.9	6.9	7.6
clma	6/6	7/8	5/5	63.8	68.0	64.5	65.9k	71.2k	53.7k	8.58	6.31	7.80	11.1	8.6	9.8
des	6/6	7/8	4/4	46.9	52.0	49.2	52.1k	54.9k	40.4k	6.31	4.74	6.63	14.7	12.8	14.8
dsip	4/5	5/6	4/4	13.2	15.5	14.4	19.4k	19.5k	16.1k	1.39	1.20	1.33	11.2	9.0	9.7
frisc	6/6	7/8	5/5	44.6	53.4	51.9	46.8k	49.3k	40.2k	4.95	4.18	4.99	12.7	10.3	11.1
i10	4/4	5/6	4/4	17.8	19.0	17.9	24.7k	24.4k	19.9k	2.65	1.78	2.26	5.8	4.3	5.0
pdc	7/8	9/10	5/5	71.9	76.6	73.6	59.9k	66.4k	47.1k	9.50	7.39	11.40	9.6	7.9	9.8
s38417	5/6	7/7	4/4	43.9	51.4	47.6	66.8k	65.9k	56.1k	4.37	3.78	4.18	6.9	6.0	6.8
s38584	5/6	6/6	4/4	45.2	51.4	48.2	69.4k	68.0k	58.1k	4.41	3.72	4.10	19.4	17.3	18.2
spla	6/6	10/10	6/6	36.4	38.1	35.6	33.7k	37.7k	25.9k	5.71	3.70	4.92	6.5	5.2	6.2
AVG				1	1.11	1.04	1	1.02	0.81	1	0.78	0.95	1	0.81	0.91

The experimental results are listed in Table 8. Compared to *BC*, *CC-G* shows about 19% average performance improvement due to faster cells and less capacitance of inter-cell nets. Most benchmarks based on *CC-G/D* are also faster than *BC*-mapped ones except *pdc* and *des*, whose *CC-G/D* based implementations are less than 5% slower than *BC*.

6. Conclusions

VeSFETs may facilitate medium volume ASIC products in advanced technology nodes because manufacturing cost can be greatly reduced by VeSFETs' absolutely regular geometry. Various canvases can be pre-manufactured and circuits can be customized by interconnect configuration. It usually requires a tremendous effort to setup an automatic design flow for a new technology. In this paper, we propose a class of canvases—*chain canvases* for which it is easy to accommodate VeSFET-based designs re-using the existing CMOS ASIC infrastructure. We show that conventional EDA tools only need slight modification to build standard cell-based ASICs on *CCs*.

Although *CCs* have 2X transistor density than *BCs*, *CC*-mapped designs may not have smaller footprint area than *BC*-mapped ones because some transistors will be wasted to remove undesired connections. Despite the shortcoming on area, *CC*-mapped standard cells are easier to route, and the capacitances of intra-cell wires are much smaller than those of *BC*-mapped ones. Thus, CC-mapped cells are faster and more power efficient than BC-mapped cells. At circuit level, CC-mapped designs also outperform *BC*-mapped ones on via usage, performance and power, at a small cost of footprint area and total wire lengths.

The results presented in this paper support the conclusion that there exists an inexpensive passage (or "poor man's" path) to build the first generation tools for VeSFET-based designs by reusing of big portion of the existing CMOS EDA tools and methodologies.

7. REFERENCES

[1] W. Maly, "Integrated Circuit Device, System, and Method of Fabrication," Pub No. US 2009/0321830, Dec 31, 2009.

[2] W. Maly, and A. Pfitzner, "Complementary Vertical Transistors," Carnegie Mellon University, CSSI Techreport, No. 08-02, 01/2008.

[3] W. Maly, N. Singh, Z. Chen, N. Shen, X. Li, A. Pfitzner, D. Kasprowicz, W. Kuzmicz, Y.-W. Lin, and M. Marek-Sadowska, "Twin gate, vertical slit FET (VeSFET) for highly periodic layout and 3D integration," in *Proceedings of MIXDES'11*, pp.145-150, 2011.

[4] W. Maly, Y.-W. Lin, and M. Marek-Sadowska, "OPC-Free and Minimally Irregular IC Design Style", in *Proceedings of DAC'07*, pp. 954 -957, 2007

[5] Y.-W. Lin, M. Marek-Sadowska, and W. Maly, "Layout Generator for Transistor-Level High-Density Regular Circuits". IEEE Trans. on CAD, vol. 39, No. 2, pp. 197-210, Feb. 2010.

[6] T. Uehara and W. Van Cleemput, "Optimal Layout of CMOS Functional Arrays," IEEE Trans. on Computers, Vo. C-30, no. 5, pp. 305-311, May 1981.

[7] Y.-W. Lin, M. Marek-Sadowska, W. Maly, A. Pfitzner, and D. Kasprowicz, "Is there always performance overhead for regular canvas?" in *Proceedings of ICCD'08*, pp. 557-562, 2008.

[8] Y.-W. Lin, M. Marek-Sadowska, and W. Maly, "On Cell Layout-Performance Relationships in VeSFET-Based, High-Density Regular Circuits". IEEE Trans. on CAD, vol. 30, No. 2, pp. 229-241, Feb. 2011.

[9] Raphael: Raphael User Guide, Synopsys, Inc. D-2010.03.

[10] Y. Suzuki, K. Odagawa, and T. Abe, " Clocked CMOS calculator circuitry," IEEE Journal of Solid State Circuits, vol. SC-8, Dec. 1973, pp. 462-469.

[11] M. Marek-Sadowska, and X. Qiu, "A study on cell-level routing for VeSFET circuits," in *Proceedings of MIXDES'11*, pp. 127-132, 2011.

[12] X. Qiu and M. Marek-Sadowska, "Can pin access limit the footprint scaling?" in *Proceedings of DAC'12*, pp. 1100-1106, 2012.

[13] J. Lou, S. Thakur, S. Krishnamoorthy, and H.S. Sheng, "Estimating routing congestion using probabilistic analysis," IEEE Trans. on CAD, vol. 21, no. 1, pp. 32-41, Jan. 2002.

[14] T. Taghavi, C. Alpert, A. Huber, Z. Li, G.-J. Nam, and S. Ramji, "New placement prediction and mitigation techniques for local routing congestion," in *Proceedings of ICCAD'10*, pp. 621-624, 2010.

[15] Berkeley Logic Synthesis and Verification Group, ABC: A System for Sequential Synthesis and Verification, Release 70930. http://www.eecs.berkeley.edu/~alanmi/abc/

[16] StarRC: StarRC User Guide, Synopsys, Inc. F-2011.12.

ISPD 2013 Expert Designer/User Session (EDS)

Organizer:

Cliff C. N. Sze
IBM Research

Moderator:

Laleh Behjat
University of Calgary

Speakers:

Nikhil Jayakumar
Juniper Networks

Atul Walimbe
Intel

Giriraj Kakol
Intel

Gregory Ford
IBM

Mark Zwolinski
University of Southampton

Harish Dangat
Samsung

ABSTRACT

We have newly introduced the expert designer/user session (EDS) to ISPD in 2013, which is tailor-made for designers and users of physical design (PD) tools. Since ISPD is the premium PD-centric symposium, it is a great opportunity for designers, tools developers and PD researchers to interact and learn from each other. The benefit of including EDS in ISPD's program is twofold.

(1) It provides a focus session and friendly environment, which stimulates technical discussion among industry and academia. Experienced PD users would have a chance to get exposed to the detail of the latest state-of-the-art research in the field of physical design.

(2) The session brings in designers' experience of PD tools, which includes opportunities and challenges from users' prospective. This would directly drive the PD research direction in the future, and tighten the bond between users, researchers and developers in the PD community. In turn, it eventually helps improving the PD tools from EDA vendors.

In this Expert Designer Session (EDS), we carefully selected and invited a list of six expert designers/users around the world to present their on-hand design and tool experience.

The topics covered include tree-mesh clock distribution, large block placement, place-and-route for custom designs, layout dependent effects, metal-based ECO, and clock aware timing closure flow.

The format of EDS consists of an oral presentation session, immediately followed by a poster session. This encourages deep discussion and professional networking between PD users, researchers and developers.

Categories and Subject Descriptors

B.7.2 [Integrated Circuits]: Design Aids

General Terms

Algorithm, Performance, Design, Reliability.

Keywords

VLSI, Microprocessor designs.

Relative Timing Driven Multi-Synchronous Design: Enabling Order-of-Magnitude Energy Reduction

Kenneth S. Stevens
Electrical and Computer Engineering
University of Utah
kstevens@ece.utah.edu

ABSTRACT

Energy efficient integrated circuit design largely stems from two very different disciplines: high level system architecture, and low level physical design. These two disciplines are tied together by the algorithms, methodologies, and related CAD which generate productivity and achieve consistency between these domains. Thus a "wine goblet model" aptly depicts energy efficient design. The architecture, which delivers the payload, is represented by the cup. Physical design is the goblet base, and the CAD and algorithms are represented by the goblet's stem. All three are essential components. Energy efficient designs often result from multi-synchronous architectures because the most efficient energy-delay point normally differs for each component in a system. The most energy-efficient circuits and physical design are based on asynchronous methodologies. Thus power optimized multi-synchronous architectures can result in an order of magnitude reduction in power at the same performance over traditional design approaches, *if* new algorithms and physical design are employed. The differentiating factor between traditional design and energy efficient systems is timing, since each component may operate at an independent frequency. A new method of representing timing for multi-synchronous systems is presented that is based on a relativistic and logical model rather than numeric based timing models. Employing relativistic timing constraints provides a significant improvement in energy efficiency, but also fundamentally changes the approach to timing driven physical design optimization. How the relative timing model results in a ten-fold improvement in power at the same performance will be demonstrated by example, as well has the challenges and opportunities it poses stem and base of the wine goblet: the algorithms, CAD, and physical design.

Categories and Subject Descriptors

B.7.2 [**Integrated Circuits**]: Design Aids – *layout, placement and routing, simulation, verification.*

Keywords

VLSI, Design, Design Automation, Energy-Efficiency, Asynchronous Design.

Bio

Kenneth S. Stevens is an Associate Professor at the University of Utah. Prior to Utah, Ken worked at Intel's Strategic CAD Lab in Hillsboro Oregon where he developed timing technology for the double frequency ALU cores, multiple input switching validation, and the design of the front end of the Pentium Processor with asynchronous circuits that operated at 3.6 GHz when the lead processor's fastest clock speed was 450 MHz. Prior to Intel, Ken was an Assistant Professor at the Air Force Institute of Technology (AFIT) in Dayton Ohio where he developed asynchronous communication chips for space applications. Ken received his Ph.D. at the University of Calgary, where he researched the verification of sequential circuits and systems. Before that he worked at Fairchild Labs for AI Research and Hewlett Packard Labs. There he developed an asynchronous circuit synthesis methodology called "burst mode", and designed and fabricated an ultra high bandwidth communication chip for distributed memory multiprocessors. He received three degrees from the University of Utah, including a B.A. in Biology, and B.S. and M.S. degrees in Computer Science. Ken has published in journals and conferences, has 13 patents, and is a Senior Member of the IEEE. He also created a successful software startup company, has developed software for the GNU project, and serves on program committees and conference chairmanships. His current research focus includes reducing energy in digital electronics through novel timing verification technology that transforms circuit timing into logical expressions, network fabric and desynchronized pipeline designs, and asynchronous circuits and systems.

Network Flow Based Datapath Bit Slicing

Hua Xiang   Minsik Cho   Haoxing Ren   Matthew Ziegler   Ruchir Puri

IBM T.J. Watson Research Center, Yorktown Heights NY 10598

email: {huaxiang, minsikcho, haoxing, zieglerm, ruchir}@us.ibm.com

ABSTRACT

In deep sub-micro designs, more functions are integrated into one chip, and datapath has become a critical part of the design. Typical datapath consists an array of bit slices. The inherent high degree regularity of datapaths is especially attractive to the placement and routing to achieve regular layout with high density and high performance. However, the current design methodology may generate inferior datapath designs because the datapath regularity cannot be well understood by the traditional design tools. In previous works, several techniques are proposed to preserve/re-identify datapath structures. However, they either restrict the datapath optimization or have little tolerance on bit slice difference.

In this work, we present a novel approach to re-identify datapath bit slices. Contrary to the previous template-based approach, we convert the bit slicing problem to the bit matching problem. Then a min-cost max-flow based algorithm is proposed to identify the main-frame of bit slices so that the datapath bit matching is achieved. An efficient two way search approach is developed to derive the full bit slices based on the bit matching results. We further improve the bit slicing solution with an iterative method. The experimental results demonstrate the effectiveness and efficiency of our approach.

Category: B.7.2 [Integrated Circuits]: Design Aids - Placement and routing; J.6 [Computer Applications]: Computer-Aided Engineering - Computer-aided design

General Terms: Algorithms, Design

Keywords: datapath, bit slicing, network flow

1. INTRODUCTION

In the deep submicron design era, the level of chip integration increases dynamically with very aggressive goals. Datapath has become a critical part of the design. A datapath comprises bit slices to obtain the necessary word size. For bit slices, they have the same or similar functions and structures. The inherent high degree regularity of datapaths is especially attractive to placers such that all gates in one bit slice can get aligned to achieve regular layouts with high performance and small areas. Fig 1 shows a datapath dominant macro. All bit slices are aligned as highlighted to produce the high density design, and this kind of placement is hard to achieve with the general placement.

However, it is well understood that the traditional design tools are not well suited for high-performance datapaths since regularity is not an apparent feature of typical digital system descriptions [15]. As a result, datapaths are either manually designed to exploit the regularity or totally ignored to be treated as ordinary logic structures. But as the complexity of the datapath structure grows, the requirement on turnaround time makes datapath layout automation inevitable. Meanwhile, the high density high performance requirement on datapaths distinguishes them from the random logic design, and has to be properly handled to achieve design closure. Therefore, it is desirable and necessary to support datapath aware design.

In literature, several techniques and algorithms have been proposed to deal with datapaths. Overall, the datapath handling can be classified into two categories. (1) [1, 7, 5] is to preserve the datapath structure from HDL throughout the synthesis process. Although this well maintains the regularity, it also limits the optimization ability since a part of a datapath cannot be processed independently. For example, a gate in one bit line can be resized for better timing. But in order to keep the regularity, all the equivalent gates in other bit lines have to be resized in the same way as well. Due to area or other constraints, the resizing optimization may fail. (2) [14, 13, 11, 18, 19, 6] are to re-identify datapath structures during/after synthesis for optimization. For all these approaches, one fundamental step is to extract datapath regularity.

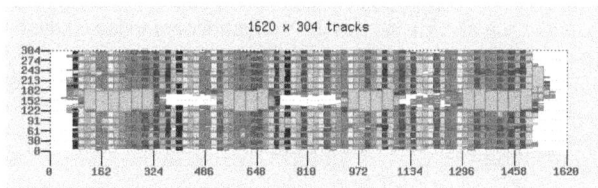

Figure 1: A datapath dominant macro with bit slices aligned.

The main idea of datapath extraction is to cover the design with either pre-defined library templates or automatically generated templates [4, 8, 9, 15, 16, 12, 17]. Template extraction requires creating all equivalent classes of the circuit graph under isomorphism, and the covering is to find the exact match against templates. However, after synthesis and technology mapping, the datapath bit lines may not be kept the same. It is also possible that not all bit lines have the same functionality. Fig 2 gives an example. In Fig 2, the datapath is from a PI vector to a PO vector with 4 bit lines. Among these four bit lines, Bit 3 ($PI(3) \rightarrow PO(2)$) have different logic function (an AND gate is connected to the PO instead of a NAND gate). For structures, these four bit slices are very close, but they are all different. Then if we apply the template-based method on datapaths, we may get some pieces with the same structure, but not the whole bit slices. On the other hand, if the placer and router can be informed with the alignment constraints on bit lines, it helps to produce high density and routing friendly design.

This motivates us to focus on bit slices instead of regularity. Typically, datapath (or datapath segment) starts from PI/latch vectors, and ends at PO/latch vectors. The names of PI/PO/latch vectors are usually featured as a string ending with an index (e.g., PO_result(0), PO_result(1) etc.), and they are seldom changed during synthesis. Besides, it is trivial to set an attribute on these PI/PO/ gates to indicate that they are vectors. Therefore, the vectors can be easily identified in the physical design stage. Once the starting/ending vectors are given, the corresponding datapath is determined. But how to derive bit slices?

Datapath Bit Slicing (DBS) Given a datapath input vector $S = (s_1, ..., s_n)$ and output vector $T = (t_1, ..., t_n)$, identify the n bit slices such that the n bit slices have similar structures.

In the design, a datapath may have multiple stages which are usually indicated by latch banks. For example, a datapath could be $PI \rightarrow LatchBank1 \rightarrow LatchBank2 \rightarrow PO$. For such datapaths, we can start from its inputs, and do bit slicing stage by stage. To simplify the presentation, we let S and T are two adjacent vectors. Furthermore, we can assume that there is no connection circles between S and T.

Although the ultimate goal is to get bit lines with similar structures, it is pretty difficult to work directly with similarity since it requires the comparison between two objects, especially the bit slice template is unknown. On the other hand, once we are able to match the starting points and the ending points, the gates between a pair of matched points can get extracted to make one bit slice. Therefore, contrary to the previous template-based approaches, we address the bit slicing problem by determining the matching between the starting bits and the ending bits so that the matching guides to produce bit slices with similar structures.

Along the datapath, if all operations are vector based, the bit lines might be clear. However, in real design, most datapaths are a combination of vector and scalar operations. Even at the VHDL/HDL level, except if users specify the bits, it is not a trivial thing to determine the slices since the index of a vector has no logic meaning, and one bit of the starting vector can have connections to multiple bits of the ending vector. As shown in Fig 2, $PI(1)$ has paths to both $PO(2)$ and $PO(4)$, and it is not straightforward to match the bits.

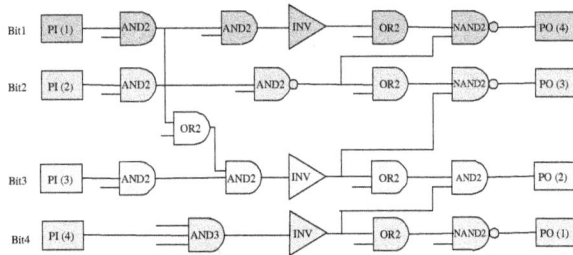

Figure 2: A datapath with 4 bit slices. The four bit slices have similar but different functions and structures.

In this work, we solve the bit slicing problem with a new approach totally different from previous methods. We present a network flow based algorithm to identify the main frame of the datapath. The motivation of this method is based on the fact that although each bit slice can be different, there may exist a main path between a starting point and its corresponding ending point. Once we are able to match the starting points and the ending points, the gates between a pair of matched points can get extracted to make one bit slice.

The main contributions of this work are:

- We develop an efficient two-way search algorithm to identify datapath related gates based on the datapath input and output vectors. Furthermore, the algorithm can be applied to identify gates in one bit slice once the starting and ending bits are given.

- We convert the datapath bit slicing problem to the bit matching problem such that the bit slices can be derived based on the matched starting/ending bits.

- We propose a totally new concept of datapath main frame, which is composed of a set of disjoint paths from the starting vector to the ending vector.

- We develop an optimal min-cost max-flow based algorithm to identify the main frame of the datapath. Based on the main frame, the bit matching solution can be easily derived.

- We present two novel techniques to create more min-cost max-flow flow solutions so that an iterative approach can be explored to further improve the datapath bit slicing solution.

The rest of the paper is organized as follows. In Section 2, we present an efficient two-way search algorithm to identify datapath related gates. In Section 3, we convert the datapath bit matching problem to the datapath main frame problem, which can be optimally solved by applying the min-cost max-flow algorithm on the constructed flow network graph. Furthermore, two novel techniques are developed to create more flow solutions. Section 4 summarizes the whole flow of our datapath bit slicing algorithm. The experimental results are presented in Section 5, and Section 6 concludes the paper.

2. TWO-WAY SEARCH EXTRACTION

For any gate related to the given datapath, one important feature is that this gate must have at least one net connection path to both the starting vector and the ending vector. Therefore, if we do tracing from one side to the other, a datapath related gate must appear in the search. This motivates us to get the Two_Way_Search_Extraction algorithm.

For each gate in the design, it is initialized with a mark 0. Then apply the breadth-first-search algorithm along the fan-out cone of S. The search stops when it hits T, a latch, a PO, or any gates that are identified as not related. For each gate appearing in the search, increase its mark to 1. Similarly, we apply another search from T to S, and increase the mark of gates in the search. If a gate has a mark 2, it means it has connections to both S and T, and this gate is accepted as a datapath related gate. The algorithm is summarized as follows.

Algorithm Two_Way_Search_Extraction(S, T)
1. Initialize each gate g in the design with $mark_g = 0$;
2.
3.   //forward search
4.   Start from S to do breath-first-search along output nets;
5.   For any gate g in the search, $mark_g ++$;
6.
7.   //backward search
8.   Start from T to do breast-first-search along input nets;
9.   For any gate g in the search, $mark_g ++$;
10.
11.   Let $G_d = (V_d, E_d)$, where
12. V_d is a set of gates whose $mark$ is 2;
13. E_d is the netlist for V_d;
14.   return G_d;

The core part of the algorithm is the breath-first-search, and its runtime is bounded by $O(p+q)$ where p is the number of gates in the design, and q is the number of nets. The initialization and the G_d generation only need traverse all gates once. Therefore, the total runtime of the extraction algorithm is $O(p+q)$.

Please note that any gate in one bit slice must have paths to both a starting bit and an ending bit. Therefore, Two_Way_Search_Extraction can also be used to identify gates in one bit slice by letting S be the starting bit and T be the ending bit. This allows us to exploit the bit matching solution to solve the datapath bit slicing problem.

3. DATAPATH BIT MATCHING

By applying Two_Way_Search_Extraction, we are able to get the underlying connection graph for the given datapath. However, the graph only shows the connectivity between two vectors, and we need to identify the bit lines from the graph.

The specialty of datapath bit lines lies in their regularity/similarity. Regularity requires equivalence between two pieces. But in real design, even if all bit lines have the same logic function, they may still get different implementation since the traditional tools have no knowledge of the datapath structure, and each bit line might be processed independently. Therefore, except bit structures are well preserved from the very beginning of the design process such as [1, 7, 5], in most cases, the datapath bit slices have similar but different structures.

However, what is similarity? Similarity draws on the comparison between two objects. So we either compare one bit line with a bit line template, or another bit line. First, bit line templates with "similar" features are even more difficult to generate than templates with "equivalence". Second, we are working on bit slicing, and the bit lines are not available yet.

On the other hand, for each bit slice, if its starting and ending bit are known, we can apply the Two_Way_Search_Extraction on these two starting/ending bits to derive the full bit slice. Therefore, we convert the bit slicing problem into the bit matching problem.

Datapath Bit Matching Given a datapath input vector $S = (s_1, ..., s_n)$ and its output vector $T = (t_1, ..., t_n)$, identify the one-to-one matching between S and T such that n bit slices extracted based on their input/output bits have similar structures.

One intuitive way to solve the bit matching problem is to enumerate all possible matching solutions between two vectors. Although datapath has limited connections between bit lines, e.g., we assume one starting vector bit has paths connecting to only two ending vector bits, the number of the total solutions can be up to $2^{n/2}$, where n is usually larger than 8, such as 32 or 64.

A bipartite graph is usually used to solve the one-to-one matching problem [10]. However, for bit slicing problems, it is not easy to setup a bipartite graph since the weight between a pair of starting and ending bits cannot be calculated independently. For example, in Fig 2, the paths $PI(1) \rightarrow PO(3)$ and $PI(3) \rightarrow PO(2)$ share two gates (one $AND2$ and one INV).

Finally, the datapath bit slicing problem is also different from partition problems. For partition, the target is to assign gates into n sets such that the union of the n sets covers all the gates with certain balancing constraints. However, for bit slicing, one fundamental difference is that not all gates need get assigned to a bit slice. For example, the OR gate between $Bit2$ and $Bit3$ in Fig 2 doesn't belong to any bit line.

In this work, we convert the datapath bit matching to an optimization problem, called Main Frame Identification (MFI) problem. The whole conversion is illustrated as Fig 3. By applying the min-cost max-flow algorithm, the datapath main frame problem can be optimally solved in polynomial time, and it is straightforward to

derive the bit matching solution from the corresponding MFI solution. Once the matching solution is obtained, we can apply the two way search algorithm to extract the gates between one starting bit and one ending bit to get one bit slice.

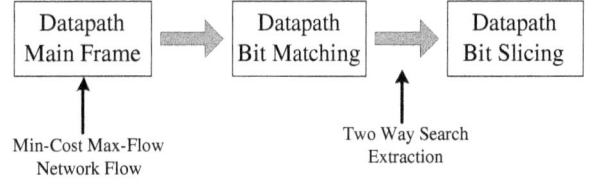

Figure 3: The datapath slicing problem is converted to datapath main frame problem.

3.1 Main Frame Identification

Now let's view the datapath from another standpoint.
Observations:
  (1) In general, all bit slices carry similar number of gates;
  (2) The connections among bit slices are limited;
  (3) All bit slices usually have at least one similar path from the input bit to the output bit, and the path is disjoint with the similar paths in other bit lines;

It is pretty hard to get bit slices directly. But it might be relatively easier to get a part of each bit slice. And if the part can help us determine the input/output bit of that slice, then the bit matching solution is obtained. This motivates us to find the main frame of a datapath, i.e., among all the similar paths from bit lines, find the longest one. Furthermore, for typical datapaths, the total number of gates from any disjoint n disjoint paths won't be larger than the total gates on these n longest paths. Therefore, we transform the concept to the following definition.

Datapath Main Frame (DMF) Given a datapath with n bits, its main frame is a set of n disjoint paths from the input to the output such that the number of datapath gates on these paths is maximized.

To identify the datapath main frame, we need to solve the Main Frame Identification (MFI) problem.

Main Frame Identification (MFI) Given a datapath input vector $S = (s_1, ..., s_n)$ and its output vector $T = (t_1, ..., t_n)$, identify n disjoint paths from S to T such that the n paths cover as many datapath gates as possible.

In MFI solutions, the n disjoint paths correspond to the n paths in n bit slices. Meanwhile, the n paths try to cover as many gates as possible. This implicitly forces to get the longest similar paths.

In this work, we present a network flow based algorithm (MFI_by_Flow) to optimally solve MFI problems. The inputs S and T are the starting and ending vectors of a datapath, respectively. $G_d = (V_d, E_d)$ is the datapath graph obtained by Two_Way_Search_Extraction algorithm, where V_d is the datapath gate set, and E_d is the datapath netlist.

Algorithm MFI_by_Flow(S, T, G_d)
1. Construct the network graph $G_f = (V_f, E_f)$ based on G_d;
2. Assign capacities U_f and costs C_f;
3. Apply min-cost max-flow algorithm on G_f;
4. Derive disjoint paths from S to T;
5. Return the datapath matching result;

Based on the datapath graph $G_d = (V_d, E_d)$, we first construct a flow network as follows.

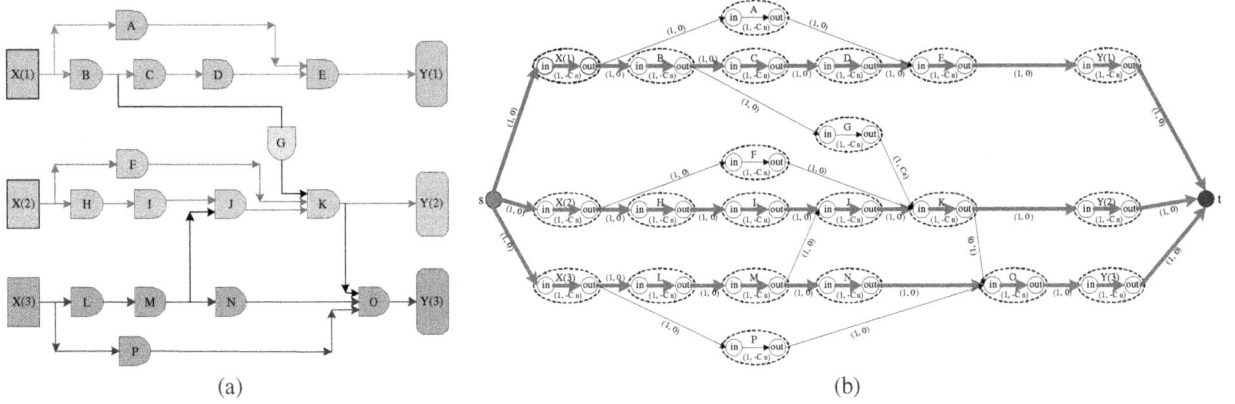

Figure 5: (a) The MFI problem has the input vector $X = \{x(1), x(2), x(3)\}$, and the output vector $Y = \{y(1), y(2), y(3)\}$. The gates/nets are obtained by applying Two_Way_Search_Extraction. (b) The corresponding flow network. And the thick lines show the flow of a min-cost max-flow solution. The three paths are $X(1) \rightarrow Y(1)$, $X(2) \rightarrow Y(2)$ and $X(3) \rightarrow Y(3)$.

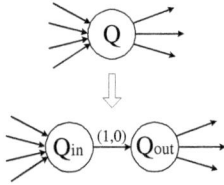

Figure 4: One node Q is split to two nodes Q_{in} and Q_{out}. In this way, the node capacity is transformed to the edge capacity.

(1) $V_f = \{s, t\} \cup V_d$, where s is the source node, and t is the sink node.

(2) $E_f = \{(s, s_i)|s_i \in S, i = 1..n\} \cup E_d \cup \{(t_i, t)|t_i \in T, i = 1..n\}$.

(3) Node Capacity: for $v \in V_d$, $U_f(v) = 1$.

(4) Edge Capacity: for $e \in E_f$, $U_f(e) = 1$.

(5) Node Cost: for $v \in V_d$, $C_f(v) = -C_n$,

    where C_n is a positive integer;

(6) Edge Cost: for $e \in E_f$, $C_f(e) = 0$.

After constructing the flow graph, the min-cost max-flow algorithm will be applied on the flow network. Since the target of MFI is to maximize the number of identified gates, each gate is assigned with a negative cost $-C_n$. Then the more gates on the flow, the less cost we get.

As we note that the classical network flow problem only assigns capacities and costs on flow edges. We handle node capacity/cost by splitting the node Q into two nodes Q_{in} and Q_{out}. One edge is added between Q_{in} and Q_{out}, and assign the node capacity and cost as the edge capacity and cost, respectively. Then all the original incoming edges are pointed to Q_{in} while all the original out-coming edges are connecting from Q_{out}. Fig 4 illustrates the idea of the node splitting.

Fig 5 shows an example of the flow network construction. In this example, the input vector is $X = (x_1, x_2, x_3)$, and the output vector is $Y = (y_1, y_2, y_3)$. Fig 5 (a) is the datapath gate graph obtained with Two_Way_Search_Extraction. Visually, we can see that there are three bit lines, and each bit line has two paths from the input bit to the output bit. Also there are some net connections among these three bit lines. Each gate in the datapath is represented by two nodes (*in* and *out*), and the edge (*in*, *out*) has a capacity 1 and cost $-C_n$. By applying the min-cost max-flow algorithm on this network, we got the flow solution as the thick lines in Fig 5 (b). Finally, by tracing the flow paths in Fig 5 (b), it is easy to derive the bit matching as $X(1) \rightarrow Y(1)$, $X(2) \rightarrow Y(2)$, and $X(3) \rightarrow Y(3)$.

The optimality of the min-cost max-flow algorithm guarantees that the maximum number of gates are identified since the gate edge cost is negative. The more gates found, the less cost.

Finding a min-cost max-flow solution in a flow network is a classical problem, and several polynomial algorithms are available [2, 3, 10]. Therefore, if the double scaling algorithm in [3] is used, the time complexity can be bounded by $O(|V_f| \cdot |E_f| log|V_f|)$ where $|V_f|$ is the number of nodes in the flow network, and $|E_f|$ is the number of edges. It is easy to see that both $|V_f|$ and $|E_f|$ are linearly bounded by the gates and nets in the datapath graph, respectively.

THEOREM 1. *The MFI_by_Flow algorithm can exactly solve the MFI problem as long as one solution exists. The algorithm runtime is bounded by $O(|V_d| \cdot |E_d| log|V_d|)$ where $|V_d|$ and $|E_d|$ are the number of gates and nets in the datapath graph which is obtained by applying Two_Way_Search_Extraction on the given input and output datapath vectors.*

Once we get the bit matching through MFI_by_Flow, the starting and ending points of one bit slice are determined. Then we can apply Two_Way_Search_Extraction again on each starting/ending bit pair to get the bit slices. As illustrated in Fig 5 (a), there are three bit lines, and each bit slice includes 5 gates. The following algorithm converts the datapath bit matching solution to datapath bit slices. M is the matching results. For each item in M, s records the starting bit in S, and t indicates the ending bit in T.

Algorithm Datapath_Bit_Matching_to_Slices(M)
1. for $i = 1$ to $|M|$
2. $Slices[i]$ = Two_Way_Search_Extraction($M[i].s, M[i].t$);
3. return $Slice$

Note that during Two_Way_Search_Extraction, some gates might belong to multiple bit slices. For example, the gates J and K in Fig 5 (a) are included in the extractions of $X(2) \rightarrow Y(2)$ and $X(3) \rightarrow Y(3)$. For this kind of gates, we first check if they are on the MFI flow paths or not. If yes, then they are assigned to the corresponding bit lines. Therefore, for the gates J and K, they are assigned to the bit slice $X(2) \rightarrow Y(2)$. Otherwise, the gates are marked to be shared by multiple bit lines so that the placer can handle them correctly.

3.2 Multiple MFI Solution Generation

NOTE: An MFI solution is not a bit slicing solution. Instead, MFI returns a bit matching solution. By applying Two_Way_Search _Extraction on each paired bits, a datapath bit slice can be derived.

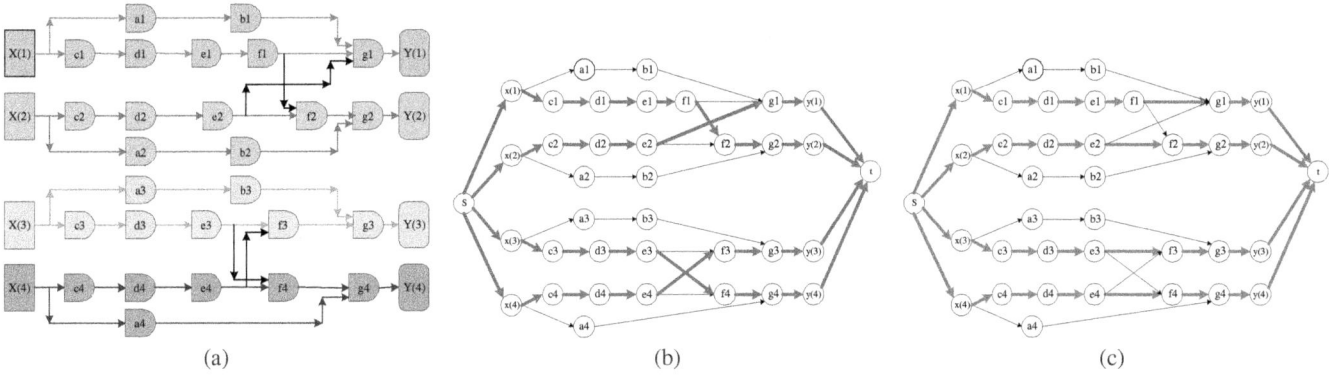

Figure 6: (a) A datapath graph with 4 bit slices. (b) and (c) are two optimal MFI solutions.

However, an optimal MFI solution does not directly map to an optimal bit slicing solution since a datapath bit slice may include multiple paths while MFI focuses only on one path between a starting bit and an ending bit. On the other hand, if there are multiple MFI solutions, then multiple bit slicing solutions can be created, and it helps to improve the bit slicing quality.

MFI_by_Flow draws on the min-cost max-flow algorithm to find an optimal MFI solution. But an MFI problem may have multiple optimal solutions, and different optimal solutions may lead to different bit slices as well. Fig 6 gives an example. As illustrated in Fig 6 (a), this datapath includes 4 bit lines. Fig 6 (b) and (c) show the optimal flows on the constructed network in MFI_by_Flow. To simplify the representation, the *in* node and *out* node are represented by one gate node. The thick lines in the flow graph are the flow results with the min-cost max-flow algorithm. Both solutions are optimal as they cover the same number of gates.

Based on the bit matching solution from Fig 6 (b), we can derive four bit slices as follows (the starting/ending gates are not included in gate counting).

$Bit1 : X(1) \rightarrow Y(2)$ with 6 gates;
$Bit2 : X(2) \rightarrow Y(1)$ with 4 gates;
$Bit3 : X(3) \rightarrow Y(4)$ with 5 gates;
$Bit4 : X(4) \rightarrow Y(3)$ with 5 gates.

On the other hand, the bit slices obtained from Fig 6 (c) are:

$Bit1 : X(1) \rightarrow Y(1)$ with 7 gates;
$Bit2 : X(2) \rightarrow Y(2)$ with 7 gates;
$Bit3 : X(3) \rightarrow Y(3)$ with 7 gates;
$Bit4 : X(4) \rightarrow Y(4)$ with 6 gates.

Obviously, the bit slicing solution with Fig 6 (c) is better since each slice has closer structures and covers more datapath gates.

To resolve this issue, we need explore more MFI solutions. In this section, we propose two methods to create more MFI solutions. One is to change the edge cost of the flow network, and the other is to build a new optimal MFI solution based on two optimal MFI solutions.

3.2.1 Flow Adjustment

Since the min-cost max-flow algorithm is a deterministic algorithm, if we want to get another optimal solution, the flow network must be changed in some way. For two optimal solutions, they must include the same number of gates. And it is also very likely that they are covering the same set of gates. For example, in Fig 6 (b) and (c), the gates on the two flow solutions are the same. Meanwhile, the number of edges in two optimal solutions is also the same since the flow results are paths from one end to the other. But the two sets of edges must be different, especially, the two sets of net edges (i.e., edges corresponding to the nets in the datapath graph) are different. In Fig 6 (c), the net between $f1$ and $g1$ is in the flow,

but it is not used in the flow solution in Fig 6 (b). This motivates us to utilize the cost of these net edges to generate different optimal MFI solutions.

Algorithm Adjust_Flow_Network_Cost $(G_f, G_d, flow)$
1.　for each edge e in the $flow$
2.　　if e corresponds to a net in G_d
3.　　then $C_f(e)+=\delta$

For each edge in the flow, if it is a net edge in the datapath graph G_d, then increase its cost with a small positive number δ ($\delta \ll C_n$). Since the edge cost is much smaller than the absolute value of the gate cost $|-C_n|$, it won't affect the solution optimality, i.e., the number of gates in the flow won't be decreased. On the other hand, after the flow network adjustment, if there is another MFI optimal solution, the min-cost max-flow algorithm will identify a new one. For example, in Fig 6, at the very beginning, the gate cost is -100, and the edge cost is 0. Suppose the first optimal MFI solution we get is as Fig 6 (b). Next we increase the edge cost to 1 for the thick edges in Fig 6 (b). When we apply the min-cost max-flow algorithm again on this flow network, this flow solution is not an optimal solution any more since its cost is $-100*28 + 1*24 = -2776$. (Totally 28 gates and 24 net edges in the flow.) While the solution in Fig 6 (c) has a total cost $-100*28 + 1*20 + 0*4 = -2780$. (Totally 28 gates and 24 net edges in the flow. But 20 net edges are in Fig 6 (b) as well.) By iteratively adjust the flow network, we are able to obtain more MFI solutions.

3.2.2 Group-Piece based Flow Creation

Draw on Adjust_Flow_Network_Cost, we are able to get multiple MFI optimal solutions. One important feature of these optimal MFI solutions is that the flow paths can be partitioned into groups such that different combinations of the flows from each group also lead to optimal MFI solutions. As shown in Fig 7 (a) and (b), if we let the first two flow paths be as one group, and the last two flow paths as another group, we get

$Group1 = \{x(1), x(2)\} \rightarrow \{y(1), y(2)\}$
$Group2 = \{x(3), x(4)\} \rightarrow \{y(3), y(4)\}$.

As we note that the optimal flow solutions on each group do not share any nodes or edges. So for each group, we could select a flow solution from different MFI solutions, and piece them together as a new MFI solution. As shown in Fig 7 (c), the flows for $Group1$ are from Fig 7 (a), while the flows for $Group2$ are from Fig 7 (b). And this new network flow is also an optimal MFI solution. With this group-piece technique, we are able to create multiple optimal MFI solutions.

The reason that we are able to get disjoint groups for most of the datapaths still relies on the fact that the connections among datapath bit slices are rather limited. Therefore, the correspond-

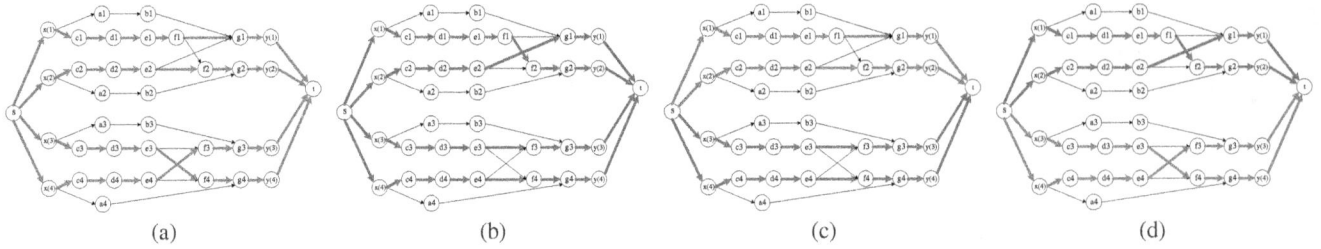

Figure 7: (a) and (b) are two MFI optimal solutions. (c) and (d) are two new MFI optimal solutions by piecing flows from (a) and (b).

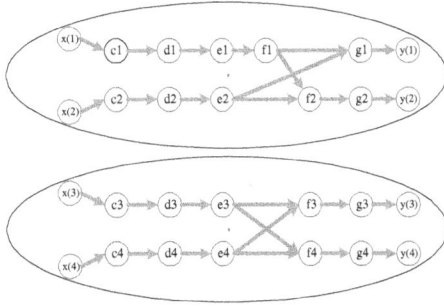

Figure 8: The two MFI solutions in Fig 7 (a) and (b) are merged into one graph. The two disjoint subgraphs are illustrated by the dash ellipses.

ing flow network graph is different from the general network graph such that each bit slice or bit slices with connections implicitly map to a group. For an ideal datapath, each bit slice corresponds to one group. On the other hand, for a general network graph, the connections among nodes may be very random and complicated, and the group partition may be hard.

Given two MFI optimal solutions, we use Group_Identification to find flow disjoint groups. In the return solution, each subgraph corresponds to one group.

Algorithm Group_Identification (*Flow*1, *Flow*2)
1. For node v in $Flow1 \cup Flow2$ excluding s and t
2.   add v to V_p;
3. For edge e in $Flow1 \cup Flow2$ excluding edges with s or t
4.   add e to E_p;
5. Find the disjoint subgraphs from $G_p = (V_p, E_p)$;
6. return disjoint subgraphs;

The main idea is to merge the two MFI solutions into one graph G_p. Then G_p includes all the flow paths from two solutions. If there are disjoint subgraphs, it means that the flow solution on each subgraph is independent on each other. Therefore, for each subgraph, we can choose the flow solution from either given MFI solutions, and the new flow solution is still an optimal MFI solution. For the two MFI solutions in Fig 7 (a) and (b), we could construct G_p as shown in Fig 8. In this graph, there are two disjoint subgraphs as illustrated by ellipses. For each subgraph, its flow solution can from Fig 7 either (a) or (b), and two new MFI solutions are obtained as Fig 7 (c) and (d). The solution of Fig 7 (c) gives the best slicing results to the problem of Fig 6 (a).

THEOREM 2. *For any two optimal solutions of a given MFI problem, another optimal solution can be obtained by identifying the disjoint groups with Group_Identification, and then for each group, copying the flow solution from either of the given MFI solutions.*

3.3 Iterative Bit Slicing Improvement

Comparing to the bit slice identification, it is relatively easier to evaluate if a datapath bit slicing solution is good or bad. Various criteria (for example, the total number of datapath gates covered by bit slices, the distribution of gate numbers of each slice etc.) can be defined to evaluate the slicing quality. If there are multiple slicing solutions, the datapath evaluation can be called to pick the best one.

In the above sections, we present methods to create multiple optimal datapath main frame solutions. This also means we are able to generate more bit slicing solutions. In this section, we propose an iterative approach to improve the datapath bit slicing solution. The algorithm is summarized as follows. *Iters* is the number of iterations. S and T are the starting and ending vectors, respectively. G_f is the network graph, and G_d is the netlist between S and T. *InitMatch* and *InitFlow* are a datapath matching solution and the corresponding network flow.

Algorithm Iter_Improve (*Iters, S, T, G_f, G_d, InitFlow, InitMatch*)
1. *InitSlices* = Datapath_Bit_Matching_to_Slices(*InitMatch*);
2. *BestCost* = Evaluate_BitSlice_Solution(*InitSlices*, G_d);
3. if *BestCost* == *PERFECT_COST*
4. then return *InitSlices*;
5.
6. *BestFlow* = *InitFlow*;
7. *BestSlices* = *InitSlices*;
8. for($i = 0; i < Iters; i++$) {
9.   Adjust_Flow_Network_Cost(G_f, G_d, *BestFlow*);
10.   *NewFlow* = min-cost max-flow on G_f;
11.   *Groups* = Group_Identification(*BestFlow*, *NewFlow*);
12.   for each group g in *Groups* {
13.   *WorkFlow* = *BestFlow*;
14.   replace *WorkFlow* in g with that in *NewFlow*;
15.   derive matching results *NewMatch* from *WorkFlow*;
16.   *NewSlices* = Datapath_Bit_Matching_to_Slices(*NewMatch*)
17.   *NewCost* = Evaluate_BitSlice_Solution(*NewSlices*, G_d);
18.   if *NewCost* == *PREFECT_COST*
19.   then return *NewSlices*;
20.   else if(*NewCost* > *BestCost*)
21.   then *BestCost* = *NewCost*;
22.   *BestFlow* = *WorkFlow*;
23.   *BestSlices* = *NewSlices*;
24.   }
25. }
26. Return *BestSlices*

Evaluate_BitSlice_Solution is to evaluate the bit slicing results. If the result is good enough, for example, we get n disjoint slices, then exit directly. In this case, the datapath slice evaluation will return the cost *PREFECT_COST*. Otherwise, improve the slicing results with iterations. For each iteration, by adjusting the flow cost with *Adjust_Flow_Network_Cost*, a new network flow solution is obtained. By combining flows from different groups, new datapath matching solutions are created. For each new matching solution, derive the slicing result and keep the best slicing solution.

4. DATAPATH BIT SLICING ALGORITHM

In this section, we summarize the whole flow of our datapath bit slicing algorithm.

The ultimate goal of this work is to identify datapath bit slices. To solve the bit slicing problem, our first step is to determine the gates along the datapath, i.e., datapath extraction. We propose a two-way search method to extract all gates between the two vectors. Please note that these gates should cover all bit slice gates, but it is not necessary that every extracted gate belongs to a certain bit. For example, in Fig 2, the *OR2* gate between *Bit3* and *Bit4* gets extracted, but it doesn't belong to any bit slice.

Then the next step is to do bit matching on the extracted datapath graph, i.e., finding the perfect matching between the starting vector S and the ending vector T. Once the starting point and the ending point are determined, the two-way search extraction algorithm can be applied again to get one bit slice.

To address the bit matching problem, we convert it to the datapath main frame problem, which can be optimally solved with a min-cost max-flow solution. From the flow solution, we can easily derive the bit matching solution.

Furthermore, the flow network may have multiple optimal min-cost max-flow solutions, while the algorithm only returns one. To address this issue, we propose two effective techniques, adjusting flow network and combining flow solutions, to generate more flow solutions so that an iterative approach can be used to produce more bit slicing solutions. Based on the datapath bit slice evaluation, the best bit slicing solution can be identified. The whole bit slicing flow is summarized as follows.

Algorithm Datapath_Bit_Slicing(S, T, Iters)
1. G_d = Two_Way_Search_Extraction(S, T);
2. (InitFlow, InitMatch) = MFI_by_Flow(S, T, G_d);
3. Slices=Iter_Improve(Iters, S, T, G_f, G_d, InitFlow, InitMatch)
4. return Slices;

Please note that the proposed algorithm is applied on each individual datapath segment instead of the whole datapath. Meanwhile, the underlying engine is the min-cost max-flow algorithm, and it can handle problems with thousands of nodes efficiently. Therefore, the proposed algorithm is capable of processing large datapaths.

5. EXPERIMENTAL RESULTS

We implemented the proposed algorithms in C++, and integrated into a physical-synthesis system. To evaluate the bit slicing solution, we define the cost as $C = \alpha \cdot \sum_{i=1}^{n} G_i - \beta \cdot \sum_{i=0}^{n} |G_i - G_{avg}|$, where G_i is the number of gates in the ith bit slice, and G_{avg} is the average of G_i. α and β are user defined parms. We let $\alpha = 1$ and $\beta = 0.1$. For this cost definition, our target is to let bit slices cover as many gates as possible. Meanwhile, the difference on slice gate numbers is minimized.

Table 1: Bit Slicing Tests on Generated Designs

Datapath	Width	Slices	MinSize	MaxSize	runtime
Test1-1	4	4	4	5	0.08s
Test2-1	4	4	4	5	0.09s
Test3-1	7	7	1	5	0.36s
Test4-1	32	32	3	8	6.5s
Test4-2	32	32	2	3	6.9s
Test4-3	32	32	2	3	8.6s

To verify the effectiveness of our algorithm, we tested our algorithm on two sets of designs. The first test set include 4 testcases. They were manually created so that the datapath bit slicing is known to us. The second set was derived from industrial designs. And the

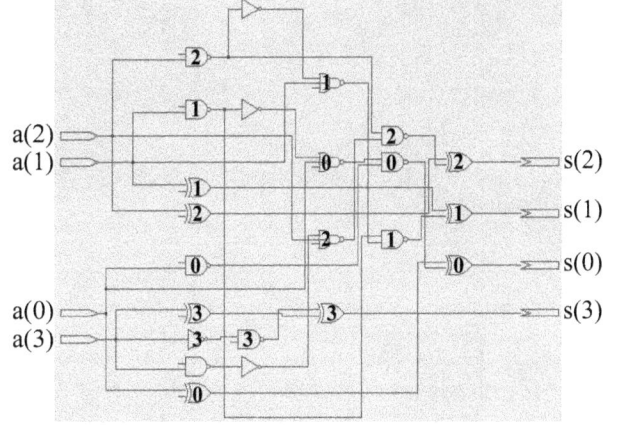

Figure 9: The bit slicing solution on $Test1 - 1$

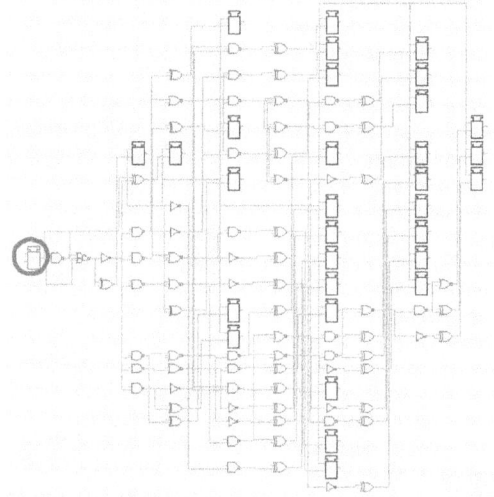

Figure 10: One design piece of $Test4 - 1$. **The gate in the circle is the starting bit of** $Bit31$, **and it has paths to** $Bit0 \sim Bit31$. **The ending bits are highlighted.**

slicing results were verified by the design team. The datapath is named as *testname − pathIndex*. For example, $Test1 - 1$ refers to the first datapath in *Test1*.

The first test set was tested on an AIX workstation ($2GHz$). For all the four designs, we are able to slice all the datapaths, and Table 1 shows the results of some datapaths. *Width* is the number of bits for a given datapath. *Slices* is the number of bit slices obtained by our algorithm. *MinSize* and *MaxSize* are the minimum and maximum value of G_i ($i = 1..n$), respectively, i.e., the minimum and maximum number of gates in one slice.

As we notice that, the number of gates of each bit slice can be different. This indicates that the logic or structure of bit lines might be different. Still our algorithm can slice the datapath as expected. On the other hand, the template-based approaches [4, 16, 17] are not able to identify all bit slices due to the bit line differences.

Test1 is a 4 bit adder, and one datapath ($a(*) \rightarrow s(*)$) is shown in Fig 9. Although the logic is pretty simple, the datapath slices are not straightforward. Each bit slice has multiple paths between the starting and the ending bits, and there are connections between bit lines (for example, $a(1)$ has paths to both $s(0)$ and $s(1)$). Furthermore, the logic and structure of $Bit3$ ($a(3) \rightarrow s(3)$) are different

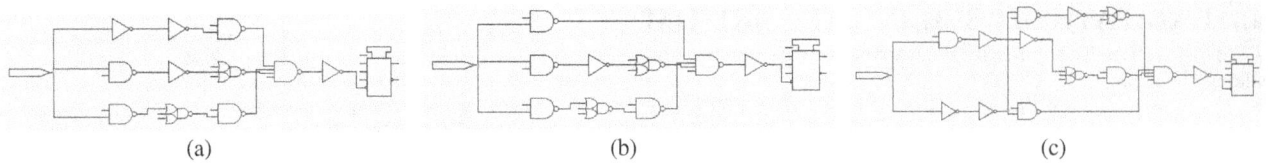

(a)　　　　　　　　　　(b)　　　　　　　　　　(c)

Figure 12: Three datapath bit lines from $Test6-1$. The three bit lines have the same logic function, but they are implemented with different numbers of gates.

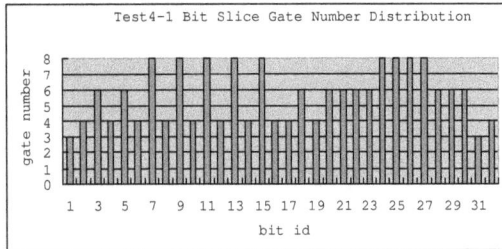

Figure 11: The bit slice gate number distribution of $Test4-1$.

from the rest. By applying our bit slicing algorithm, we can identify these four bit slices. The numbers on gates indicate which bit slice a gate belongs to. For this datapath, there are 4 gates that do not have numbers with them. This means that they do not belong to any bit slice. Therefore, the traditional partition algorithm won't work for this kind of bit slicing problems.

Among these four tests, $Test4$ has a very complicated datapath since one bit line has heavy connections to other bits. Fig 10 shows one piece of $Test4-1$. This datapath is $Latch \rightarrow Latch$. The gate in the circle is the starting bit of $Bit31$, and it has paths to $Bit0 \sim Bit31$. Furthermore, the gates in each bit slice are different too. Fig 11 shows the distribution of bit slice gate numbers. The number of gates in one slice ranges from 3 to 8. Still, the algorithm can identify the 32 bit lines.

The second test set was tested on a linux workstation ($2.8GHz$). Table 2 summarizes the test results of the second test set. For all these datapaths, our algorithm can find the right slicing solution very quickly. Figure 12 shows three bit lines from $Test6-1$. Logically, these three bit lines are the same, but they are implemented with a different number of gates. The highlighted paths are the MFI paths of the three bit lines. The highlight paths of Figure 12 (a) and (b) have the same logic gates, while (c) has two more consecutive inverters. For this case, although some logic calculation (e.g., double inverter removal) could help to identify the datapath bit slices, our approach does not require any logic analysis and calculation.

Table 2: Bit Slicing Tests on Industrial Designs

Datapath	Width	Slices	MinSize	MaxSize	runtime
Test5-3	8	8	10	12	0.55s
Test5-4	64	64	8	10	2.58s
Test5-1	10	10	31	37	0.68s
Test5-2	8	8	6	14	0.57s
Test5-5	16	16	13	17	1.45s
Test6-1	64	64	9	13	0.28s
Test6-2	32	32	4	8	0.17s
Test6-3	64	64	5	7	0.22s
Test6-3	64	64	5	5	0.24s
Test6-4	56	56	6	6	0.19s
Test6-4	56	56	4	6	0.20s
Test7-1	56	56	3	7	0.18s

6. CONCLUSION

In this work, we present a novel approach to identify datapath bit slices. Contrary to the previous template-based approach, we convert the bit slicing problem to the bit matching problem so that bit slices with relatively large difference can still be identified. In order to solve the datapath bit matching, a min-cost max-flow based algorithm is proposed to identify the main-frame of bit slices which indicates the longest similar paths in each bit slice. To further improve the bit slicing solution, an iterative approach is adopted drawing on two new techniques to create more bit matching solutions. Finally, an efficient two way search approach is developed to derive the full bit slices based on the bit matching results. The experimental results demonstrate the effectiveness and efficiency of our approach.

7. REFERENCES

[1] Synopsys Module Compiler User Manual.
[2] R. K. Ahuja, A. V. Goldberg, J. B. Orlin, and R. E. Targan, Finding minimum-cost flows by double scaling, Mathematical programming 53, pp. 243-266, 1992.
[3] R. K. Ahuja, T. L. Magnanti, and J. B. Orlin, Network Flows, Prentice Hall, 1993.
[4] S. R. Arikati and R. Varadarajan, A Signature based Approach to Regularity Extraction, in Proc. Intl. Conf. on Computer Aided Design, 1997.
[5] T. Chan, A. Chowdhary, B. Krishna, A. Levin, G. Meeker, and N. Sehgal, Challenges of CAD Development of Datapath Design, Intel Technology Journal Q1, 1999.
[6] S. Chou, M-K Hsu, Y-W Chang, Structure-aware placement for datapath-intensive circuit designs, in Proc. DAC, 2012.
[7] A. Chowdhary, and R. Gupta, A Methodology for Synthesis of Data Path Circuits, IEEE Design & Test of Computers, 2002.
[8] A. Chowdhary, S. Kale, P. Saripella, N. Sehgal, and R. Gupta, A General Approach for Regularity Extraction in Datapath Circuits, in Proc. Intl. Conf. on Computer-Aided Design, Nov, 1998.
[9] M.R. Corazao, M.A. Khalaf, L.M. Guerra, M. Potkonjak, and J.M. Rabaey, Performance optimization using template mapping for datapath-intensive high-level synthesis, In IEEE Transactions on CAD, pp. 877-887, Aug, 1996.
[10] T. Cormen, C. Leiserson, R. Rivest, and C. Stein, Introduction to Algorithms, the MIT press.
[11] S. Das, and S.P. Khatri, A Regularity-Driven Fast Gridless Detailed Router for High Frequency Datapath Designs, ISPD, 2001.
[12] S. Hassoun, and C. McCreary, Regularity Extraction Via Clan-Based Structural Circuit Decomposition, in Proc. Intl. Conf. on Computer-Aided Design, Nov, 1999.
[13] T. Kutzschebauch, Efficient Logic Optimization Using Regularity Extraction, in Proc. Intl. Conf. on Computer Design, 2000.
[14] T. Kutzschebauch, and L. Stok, Regularity Driven Logic Synthesis, in Proc. Intl. Conf. on Computer-Aided Design, 2000.
[15] D. S. Rao and F. J. Kurdahi, Partitioning by Regularity Extraction, in Proc. Design Automation Conf., 1992.
[16] D. S. Rao and F. J. Kurdahi, On Clustering for Maximal Regularity Extraction, IEEE Trans. on Computer Aided Design, vol 12, 1993.
[17] A. P. E. Rosiello, F. Ferrandi, D. Pandini, and D. Sciuto, A Hash-based Approach for Functional Regularity Extraction during Logic Synthesis, IEEE Computer Society Annual Symposium on VLSI, 2007.
[18] S. I. Ward, M. Kim, N. Viswanathan, Z. Li, C. Alpert, E. E. Swartzlander, Jr., D. Z. Pan, Keep it Straight: Teaching Placement how to Better Handle Designs with Datapaths, in Proc. ISPD, 2012.
[19] S. Ward, D. Ding, D. Z. Pan, PADE: A High-Performance Placer with Automatic Datapath Extraction and Evaluation through High-Dimensional Data Learning, in Proc. DAC, 2012.

FF-Bond: Multi-bit Flip-flop Bonding at Placement

Chang-Cheng Tsai[1], Yiyu Shi[2], Guojie Luo[3], and Iris Hui-Ru Jiang[1]

[1]Dept. of Electronics Engineering and Inst. of Electronics, National Chiao Tung University, Hsinchu 30010, Taiwan
[2]Dept. of ECE, Missouri University of Science and Technology, Rolla, MO 65409, US
[3]Center for Energy-Efficient Computing and Applications (CECA), Peking University, Beijing 100871, P.R. China

shrimpcct@gmail.com; yshi@mst.edu; gluo@pku.edu.cn; huiru.jiang@gmail.com

ABSTRACT

Clock power contributes a significant portion of chip power in modern IC design. Applying multi-bit flip-flops can effectively reduce clock power. State-of-the-art work performs multi-bit flip-flop clustering at the post-placement stage. However, the solution quality may be limited because the combinational gates are immovable during the clustering process. To overcome the deficiency, in this paper, we propose multi-bit flip-flop bonding at placement. Inspired by ionic bonding in Chemistry, we direct flip-flops to merging friendly locations thus facilitating flip-flop merging. Experimental results show that our algorithm, called FF-Bond, can save 27% clock power on average. Compared with state-of-the-art post-placement multi-bit flip-flop clustering, FF-Bond can further reduce 14% clock power.

Categories and Subject Descriptors

B.7.2 [INTEGRATED CIRCUITS]: Design Aids – *placement and routing*

General Terms

Algorithms, Performance, Design.

Keywords

Multi-bit flip-flops, placement, clock power, timing.

1. INTRODUCTION

Clock power has become the main source of chip power in modern IC design [1]. As revealed by [2][3][4], relocating flip-flops benefits clock network synthesis. As shown in Figure 1, compared with single-bit flip-flops, multi-bit flip-flops (MBFFs) present a smaller load on the clock network due to the shared clock logic in the cell [5]. Thus, replacing flip-flops with MBFFs can effectively reduce both the clock network power and the MBFF power consumption. However, the signal wirelength may somewhat increase which may not be acceptable or lead to an increase of power consumption on timing critical paths. Thus, use of MBFFs requires ensuring sufficient timing slacks to avoid impacting timing critical paths.

Due to the lack of physical information before the placement stage, state-of-the-art work handles MBFF clustering at the post-placement stage, e.g., [6][7][8][9][10]. In order not to sacrifice timing, most of these works model the movable regions of flip-flops by an intersection graph. A clique of a proper size in the

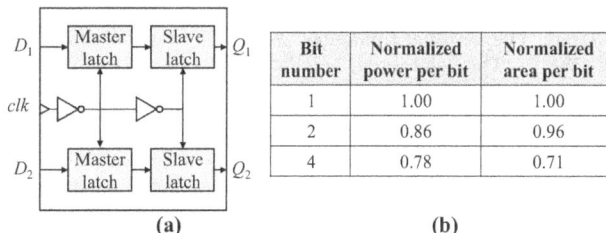

Figure 1. Multi-bit flip-flop. (a) A dual-bit flip-flop, where the inverter chain is shared. (b) Power and area of the MBFF library.

Bit number	Normalized power per bit	Normalized area per bit
1	1.00	1.00
2	0.86	0.96
4	0.78	0.71

intersection graph corresponds to an MBFF. Yan and Chen form MBFFs from largest maximal cliques in [6]. Chang *et al.* present a progressive window-based clustering method in [7]. Wang *et al.* allocate MBFFs extracted from a randomly sampled subset of maximal cliques in [8]. Jiang *et al.* encode the intersection graph by interval graphs to identify mergeable flip-flops in [9]. Liu *et al.* propose a bottom-up merging method in [10]. Among these works, [9] delivers the most power efficient result.

However, the combinational gates are immovable during the post-placement MBFF clustering scheme. The clustering flexibility and quality are thus limited. To break this limitation, in this paper, we perform MBFF bonding at placement.

A possible solution is to directly integrate placement and post-placement MBFF clustering together. These two tasks are sequentially applied at each iteration. Nevertheless, if doing so, the movement of flip-flops is constrained by the placement at the current iteration and may oscillate among iterations.

In contrast, inspired by ionic bonding in Chemistry [11], we guide flip-flops to move towards merging friendly locations at the global placement stage without sacrificing timing. An ionic bond is formed when the atom of an element releases some of its electron(s) and the atom of another element then captures the electron(s) to attain a stable electron configuration. (see Figure 2(a)) We devise a flip-flop bonding scheme so that flip-flops are

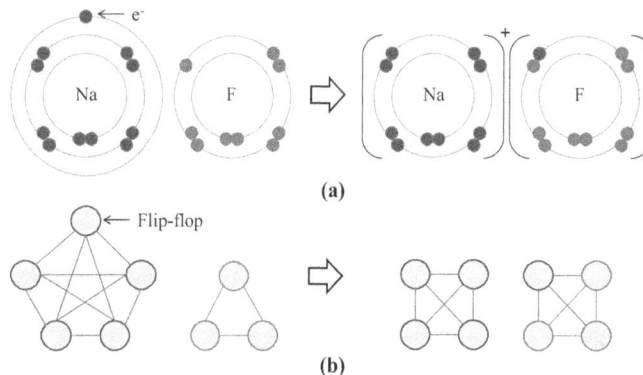

Figure 2. (a) Ionic bonding: Na + F→ NaF. (b) Flip-flop bonding.

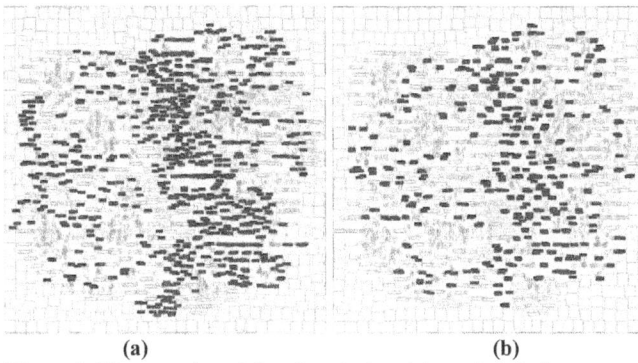

Figure 3. The snapshot right after timing-driven global placement of s38417. (The solid squares indicate single-bit flip-flops.) (a) Without flip-flop bonding. (Number of resulting 4-/2-/1-bit flip-flops: 35/252/237.) (b) With flip-flop bonding. (Number of resulting 4-/2-/1-bit flip-flops: 159/105/35.)

moved to merging friendly locations. For example, according to the MBFF library given in Figure 1(b), a four-bit flip-flop is most power efficient. Thus, a clique of size 4 in the intersection graph is considered perfect. Via flip-flop bonding, we release flip-flops from an oversized clique (larger than 4) to an undersized clique (less than 4). (see Figure 2(b))

In this paper, we propose an MBFF bonding at placement algorithm, called FF-Bond. To demonstrate our flow, we develop a net-based timing-driven placer [12]. The wirelength-driven placement kernel is based on an analytical placement method proposed in [13]. Rather than incorporating an approximate delay model into the placer, we tune the net weights by the timing slacks computed by a signoff timing engine for more accurate timing information. By introducing a flip-flop bonding force, we guide each flip-flop to a merging friendly location. (see Figure 3) Consequently, after timing-driven global placement with flip-flop bonding, flip-flops can easily be merged together thus reducing power. Legalization and detailed placement are then performed to remove overlap and incrementally refine the placement result.

Experimental results show that FF-Bond can save 27% clock power on average. Compared with state-of-the-art post-placement MBFF clustering, FF-Bond can further reduce 14% clock power. The remainder of this paper is organized as follows. Section 2 introduces post-placement MBFF clustering and gives the problem formulation. Section 3 details our MBFF bonding at placement algorithm, FF-Bond. Section 4 lists experimental results. Finally, Section 5 gives a conclusion.

2. PRELIMINARIES

In this section, we introduce post-placement MBFF clustering and give the problem formulation.

2.1 Post-Placement MBFF Clustering

The post-placement MBFF clustering problem is that given a placed design, an MBFF library, timing slacks, and placement density constraints, replace flip-flops with MBFFs such that the power is minimized and the timing and placement density constraints are satisfied.

As mentioned in Section 1, INTEGRA proposed in [9] delivers the most power efficient result among prior works. We take INTEGRA as an example to demonstrate post-placement MBFF clustering.

First of all, timing analysis reports the timing slacks of the fanin/fanout pin of each flip-flop. Based on a delay-wirelength

conversion, the movable region of each flip-flop without hurting timing is obtained. As shown in Figure 4(a), the fanin and fanout slacks are converted to diamonds. The overlap region of these diamonds is the feasible region of a flip-flop. Figure 4(b) illustrates the extracted feasible regions of flip-flops for a sample design. The corresponding intersection graph is constructed as shown in Figure 4(c). If the feasible regions of several flip-flops overlap (i.e., a clique in the intersection graph), these flip-flops can form an MBFF.

As shown in Figure 4(d), INTEGRA applies coordinate transformation and encodes the intersection graph by two interval graphs. Two sequences are used to record the starting (type s) and ending (type e) x'-/y'-coordinates of feasible regions in ascending order. It is shown that all maximal cliques can be extracted at decision points (the 'se' patterns in the sorted x'-sequence). {1, 2, 4} and {1, 3, 4} are found at the first decision point; {3, 4, 5, 6} is found at the second one; {4, 5, 6, 7, 8} is found at the third one. INTEGRA scans the x'-sequence and generates MBFFs at decision points. The clustering result is shown in Figure 4(e).

2.2 Problem Formulation

As mentioned in Section 1, the flexibility and solution quality of post-placement MBFF clustering are limited. To overcome the deficiency, in this paper, we perform MBFF bonding at placement.

The problem formulation is described as follows.

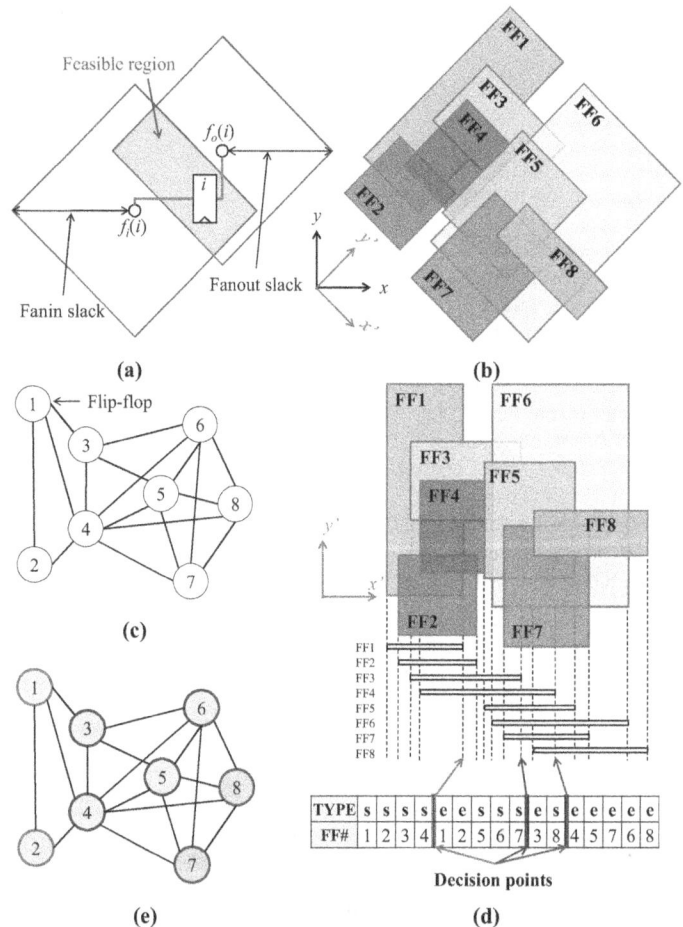

Figure 4. Post-placement MBFF clustering. (a) Feasible region. (b) Feasible region extraction. (c) Intersection graph. (d) INTEGRA. (e) MBFF clustering result: {1, 2}, {3, 4, 5, 6}, {7, 8}.

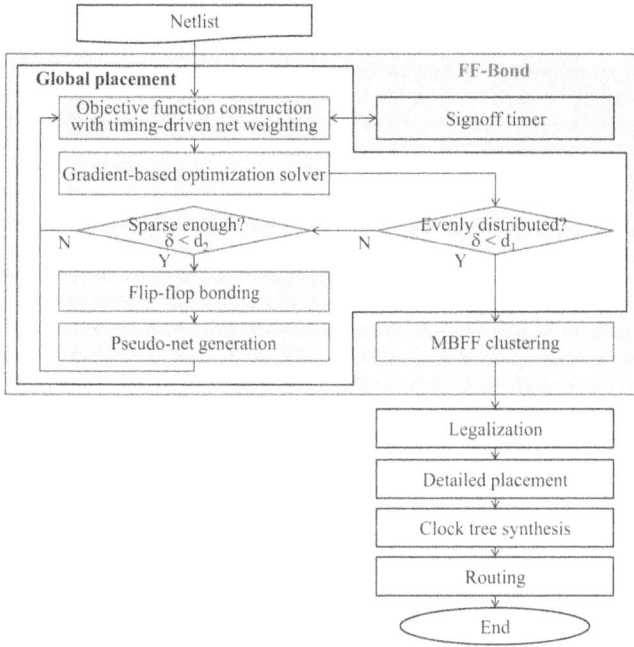

Figure 5. The overview of FF-Bond.

The MBFF Bonding at Placement Problem: Given a netlist, an MBFF library, timing constraints, and placement density constraints, find a placement and replace flip-flops with MBFFs such that the power is minimized and the timing and placement density constraints are satisfied.

3. OUR ALGORITHM—FF-BOND

In this section, we propose the MBFF bonding at placement algorithm, FF-Bond.

3.1 Overview

It can be seen that if flip-flop 2 in Figure 4 moves towards flip-flop 3, their feasible regions may overlap, and the clustering result will be improved. (Two four-bit flip-flops {1, 2, 3, 4} and {5, 6, 7, 8} can be formed.) Therefore, we propose FF-Bond to guide flip-flops towards merging friendly locations at the global placement stage without sacrificing timing.

Figure 5 shows the overview of FF-Bond. To demonstrate our flow, we develop a net-based timing-driven placer [12]. The wirelength-driven placement kernel is based on an analytical placement method, mPL5, proposed in [13]. Instead of using an approximate delay model in the placer, the net weights are adjusted according to the timing slacks computed by a signoff timing engine for more accurate timing information. By introducing a flip-flop bonding force, we guide each flip-flop to a merging friendly location. After timing-driven global placement with flip-flop bonding, flip-flops are merged. Legalization and detailed placement are then applied to remove overlaps and refine the placement. Clock network synthesis and routing are finally performed.

3.2 Timing-Driven Placement

To demonstrate our flow, we develop a timing-driven placer based on a pure wirelength-driven placement kernel mPL5 [13] and the slack-based net-weighting technique [14]. Pure wirelength-driven global placement is applied only for the first iteration, while timing-driven global placement is applied for the subsequent iterations. We shall introduce these two techniques in this subsection.

3.2.1 Wirelength-driven placement kernel
The pure wirelength-driven placement kernel is based on an analytical placement method, mPL5, proposed in [13]. A netlist is modeled by a hypergraph $H=(V, E)$, where V denotes the set of cells and hyperedges in E represent nets. (x_i, y_i) represents x-/y-coordinates of cell i. First of all, the placement region is divided into $m \times n$ non-overlapping uniform bins. The following constrained minimization problem is considered.

$$\min W(x, y)$$
$$\text{s.t.} D_{ij} = K, 1 \le i \le m, 1 \le j \le n, \qquad (1)$$

where $W(x, y)$ is the wirelength function defined by half-perimeter wirelength (HPWL), D_{ij} means the average density of bin B_{ij}, and K is the target density computed by the total cell area divided by the area of the placement region. The objective function and constraints are not differentiable. The wirelength function is smoothed by log-sum-exp approximation [15].

$$\widehat{W}(x, y) = \eta \sum_{e \in E} \left(\log \sum_{v_k \in e} \exp\left(\frac{x_k}{\eta}\right) + \log \sum_{v_k \in e} \exp\left(\frac{-x_k}{\eta}\right) + \right.$$
$$\left. \log \sum_{v_k \in e} \exp\left(\frac{y_k}{\eta}\right) + \log \sum_{v_k \in e} \exp\left(\frac{-y_k}{\eta}\right) \right). \qquad (2)$$

Furthermore, the inverse Laplace transformation is applied to smooth the density function.

$$\min \widehat{W}(x, y)$$
$$\text{s.t.} \psi_{ij} = \overline{K}, 1 \le i \le m, 1 \le j \le n. \qquad (3)$$

Via Lagrange multipliers, the density constraint is converted to a penalty into the objective function.

$$L(x, y, \lambda) = \widehat{W}(x, y) + \sum_{i,j} \lambda_{ij}(\psi_{ij} - \overline{K}). \qquad (4)$$

A gradient-based optimization solver is then applied to solve the nonlinear program.

3.2.2 Slack-based net weighting
We adopt slack-based net weighting since this timing-driven placement approach has low computational complexity and high flexibility [12]. To reflect timing criticalities, we adjust net weights at each iteration according to the timing slacks. Instead of incorporating an approximate delay model into the placer, we rely on a signoff timing engine. We assign negative slack nets with larger net weights than positive slack nets. Thus, the placement kernel tends to shorten the negative slack nets to resolve timing violations. The net weight at an iteration is defined as follows [14].

$$net\ weight = \left(1 - \frac{slack}{T_{clk}}\right)^{\alpha}, \qquad (5)$$

where T_{clk} is the clock period for a particular net, and $\alpha > 1$ is the criticality exponent to emphasize critical nets. At the first iteration, slack is set to 0 for pure wirelength-driven global placement.

3.3 Flip-flop Bonding
In this subsection, we shall detail the flip-flop bonding mechanism to guide flip-flops towards merging friendly locations.

Consider the possible at-placement MBFF merging method mentioned in Section 1, where placement and post-placement MBFF clustering are directly integrated together. If doing so, the movement of flip-flops is guided by the post-placement MBFF clustering result according to the current placement. This guidance does not encourage orphan flip-flops to merge with others and may oscillate among iterations.

In contrast, we devise a flip-flop bonding mechanism inspired by ionic bonding in Chemistry [11]. For example, consider two maximal cliques of size 5 and 3. Based on the MBFF library given in Figure 1(b), post-placement MBFF clustering may generate one four-bit flip-flop, one dual-bit flip-flop, and two single-bit flip-flops (orphans). As illustrated in Figure 2(b), if one flip-flop in the maximal clique of size 5 is attracted to the maximal clique of size 3, we may have two four-bit flip-flops instead. Hence, by introducing a flip-flop bonding force, we direct each flip-flop towards a merging friendly location, thus forming more larger-bit flip-flops.

'Given an MBFF library, the bit number of the most power efficient cell is considered as the perfect clique size. (The most power efficient flip-flop cell has the lowest normalized power per bit.) Hence, all extracted maximal cliques are classified into oversized, perfect, and undersized cliques accordingly. (e.g., Figure 6) An oversized clique can form at least one perfect-sized clique and possibly leave several single-bit flip-flops (that we try to avoid). A perfect-sized clique is desired. An undersized clique is to attract flip-flops to form a perfect-sized clique.

Flip-flop bonding tries to bond flip-flops into perfect-sized cliques. The priority of processing maximal cliques is in the following order: perfect, undersize, oversize. Perfect-sized cliques are preserved first. An investigated undersized/oversized clique selects the most adjacent flip-flops in a specified search region to form a target-sized clique. (The search region and adjacency are defined later.) Undersized cliques are handled in descending order of clique size. The target size of an undersized clique means the bit number defined in the MBFF library that is larger than and nearest to the investigated clique size. Similarly, the target size of an oversized clique is the flip-flop configuration that is larger than, nearest to, and more power efficient than the investigated clique size. For example, considering the MBFF library given in Figure 1, 2 is the target size for 1; 4 is for 2 and 3; 6 is for 5; 8 is for 6 and 7. Please note that our flip-flop bonding mechanism is general, not limited to the MBFF library given in Figure 1(b).

Figure 7 demonstrates a flip-flop bonding example. As shown in Figure 7(b), first of all, all maximal cliques are extracted based on the method presented in [9]. Figure 7(c) shows the clusters based on our flip-flop bonding strategy, where the processing order is indicated by the number beside each cluster. After the flip-flop bonding force is applied (see Section 3.4), flip-flops in each cluster are moved to each other, thus facilitating MBFF merging.

In some cases, maximal cliques may overlap. Basically, we apply the same flip-flop bonding strategy. For cliques of the same size, the clique with most independent flip-flops is processed first. An independent flip-flop means a flip-flop exists in exactly one maximal clique. Figure 7(e) shows an example with overlapping maximal cliques, while Figure 7(f) shows the resulting bonding clusters. The processing order of cliques of size 3 is indicated by the number beside each cluster. After the first two cliques of size 3 are processed, the third one has no independent flip-flops, and thus this clique is skipped.

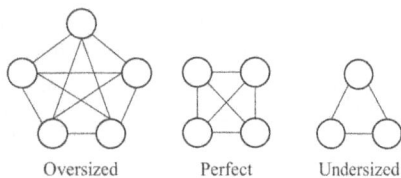

Figure 6. Clique sizes. For the MBFF library given in Figure 1, oversize means clique size > 4; perfect size means clique size = 4; undersize means clique size < 4.

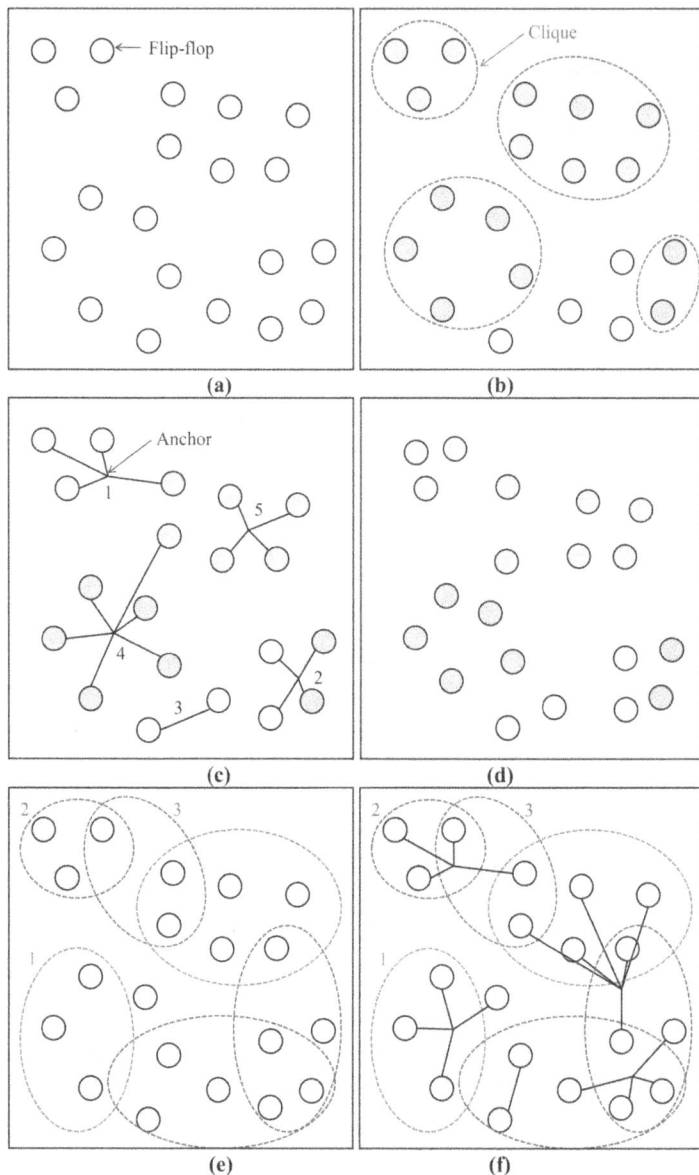

(a) (b) (c) (d) (e) (f)

Figure 7. Flip-flop bonding. (a) Flip-flops before flip-flop bonding. (b) Maximal cliques. (c) Flip-flop bonding clusters. (d) Flip-flops after flip-flop bonding. (e) Maximal cliques overlap. (f) Flip-flops after flip-flop bonding.

For the flip-flop bonding strategy, we define a search region and adjacency. The search region prevents flip-flops from attracting distant flip-flops. The adjacency reflects the physical distance and timing information. Let (x_C, y_C) denote the average x-/y-coordinates of clique C and (x_i, y_i) denote the x-/y-coordinates of flip-flop i. Assume that the fanin and fanout slack of flip-flop i is $s_{fi}(i)$ and $s_{fo}(i)$, respectively. The adjacency between clique C and flip-flop i is defined as follows.

$$A(C,i) = |x_C - x_i| + |y_C - y_i| - \varepsilon_i \left(s_{fi}(i) + s_{fo}(i) \right),$$
$$\varepsilon_i = \begin{cases} -\infty, & \text{if } s_{fi}(i) < 0 \text{ or } s_{fo}(i) < 0 \\ \tau, & \text{otherwise} \end{cases} \tag{6}$$

τ is the delay-wirelength conversion parameter used in Section 2.1.

3.4 Pseudo-Net Generation

After flip-flop bonding, we introduce a flip-flop bonding anchor for each flip-flop cluster. Each flip-flop within a cluster is linked

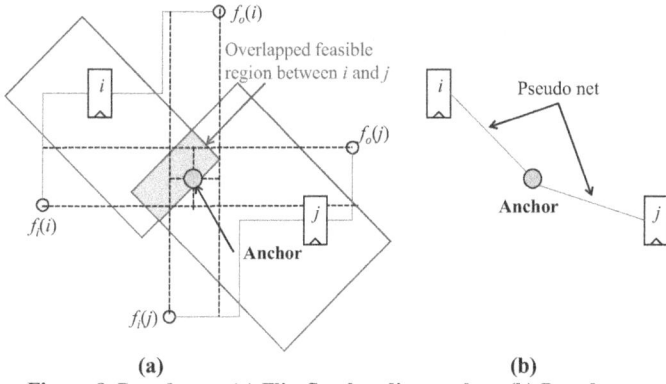

Figure 8. Pseudo net. (a) Flip-flop bonding anchor. (b) Pseudo net.

to the anchor using a two-pin pseudo net that is assigned a high net weight. Within FF-Bond, these two-pin nets are viewed as flip-flop bonding attractions. To emphasize the attractions, their weights should be greater than the default net weight for a positive slack net, say 5X in our experiments.

The anchor is set to the desired location of the potentially formed MBFF. The post-placement MBFF clustering methods place MBFFs within feasible regions due to timing constraints. However, at global placement, cells can flexibly be moved and timing is still maintained. Therefore, the anchor is set to the (center of) median of all fanin/fanout x-/y-coordinates (for signal wirelength consideration). Figure 8 shows a flip-flop cluster with two flip-flops. Figure 8(a) illustrates the anchor introduced, while Figure 8(b) shows the generated pseudo nets.

3.5 The Condition to Apply Flip-flop Bonding

At the early iterations of global placement, cells strongly overlap. While the gradient-based optimization solver computes Lagrange multipliers, cells are gradually moved towards optimal locations. During this process, the overlap among cells is iteratively reduced. Finally, when the amount of overlap is small enough, the optimizer stops. Because the placement at early iterations is quite different from the final result, flip-flop bonding is applied when cells are sparse enough.

We use an overlap index to control the global placement flow. The overlap index δ is defined by the total overlap cell area divided by the total cell area. During the global placement optimization, cells are gradually spread out, and thus δ decreases iteration by iteration. When $\delta < d_2$, flip-flop bonding is applied. d_2 is a user-specified parameter. The greater d_2, the earlier iteration flip-flop bonding is applied (the potentially larger flexibility to merge flip-flops). Later, in our experiments, d_2 is set to 0.5.

The timing-driven incremental placement and flip-flop bonding are repeated until cells are evenly distributed, $\delta < d_1$. d_1 is a user-specified parameter. Usually, d_1 is very small such that the overlap of cells is small and the density constraint is satisfied. Later, in our experiments, d_1 is set to 0.1.

4. EXPERIMENTAL RESULTS

FF-Bond was implemented in the C++ programming language on a Linux workstation with an Intel Xeon 2.4GHz CPU and 16GB memory. Experiments are conducted on the circuits from IWLS 2005 benchmark [16]. The MBFF library is designed by [17]. (see Figure 1) These circuits are synthesized and legalized by state-of-the-art commercial tools [18][19] based on UMC 55nm technology. The signoff timing engine is [20]. To test the effectiveness, FF-Bond is compared with two representative flows:

Post-placement MBFF clustering (PMC) performs timing-driven global placement followed by post-placement MBFF clustering. Interleaving placement and post-placement MBFF clustering (IMC) interleaves timing-driven global placement and post-placement MBFF clustering. The post-placement MBFF clustering method used in our experiments is based on [9] because of its superior power efficiency. The parameters in FF-Bond used in our experiments are set as follows: $d_1 = 0.1$, $d_2 = 0.5$, $\alpha = 1.2$, the search region is bounded by 20% of chip dimension, and the net weight of a pseudo net is 5X the default value for a positive slack net.

Table 1 compares the three flows on flip-flop power and generated MBFFs. FF-Bond obtains the best power efficiency among the three flows. 'FF power ratio' means the total power of generated flip-flops divided by the power of using only single-bit flip-flops. Without timing consideration, the lower bound of FF power ratio is 0.78. (All are four-bit flip-flops.) The FF power ratio of FF-Bond is very close to the lower bound. Moreover, the main constituents of formed flip-flops of FF-Bond are four-bit flip-flops (compared with single-bit flip-flops for PMC, and dual-bit flip-flops for IMC). These results show that flip-flop bonding indeed effectively guides flip-flops to merging friendly locations. Although FF-Bond results in longer wirelength, the increased wirelength induces less than 1% chip power increase in our experiments (because data signals do not always toggle in every cycle). Hence, the tradeoff between signal power and clock power is good. Figure 9 shows the global placement and MBFF merging results of s38417 of the three flows.

Table 2 compares the three flows on clock power (including clock network and MBFFs). The clock network is synthesized based on [21]. FF-Bond obtains the lowest clock power among the three flows. The fewer clock sinks, the simpler clock network. Hence, FF-Bond saves 19% flip-flop power and saves even more in terms of clock power. For s38584, IMC consumes slightly lower clock power than FF-Bond because of smaller clock buffers used, but FF-Bond still achieves fewer clock sinks. FF-Bond can totally save 27% clock power on average. Compared with post-placement MBFF clustering, FF-Bond can further reduce 14% clock power.

Table 3 compares PMC and FF-Bond on timing slacks. The timing slack of each endpoint means the worst slack over all paths ending at this endpoint. 'Worst slack' represents the worst timing slack over all endpoints, while 'Average slack' means the average. Because we consider timing during FF-Bond, the slacks of the two flows are quite similar.

Table 4 shows the impact of d_2 on the MBFF bonding results of FF-Bond. For a smaller d_2, flip-flop bonding starts at later iterations but does not guide flip-flops well because the low flexibility of moving flip-flops. In contrast, for a larger d_2, flip-flop bonding starts from earlier iterations but results in longer wirelength because distant flip-flops are attracted.

Table 5 and Figure 10 show the impact of the search region on flip-flop power, slack, and wirelength. For s38417, when the search region is set to 0.2X chip dimension, the balance between flip-flop power, slack, and wirelength is good.

Tables 1–5 show that FF-Bond is promising to merge flip-flops.

5. CONCLUSION

Applying MBFFs can effectively reduce clock power. Unlike state-of-the-art work performed MBFF clustering at the post-placement stage, in this paper, we did MBFF bonding at placement. Inspired by ionic bonding in Chemistry, we directed flip-flops to merging friendly locations. Experimental results showed that FF-Bond can save 27% clock power on average. Compared with post-placement MBFF clustering, FF-Bond can further reduce 14% clock power. Future work includes MBFF bonding with routability consideration.

6. REFERENCES

[1] L.-T. Wang, Y.-W. Chang, and K.-T. Cheng (editors). *Electronic Design Automation: Synthesis, Verification, and Test*, Elsevier/Morgan Kaufmann, 2009.

[2] Y.-S. Cheon, P.-H. Ho, A. B. Kahng, S. Reda and Q. Wang. Power-aware placement. In *Proc. Design Automation Conf. (DAC)*, pp. 795–800, 2005.

[3] D. Papa, N. Viswanathan, C. Sze, Z. Li, G.-J. Nam, C. Alpert, I.L. Markov. Physical synthesis with clock-network optimization for large systems on chips. *IEEE Micro*, vol. 31, no. 4, Apr. 2011, pp. 51–62.

[4] D.-J. Lee and I. L. Markov. Obstacle-aware clock-tree shaping during placement. *IEEE Trans. Computer-Aided Design (TCAD)*, vol. 31, no. 2, Feb 2012, pp. 205–216.

[5] 2010 CAD contest of Taiwan, http://cad contest.ee.ntu.edu.tw/cad10/Problems/B1_Faraday_091223_MultiBitFF.pdf.

[6] J.-T. Yan and Z.-W. Chen. Construction of constrained multibit flipflops for clock power reduction. In *Proc. Int. Conf. Green Circuits Syst. (ICGCS)*, pp. 675–678, 2010.

[7] Y.-T. Chang, C.-C. Hsu, M. P.-H. Lin, Y.-W. Tsai, and S.-F. Chen. Post-placement power optimization with multi-bit flip-flops. In *Proc. Int. Conf. on Computer-Aided Design (ICCAD)*, pp. 218–223, 2010.

[8] S.-H. Wang, Y.-Y. Liang, T.-Y. Kuo, and W.-K. Mak. Power-driven flip-flop merging and relocation. In *Proc. Int. Symp. on Physical Design (ISPD)*, pp. 107–114, 2011.

[9] Iris H.-R. Jiang, C.-L. Chang, Y.-M. Yang. INTEGRA: Fast multibit flip-flop clustering for clock power saving. *IEEE Trans. Computer-Aided Design (TCAD)*, vol. 31, no. 2, Feb 2012, pp. 192–204. Also see *Proc. Int. Symp. on Physical Design (ISPD)*, pp. 115–121, 2011.

[10] S. S.-Y. Liu, C.-J. Lee and H.-M. Chen. Agglomerative based flip-flop merging for power optimization. In *Proc. Design, Automation & Test in Europe Conference & Exhibition (DATE)*, pp. 1391–1396, 2012.

[11] Ionic and covalent bonds. http://chemwiki.ucdavis.edu/.

[12] D. Z. Pan, B. Halpin, and H. Ren. Timing driven placement. Handbook of Algorithms for VLSI Physical Automation, CRC Press, 2007.

[13] T. Chen, J. Cong, and K. Sze. Multilevel generalized force-directed method for circuit placement. In *Proc. Int. Symp. on Physical Design (ISPD)*, pp. 185–192, 2005.

[14] A. Marquardt V. Betz and J. Rose. Timing driven placement for FPGAs. In *Proc. Int. Symp. on Field Programmable Gate Arrays (FPGA)*, pp. 203–213, 2000.

[15] W. C. Naylor, R. Donelly, and L. Sha. Non-linear optimization system and method for wire length and delay optimization for an automatic electric circuit placer. *U.S. Patent 6301693*, Oct. 9, 2001.

[16] IWLS 2005 benchmarks. http://iwls.org/iwls2005/benchmarks.html.

[17] Faraday Technology Corp.

[18] Design Compiler. Synopsys, Inc.

[19] SoC Encounter. Cadence Design Systems, Inc.

[20] Prime Time. Synopsys, Inc.

[21] Y.-C. Chang, C.-K. Wang and H.-M. Chen. On constructing lower power and robust clock tree via slew budgeting. In *Proc. Int. Symp. on Physical Design (ISPD)*, pp. 129–136, 2012.

Table 3. Slack Comparison.

Circuit Name	Clock period (ns)	PMC		FF-Bond	
		Worst slack (ns)	Average slack (ns)	Worst slack (ns)	Average Slack (ns)
s13207	1.5	0.042	0.580	0.041	0.579
s15850	1.8	0.158	0.336	0.154	0.334
s38417	2.0	0.164	1.049	0.163	1.047
s38584	2.0	0.217	0.871	0.209	0.869
b17	3.0	0.122	1.272	0.128	1.269
b19	2.7	0.112	1.278	0.109	1.273

Table 5. FF-Bond: Search Region vs. Flip-flop Power, Wirelength, and Slack. (s38417)

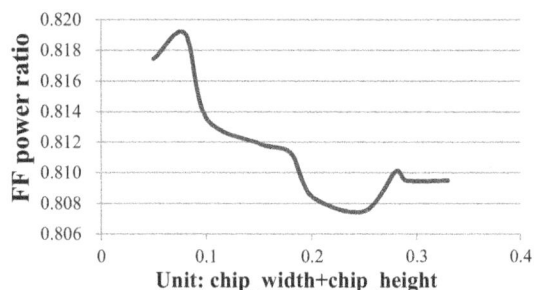

Search region (Unit: chip_width+chip_height)	FF power ratio	HPWL	Worst slack (ns)
0.05	0.817	4.89E+07	0.163
0.08	0.819	4.98E+07	0.163
0.10	0.814	5.07E+07	0.164
0.15	0.812	5.23E+07	0.163
0.18	0.811	5.34E+07	0.164
0.20	0.808	5.41E+07	0.163
0.25	0.807	5.41E+07	0.163
0.28	0.810	5.41E+07	0.163

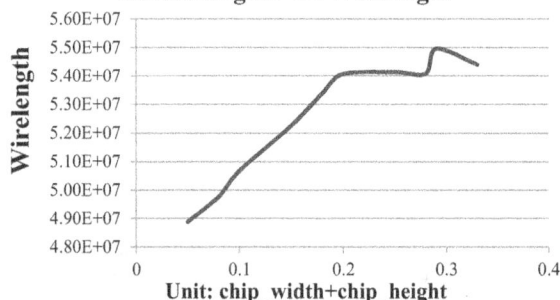

Figure 10. FF-Bond: Search region analysis. (s38417)

Table 1. Flip-flop Power Comparison.

Circuit Name	#Flip-flops	PMC			IMC			FF-Bond		
		#MBFFs 4-/2-/1-bit	FF power ratio	HPWL	#MBFFs 4-/2-/1-bit	FF power ratio	HPWL	#MBFFs 4-/2-/1-bit	FF power ratio	HPWL
s13207	212	8/57/66	0.892	4.569E+06	23/51/18	0.837	4.975E+06	35/31/10	0.814	5.344E+06
s15850	128	10/29/30	0.868	2.117E+06	18/23/10	0.826	2.869E+06	23/15/6	0.809	2.903E+06
s38417	881	35/252/237	0.885	4.599E+07	105/179/103	0.838	4.789E+07	159/105/35	0.808	5.406E+07
s38584	1069	46/291/303	0.886	5.992E+07	96/282/121	0.847	6.213E+07	203/116/25	0.803	6.926E+07
b17	1068	53/264/328	0.887	1.346E+08	137/201/118	0.834	1.363E+08	196/124/36	0.806	1.470E+08
b19	4384	378/886/1100	0.868	7.187E+08	593/742/528	0.834	7.267E+08	851/425/130	0.802	8.023E+08
Avg. ratio	-	0.21/0.91/1.00	0.881	0.85	1.20/2.05/1.00	0.836	0.92	5.33/3.33/1.00	0.807	1.00

Table 2. Clock Power Comparison. (Flip-flops and Clock Networks)

Circuit Name	Without MBFF clustering				PMC				IMC				FF-Bond			
	Total Cap. (pF)	Sinks	Buffer	Wire	Total Cap. (pF)	Sinks	Buffer	Wire	Total Cap. (pF)	Sinks	Buffer	Wire	Total Cap. (pF)	Sinks	Buffer	Wire
s13207	1.333	48.5%	36.4%	15.1%	1.223	46.8%	39.7%	13.5%	1.094	49.6%	38.8%	11.7%	1.056	49.8%	40.2%	10.0%
s15850	0.901	43.3%	47.1%	9.6%	0.837	40.9%	50.6%	8.5%	0.806	39.6%	52.6%	7.8%	0.799	39.5%	53.1%	7.4%
s38417	5.051	53.2%	31.6%	15.2%	4.113	57.9%	26.8%	15.3%	3.884	58.2%	27.4%	14.5%	3.711	58.5%	28.6%	12.9%
s38584	6.100	53.5%	28.9%	17.6%	5.352	54.3%	29.9%	15.8%	4.576	60.6%	24.1%	15.3%	4.870	53.9%	32.8%	13.3%
b17	6.273	51.9%	28.1%	20.0%	5.574	51.8%	28.7%	19.5%	5.241	51.9%	30.5%	17.6%	4.513	58.2%	26.3%	15.5%
b19	25.611	52.2%	26.8%	21.0%	22.081	52.4%	28.1%	19.5%	19.410	57.4%	23.9%	18.7%	18.277	58.7%	24.5%	16.8%
Avg. Ratio	1.00	-	-	-	0.87	-	-	-	0.77	-	-	-	0.73	-	-	-

Table 4. FF-Bond: Flip-Flop Power and Wirelength Comparison under Different d_2.

Circuit Name	#Flip-flops	d_2=0.3			d_2=0.5			d_2=0.7		
		#MBFFs 4-/2-/1-bit	FF power ratio	HPWL	#MBFFs 4-/2-/1-bit	FF power ratio	HPWL	#MBFFs 4-/2-/1-bit	FF power ratio	HPWL
s13207	212	27/41/22	0.834	5.243E+06	35/31/10	0.814	5.344E+06	38/25/10	0.809	5.518E+06
s15850	128	20/20/8	0.819	2.620E+06	23/15/6	0.809	2.903E+06	22/16/8	0.814	2.895E+06
s38417	881	171/85/27	0.802	5.258E+07	159/105/35	0.808	5.406E+07	164/89/47	0.808	5.569E+07
s38584	1069	194/135/23	0.805	6.861E+07	203/116/25	0.803	6.926E+07	192/134/33	0.807	6.944E+07
b17	1068	186/135/23	0.809	1.460E+08	196/124/36	0.806	1.470E+08	202/116/28	0.803	1.485E+08
b19	4384	847/427/142	0.803	7.927E+08	851/425/130	0.802	8.023E+08	851/431/118	0.802	8.037E+08
Avg. ratio	-	4.99/3.44/1.00	0.812	0.97	5.33/3.33/1.00	0.807	1.00	5.04/3.04/1.00	0.807	1.01

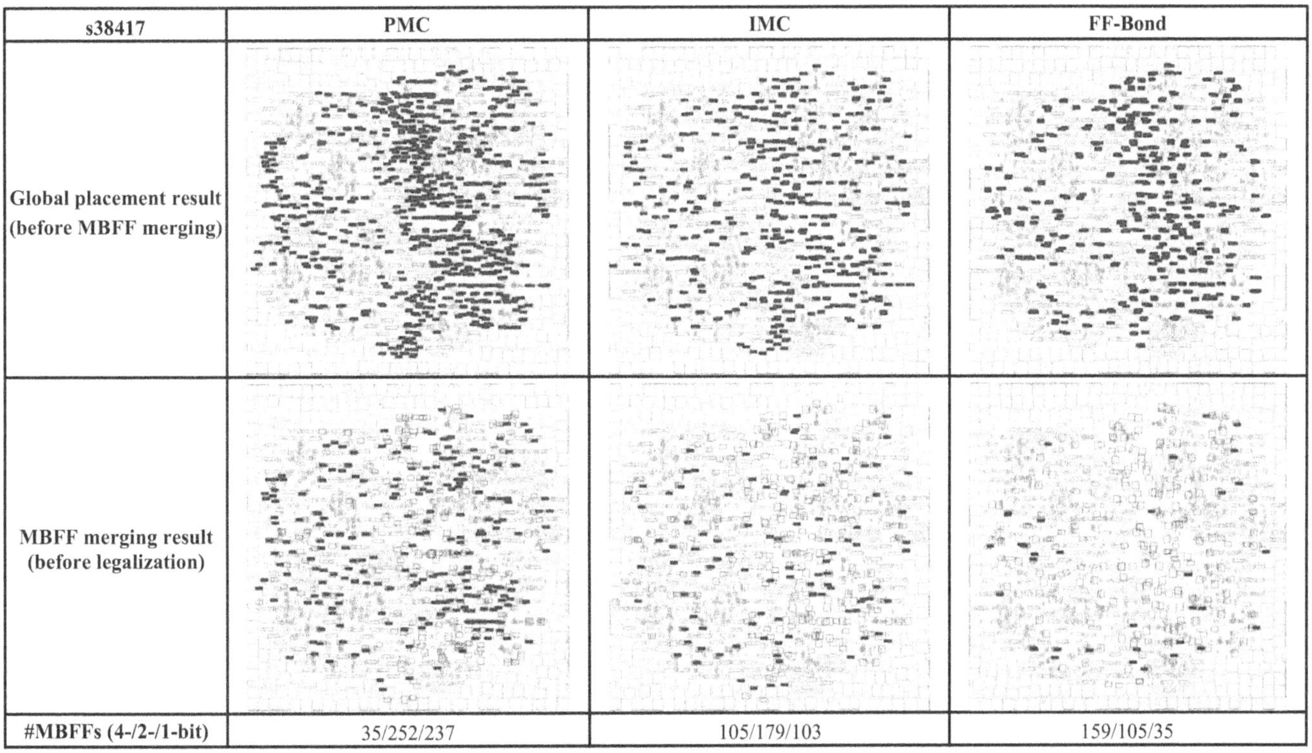

s38417	PMC	IMC	FF-Bond
Global placement result (before MBFF merging)			
MBFF merging result (before legalization)			
#MBFFs (4-/2-/1-bit)	35/252/237	105/179/103	159/105/35

Figure 9. Global placement and MBFF merging results of s38417. Grey boxes indicate combinational cells and IOs. Solid boxes indicate single-bit flip-flops, while dark squares indicates MBFFs.

Buffer Sizing for Clock Networks Using Robust Geometric Programming Considering Variations in Buffer Sizes

Logan Rakai, Amin Farshidi, Laleh Behjat, David Westwick
University of Calgary
2500 University Dr NW
Calgary, Alberta, Canada
{lmrakai, afarshid, laleh, dwestwic}@ucalgary.ca

ABSTRACT

Minimizing power and skew for clock networks are critical and difficult tasks which can be greatly affected by buffer sizing. However, buffer sizing is a non-linear problem and most existing algorithms are heuristics that fail to obtain a global minimum. In addition, existing buffer sizing solutions do not usually consider manufacturing variations. Any design made without considering variation can fail to meet design constraints after manufacturing. In this paper, first we proposed an efficient optimization scheme based on *geometric programming* (GP) for buffer sizing of clock networks. Then, we extended the GP formulation to consider process variations in the buffer sizes using *robust optimization* (RO). The resultant variation-aware network is examined with SPICE and shown to be superior in terms of robustness to variations while decreasing area, power and average skew.

Categories and Subject Descriptors

I.6.5 [**Model Development**]: [Modeling Methodologies]; J.6 [**Computer-Aided Engineering**]: [Computer-Aided Design (CAD)]

General Terms

Algorithms, Design

Keywords

Clock network, Robust optimization, Geometric programming

1. INTRODUCTION

Modern clock networks comprise a significant amount of wire and area in integrated circuits and consume a large portion of power [36]. Buffers are an integral part of a clock network added to maintain the signal integrity. The goal of the buffer insertion step is to insert buffers to meet the skew and slew constraints with an appropriate safety margin [27]. Buffers inserted in the clock network consume a large amount of power and it is necessary to keep the increase in power consumption of the network as small as possible, while maintaining low skew. Hence, the buffers are sized to optimize the performance of the network. The current buffer sizing formulations result in optimization problems which are hard to solve. In addition, under process variation, the sizes of these buffers can vary greatly, affecting the predictability of the performance and lowering the yield of the design.

In this paper, we propose a low power solution to the *buffer sizing problem*, where the clock tree is designed and buffers are already inserted. The main contributions of this paper are:

- **Developing a GP formulation** for the clock network buffer sizing problem, where the non-convex buffer sizing problem is transformed into a series of *convex optimization* problems that can be solved efficiently.

- **Efficient variation modelling** in the buffer sizing formulation using *spatial correlation* of process variations instead of the worst-case variation scenarios, resulting in more area-efficient designs with minimal degradation in quality.

- **Applying a robust optimization** technique to solve the GP formulation and find a robust solution.

In all the experiments, **SPICE** is used to calculate skew, power and producing reliable and accurate results. Other properties of our work are its flexibility in extending the formulation to include other network components and constraints such as wire widths, and offering designers the ability to optimally trade area for skew.

In summary, the paper illustrates that by performing theoretical work, e.g. converting buffer sizing problem to GP format, a designer can produce superior results and non-convex problems that could previously only be solved by heuristics can be solved using efficient convex optimization techniques and obtain considerable power savings. The second major contribution of this paper is showing that process variation can be very efficiently included in the design process using state-of-the-art optimization technique, robust optimization, to greatly improve the solution quality.

The rest of this paper is organized as follows: In Section 2, previous works on clock networks are presented. In Section 3, the proposed GP and Robust GP (RGP) models are discussed. Model examination and validation are given in Section 4. Conclusions are provided in Section 5.

2. BUFFER SIZING PRELIMINARIES

A clock network can be designed using meshes [14] or trees [19]. Mesh networks reduce variability [20, 32], but are power hungry and costly. Classical tree structures for clock networks vary from H-Trees [6], and the method of means and medians [17], to deferred-merge embedding [11]. Trees are less power hungry than meshes, but harder to design in the presence of process variations. Trees with cross links which are hybrid networks combining aspects of trees and meshes, have been proposed to reduce skew variability. In [29], a set of rules is given for where cross links would be most effective in reducing variability in a clock tree. This work was refined by adding control over the distribution of links to prevent excessive linking in small areas [30]. A method which initially inserts links between each sink and its nearest neighbors and then uses a rule-based refinement to remove unnecessary links is presented in [24]. In [26], the authors show that cross links have limited success when inserted in the tree sinks and propose to insert links higher up in the tree. In Contango3.0 [22], it is proposed to generate clock networks by producing several auxiliary-trees and fusing them to achieve more tolerance to variations.

Buffer insertion is required in clock trees to prevent signal degradation and satisfy slew constraints. Process technology provides a library of available buffers with varying properties, such as transistor sizes, resistance and capacitance. In a typical buffer sizing problem, the number and locations of buffers are set, and the widths and lengths of buffers are determined while minimizing maximum delay or skew.

A dynamic programming method, known as van Ginneken's algorithm has been proposed in [34], which for a given routing tree and set of candidate buffer locations, will find the optimal buffer sizes to minimize the maximum delay over the tree using the Elmore delay model. Extensions of the algorithm have included slew and power constraints while maintaining optimality [23]. A sequential linear programming technique for sizing buffers and wires considering skew and power supply variation using first order Taylor expansion is proposed in [35]. Works presented in [18, 21] attempt to perform buffer insertion in the presence of blockages.

The problem of wire sizing and buffer insertion has been well studied for general signal nets [4, 13, 31] and closed form solutions for optimal buffer and wire sizing are available for a single wire segment [12]. However, these solutions are aimed at minimizing maximum delay, not skew or power.

As power consumption in a clock tree is the most important design factor, in this paper, a buffer sizing formulation that will minimize the power, subject to clock skew and slew constraints is considered specifically in this paper. As area is well correlated with power consumption [31], this problem is expressed mathematically as:

$$\min. \ \text{Area}(\mathbf{x}) = \sum_{b \in \mathbf{B}} x_{w_b} x_{l_b} \tag{1}$$

$$s.t. \ \max\{d_i(\mathbf{x}) - d_j(\mathbf{x})\} \leq t_{\text{skew}}, \ \forall i, j \in \mathbf{S}, \ i \neq j \tag{2}$$

$$\text{slew}_k(\mathbf{x}) \leq t_{\text{slew}}, \ \forall k \in \mathbf{N} \tag{3}$$

$$x_{l_b} \geq l_{min}, \ \forall b \in \mathbf{B} \tag{4}$$

$$x_{w_b} \geq w_{min}, \ \forall b \in \mathbf{B}, \tag{5}$$

where $[x_{w_b}, x_{l_b}]$ are the width and length of buffer b belonging to the set of all buffers, $\mathbf{B}$, in the clock tree. It should be mentioned that in most digital circuits x_{l_b} is equal to the l_{min}. In this paper we have x_{l_b} as a variable to keep the generality of the formulation, Variable $\mathbf{x}$ is the vector representing the buffer widths and lengths of all the buffers in the clock tree and Area($\mathbf{x}$) is the total area of the buffers. $d_i(\mathbf{x})$ is the delay from the source to clock sink i in the set of all sinks $\mathbf{S}$ and target clock skew is represented by t_{skew}. The slew at each node k in the set of all nodes of the clock tree, $\mathbf{N}$, is shown by $\text{slew}_k(\mathbf{x})$. Maximum allowed slew in the clock tree is represented by t_{slew} and l_{min} and w_{min} are the minimum length and width of a buffer. This formulation is not convex, hence, the solution is very sensitive to the initial solution and the methodology used for solving the problem.

Process variations are unavoidable and undeniable at current technology nodes. Fabricating circuits considering only nominal values can lead to a large percentage of dies failing to meet timing specifications. One method for considering variation is to explicitly consider a finite number of design corners [33]. However, the number of corners grows exponentially with the number of variables making it challenging to choose a proper set of corners. In [25], a method for clock tree routing in the presence of process variations is proposed. This method introduces spatial variations only in wire width and not in the buffer sizes. However, it uses Elmore model for delay evaluations and lacks validation of the results using SPICE simulation. In [16], a statistical technique is proposed for robust clock tree routing. The results show significant improvement in average skew. This method however loses some robustness as statistical methods accept some failures while worst-case robust techniques are superior in providing tighter control over the process. In addition, all the variations are claimed to have Gaussian distribution but experimental results are not provided for verification. In [15], process variations are considered in the buffer and wire sizing problem which is formulated as a power-constrained skew minimization problem. Sequential programming is then employed to solve the robust optimization problem. However, the buffer and wire sizes are continuous variables and the results after discretization are not reported. In addition, only Elmore model is used for delay calculations and SPICE simulation results are not available.

In [22], the effects of supply voltage variations are considered by treating the delay of each path as a random variable. The proposed technique achieves a fine trade-off between robustness and power efficiency where the capacitance of the resultant clock network increases by not more than 60%, and the clock tree is still twice as efficient in power compared to a clock mesh structure. In [10], a technique is proposed to produce robust clock trees by budgeting the power consumption between the top and bottom level trees. The empirical results show that this technique is more efficient in decreasing power consumption and runtime when compared to the existing cross link techniques. However, this technique only considers the effects of variations in supply voltage and does not account for the variations in sizes of buffers and wires.

It has been shown that the problem of minimizing logic gate area while meeting maximum delay with process variation affecting the width and length of gates can be formulated as a geometric program [31]. However, the same technique is not applicable to clock network design where clock skew is an additional concern.

3. BUFFER SIZING UNDER VARIATION

In this section, a new algorithm for buffer sizing of clock networks under variation using GP and robust optimization is presented. The first major contribution of the proposed algorithm is converting the buffer sizing problem into GP format. The advantage of the proposed GP formulation is that the problem can be solved using convex optimization techniques, where all local optima are global optima. Hence, finding the global optimum can be done more efficiently and in lower runtime. The transformation into GP format is not a trivial task, however, once the formulation is converted, it is shown that the optimal solution can be easily obtained.

Nominal solutions do not consider process variations that occur during fabrication where the sizes of buffers can change by up to 25% of their nominal sizes [31]. The second major contribution of this paper is to include process variations as part of the buffer size optimization using a state-of-the-art optimization technique, robust optimization. The solution to the robust GP problem is in essence the optimal solution to the scenario where buffer sizes are uncertain. Therefore the robust solution guarantees the quality of the solution, ensures that the constraints of the problem are met, and is independent of the underlying distribution. In the rest of this section, the algorithm outline and the two major contributions are discussed.

3.1 Algorithm Outline

The proposed algorithm is designed to further improve the power consumption of a clock network. Hence, the algorithm takes an already designed clock network and reduces its power consumption by performing buffer sizing under process variation. The inputs to the proposed algorithm are t_{skew} and a clock network including buffers, sinks and net segments. The algorithm performs buffer sizing for this clock network with the objective of minimizing the buffer area, or in essence power consumption, while meeting t_{skew} under process variation in lengths and widths of the buffers. Skew and power are calculated using SPICE during network validation. The output of the algorithm is a clock network sized considering process variations.

Algorithm: Variation-Aware Buffer Sizing

Input: Clock network, t_{skew}
Output: Optimal buffer sizes under variation

Phase 1. Optimal Nominal Sizing
  1.a Perform GP relaxation
  1.b While t_{skew} not reached
      Solve GP
  1.c Calculate covariance matrix
  1.d Calculate average skew using
      Algorithm Network Validation
Phase 2. Variation-Aware Sizing
  2.a Formulate Robust GP (RGP)
  2.b While t_{skew} not reached
      Solve robust GP
  2.c Calculate average skew of the robust network
      using Algorithm Network Validation

Figure 1: Outline of the variation-aware buffer sizing algorithm.

The proposed algorithm, as shown in Figure 1, includes two major phases. In the first phase, *Optimal Nominal Sizing*, the non-convex buffer sizing problem formulation presented in (1) is relaxed into a sequence of convex GP problems. Convex optimization techniques are used to solve large-scale problems in low runtime and obtain global optima. The sequence of relaxed GP problems are solved until t_{skew} is reached and nominal values of buffer sizes are obtained. Details of the GP relaxation and t_{skew} calculations are given in Section 3.2.

After obtaining the GP nominal solutions, skew and power have to be measured using SPICE and the effects of process variations must be considered. This means that random defects have be added to buffer sizes to simulate real life situations. As there is significant uncertainty in the variation and defects introduced by the fabrication process, Monte Carlo simulation is used to introduce defects and extract timing information. The proposed Network Validation algorithm shown in Figure 2 is used to add defects to buffer sizes, recalculate skew and measure power using SPICE. In

---
Algorithm: Network Validation

Input: Clock network, covariance matrix, P
Output: Skew distribution
---
while # simulations < max simulations required
  Introduce spatially-correlated defects
  Extract timing and power information using SPICE
end

Figure 2: Network validation algorithm

this algorithm, first, a covariance matrix, P, representing the intra-die correlation between buffer size variations is obtained. Intra-die correlation means that if two buffers are located close to each other, the variation in their sizes will be closer than the buffers that are located at a distance. P is calculated using the grid-based model described in [9]. Then, Monte-Carlo simulations during which spatially correlated defects are introduced and then timing and power information are extracted using SPICE are performed.

In the second phase of the proposed algorithm, *Variation-Aware Sizing*, the nominally-sized network from the first phase is used as an initial solution. The covariance matrix is obtained and used in Algorithm Network Validation to represent uncertainty and a new GP which considers spatially-correlated process variations is formulated. The resultant problem is called a Robust GP (RGP) formulation which is solved using robust optimization (RO) techniques until the t_{skew} is reached. The steps developed to formulate the RGP are discussed in Section 3.3.

3.2 Phase 1: Optimal Nominal Sizing

The formulation in (1) is non-convex and non-linear, making it a very hard problem to be solved using optimization techniques. In this section, we propose techniques to transform it into a series of convex programming problems in GP format and show that the new GP problem can be solved efficiently and provide superior solutions.

3.2.1 Geometric Programming

A GP is an optimization problem in the general form:

$$\begin{aligned} \min. \quad & h_0(\mathbf{z}) \\ s.t. \quad & h_i(\mathbf{z}) \leq 1, \; i = 1, ..., p \\ & g_i(\mathbf{z}) = 1, \; i = 1, ..., m. \end{aligned} \quad (6)$$

In this formulation, $\mathbf{z}$ is the nominal value of the variables,

$g_1(\mathbf{z}), \ldots, g_m(\mathbf{z})$ are monomials, and $h_0(\mathbf{z}), h_1(\mathbf{z}), \ldots, h_p(\mathbf{z})$ are posynomials. A monomial is a generic function $m : \Re^n \to \Re$ of the form of $m(\mathbf{z}) = \alpha z_1^{a_1} z_2^{a_2} \cdots z_n^{a_n}$ where $\alpha > 0$ and the exponents, $a_1, \ldots, a_n \in \Re$, are real. A posynomial is a summation of monomials where all the coefficients of the summation are positive.

The problem formulated in (6) is not convex in general, but can be transformed into a convex problem. By defining an associate variable $y_i = log(z_i)$ and $\beta = \log \alpha$, the monomial is rewritten as: $m(\mathbf{y}) = e^\beta e^{a_1 y_1} \cdots e^{a_n y_n} = e^{\mathbf{a}^T \mathbf{y} + \beta}$. Similarly, for a posynomial h_i, the transformed function has the form $\sum_{j=1}^{J_i} e^{\mathbf{a_{ij}}^T \mathbf{y} + \beta_{ij}}$, where J_i is the number of monomials in posynomial h_i, and $\mathbf{a_{ij}} \in \Re^n, i = 1, \ldots, m$ contain the posynomial exponents. The optimization problem associated with the transformation has the form:

$$
\begin{aligned}
\text{min.} \quad & \sum_{j=1}^{J_0} e^{\mathbf{a_{0j}}^T \mathbf{y} + \beta_{0j}} \\
\text{s.t.} \quad & \sum_{j=1}^{J_i} e^{\mathbf{a_{ij}}^T \mathbf{y} + \beta_{ij}} \le 1, \ i = 1, \ldots, p \\
& e^{\mathbf{g_i}^T \mathbf{y} + \beta_i} = 1, \ i = 1, \ldots, m,
\end{aligned}
$$

where $\mathbf{g_l} \in \Re^n$ contains the equality constraint coefficients. This form of optimization can be transformed into a convex problem by taking the logarithm resulting in:

$$
\begin{aligned}
\text{min.} \quad & \mathbf{lse}(A_0 \mathbf{y} + b_0) \\
\text{s.t.} \quad & \mathbf{lse}(A_i \mathbf{y} + b_i) \le 0, \ i = 1, \ldots, m \\
& G\mathbf{y} + \beta = 0,
\end{aligned}
$$

where $\mathbf{lse}$ is the log-sum-exp function: $\mathbf{lse}(\mathbf{y}) = log(e^{y_1} + \ldots + e^{y_k})$. GPs in convex form can be solved using any non-linear convex minimization technique [8].

3.2.2 Proposed GP Relaxation

In this section, the proposed techniques to relax the buffer sizing problem (1) to GP format are presented. Each one of the lines in (1) to (5) is studied and if not in a posynomial or monomial format, techniques to turn them in a posynomial or monomial are proposed.

Objective function (1): The objective, $\sum_{b \in \mathbf{B}} x_{w_b} x_{l_b}$, is the sum of buffer areas which is already a posynomial.

Skew Constraints (2): Each skew constraint, $\max\{d_i - d_j\} \le t_{\text{skew}} \ \forall s_i \in \mathbf{S}$, involves a delay calculation, maximum operator, and negative coefficients. Each one of the mentioned operations must be changed or relaxed so that (2) conforms to a posynomial format. The proposed techniques for changing these operations are explained in the following.

Delay Calculations: In order to accurately calculate delay, SPICE needs to be used. However, using SPICE is not practical when solving large-scale non-linear optimization problems. The delay model used in [31], which gives a *reasonable approximation* for the delay and is a posynomial, is used in this paper. All delay calculations are then verified using SPICE.

Negative coefficients: To eliminate the negative coefficient of (2), the skew constraint is first relaxed by replacing $d_j(\mathbf{x})$ with the minimum sink delay, d_{min}. As $d_i(\mathbf{x}) - d_{min} \ge d_i(\mathbf{x}) - d_j(\mathbf{x})$, this relaxation does not affect feasibility. This relaxation may be big for some sinks. However, as will be discussed in Section 3.2.3, it is used in an iterative scheme and d_{min} is updated in each iteration. Therefore, the target skew is eventually reached. Both sides of the relaxed constraint are added by d_{min} that leads to $\max\{d_i(\mathbf{x})\} \le t_{\text{skew}} + d_{min}$.

Maximum operator: The maximum operator in the skew constraint is replaced with $\mu \le t_{\text{skew}} + d_{min}$, and a set of constraints $d_i(\mathbf{x}) \le \mu, \ \forall s_i \in \mathbf{S}$, where μ is a dummy variable. All the new constraints are monomial constraints and together act as the maximum operator.

Slew Constraints (3): In the slew model in (3), the slew at a node in the tree is proportional to the corresponding delay from the previous upstream buffer output to the node [5]:

$$slew_k(\mathbf{x}) = ln(9) \times d_k^s(\mathbf{x}),$$

where $d_k^s(\cdot)$ is the delay from the output of the previous upstream buffer to the node. The slew calculations are derived from the already calculated components of delays, which is of importance as it saves runtime. Since the delay equation is a posynomial, the resultant slew equations are in the standard GP format. To ensure clock signal quality, one constraint is added for each buffer and sink input. As intermediate nodes, i.e. nodes at wire branches, have lower slew than the buffer or sink inputs, as long as the slew constraints are satisfied at buffer and sink inputs, the slew constraints are satisfied at intermediate circuit nodes as well. Therefore, the constraints are as follows:

$$slew_k(\mathbf{x}) slew_{\max}^{-1} \le 1, \quad \forall k \in B \cup S,$$

where, $slew_i(\mathbf{x})$ is the slew calculated for a sink or buffer input i and $slew_{\max}$ is the maximum slew allowed.

Minimum length (4) and minimum width (5): The minimum length, $l_{min} \le x_{l_b}$, and minimum width, $w_{min} \le x_{w_b}$, constraints can be easily transformed by rewriting them as $l_{min} x_{l_b}^{-1} \le 1, \ b \in \mathbf{B}$ and $w_{min} x_{w_b}^{-1} \le 1, \ b \in \mathbf{B}$.

After the proposed transformations, specially for (2), we have transformed the buffer sizing problem to an optimization problem in the standard GP format:

$$
\begin{aligned}
(GP) \quad \text{min.} \quad & \text{Area}(\mathbf{x}) = \sum_{b \in \mathbf{B}} x_{w_b} x_{l_b} \\
\text{s.t.} \quad & (t_{\text{skew}} + d_{min})^{-1} \mu \le 1 \\
& slew_k(\mathbf{x}) slew_{\max}^{-1} \le 1, \forall k \in B \cup S \\
& d_i(\mathbf{x}) \mu^{-1} \le 1, \ \forall i \in \mathbf{S} \\
& l_{min} x_{l_b}^{-1} \le 1, \ b \in \mathbf{B} \\
& w_{min} x_{w_b}^{-1} \le 1, \ b \in \mathbf{B}.
\end{aligned}
$$

3.2.3 Iterative GP Solution

Since the skew constraints $d_i - d_j \le t_{\text{skew}}$ are relaxed to $d_i - d_{min} \le t_{\text{skew}}$, the calculated minimum delay may not be the minimum delay after solving the GP. In particular, if the minimum delay decreased, another iteration is required with d_{min} updated to the new calculated minimum delay. Although it is possible for the minimum delay to decrease below d_{min}, in the experiments in this paper, the final solutions were always reached after a single iteration.

To validate the solution, once a nominal value is obtained using the proposed GP model, the delay including process variations of up to $\pm 20\%$ of the nominal values for lengths and $\pm 25\%$ for widths [31], is obtained using Monte Carlo simulations and the skew is calculated using delays from SPICE.

If the GP becomes infeasible, zero skew algorithms, such as [11], can be used to initialize the network and any t_{skew} of interest can be achieved provided the slew constraints are satisfied. If a zero skew network is made up of only minimum size buffers, GP will not modify the network and Phase 1 can be skipped.

3.3 Phase 2: Variation-Aware Sizing

The buffer sizing problem in (1) or in GP format of (3.2.2) is a deterministic optimization problem. However, once process variations are included in the formulation, variables and parameters of the problem can become non-deterministic. This means the optimal solution of (3.2.2) can become suboptimal. RO techniques have been successfully applied to problems where variability in the values of parameters exist [7]. The robust version of (3.2.2) for the buffer sizing problem is formulated as:

$$\begin{aligned} \min. \quad & \mathbf{lse}(A_0\mathbf{y} + b_0) \\ \text{s.t.} \quad & \mathbf{lse}(A_i\tilde{\mathbf{y}}(\tilde{u}) + b_i) \le 0, \ i = 1, \dots, m, \end{aligned}$$

where $\tilde{u} \in U$ is the vector of uncertain variables that belongs to the uncertainty set U, representing the uncertainty in the buffer sizes. As meeting the skew constraints is the most important part of the problem, the variations are not considered in the objective function. To be able to solve the uncertain problem, a worst case robust GP in the following form is solved by using the supremum operator sup:

$$\begin{aligned} \min. \quad & \mathbf{lse}(A_0\mathbf{y} + b_0) \quad &(7) \\ \text{s.t.} \quad & \sup_{\tilde{u}\in U} \mathbf{lse}(A_i\tilde{\mathbf{y}}(\tilde{u}) + b_i) \le 0, \ i = 1, \dots, m, \end{aligned}$$

where the supremum value of the constraints for all uncertain values is bounded by 0. If the supremum value of the constraint is in the uncertainty set, then it can be replaced by the max function. To solve this problem using existing optimization techniques, an uncertainty model needs to be developed so that the uncertain variables can be replaced in the formulation. In this paper, the robust GP problem is formulated and solved using the model in (7) and the ellipsoid uncertainty model: $U = \{u + P^{1/2}v \mid ||v||_2 \le 1, \quad v \in \Re^{2|B|}\}$, where, u is the nominal value of the variables, P is the covariance matrix of variations in length and width, and v is any vector with norm less than one. The variations are based on a worst-case ellipsoidal model where the variations are considered to be spatially-correlated to agree with manufacturing data [31]. Other variations can also be modelled in the robust solution. The experimental results show that when the clock network power is calculated using SPICE, there is a dramatic decrease in the total power.

In the rest of the section, we first explain how variations are added to the original buffer sizing problem of (3.2.2). Then, we apply the GP relaxation to the new problem and solve it.

3.3.1 Proposed Robust GP (RGP) for Buffer Sizing

The first constraints that need to be modified to include variations are the skew constraints. First, the skew constraints are combined to eliminate the dummy variable μ, then the variable $\tilde{\mathbf{x}}$ is replaced by: $\tilde{\mathbf{x}} = \mathbf{x} + \delta\mathbf{x}$. The new skew constraints are: $d_i(\mathbf{x} + \delta\mathbf{x}) \le (t_{\text{skew}} + d_{min}), \ \forall i \in \mathbf{S}$ where, d_{min} is the infimum of the random variable related to the minimum delay. The delay function $d_i(\mathbf{x}+\delta\mathbf{x})$ is approximated using first order Taylor series: $d_i(\mathbf{x} + \delta\mathbf{x}) = d_i(\mathbf{x}) + \nabla_x^T d_i(\mathbf{x})\delta\mathbf{x}$. This approximation is an affine model with $d_i(\mathbf{x})$ representing the nominal value and $\nabla_x^T d_i(\mathbf{x})\delta\mathbf{x}$ representing the variation. The variational term, $\nabla_x^T d_i(\mathbf{x})\delta\mathbf{x}$, describes how small changes in the lengths and width will affect the delay. Substituting the Taylor series approximation,

the constraint becomes:

$$d_i(\mathbf{x}) + \nabla_x^T d_i(\mathbf{x})\delta\mathbf{x} \le (t_{\text{skew}} + d_{min}), \ \forall i \in \mathbf{S}. \quad (8)$$

To solve (8) which includes the new skew constraints, its worst case, i.e., sup, should be considered.

The uncertainty vector, $\delta\mathbf{x}$ is bounded within the ellipsoid defined by the covariance matrix: $\delta\mathbf{x} = P^{1/2}v$, $||v|| \le 1$. The term $\nabla_x^T d_i(\mathbf{x})$ is split into positive terms, ϕ_1, and negative terms, ϕ_2. Hence, the variational term, $\nabla_x^T d_i(\mathbf{x})\delta\mathbf{x}$, is equal to: $\phi_1^T P^{1/2}v + \phi_2^T P^{1/2}v$, $||v|| \le 1$. Only the maximum variational term is of importance, and $\phi_1^T P^{1/2}v + \phi_2^T P^{1/2}v$, $||v|| \le 1$ is replaced by $\max_{||v||\le 1}(< P^{1/2}\phi_1, v > + < P^{1/2}\phi_2, v >)$, where $< a, b >$ is the inner product of a and b. The inequality constraint becomes:

$$d_i(\mathbf{x}) + \max_{||v|| \le 1}(< P^{1/2}\phi_1, v > + < P^{1/2}\phi_2, v >) \le t_{\text{skew}} + d_{min}.$$

To remove the max operator, robust variables r_1 and r_2 are defined as: $r_1 = ||P^{1/2}\phi_1||$, $r_2 = ||P^{1/2}\phi_2||$, hence: $\phi_1^T P\phi_1 r_1^{-2} \le 1$ and $\phi_2^T P\phi_2 r_2^{-2} \le 1$. Using $||v|| \le 1$, we will have:

$$\max_{||v|| \le 1}(< P^{1/2}\phi_1, v > + < P^{1/2}\phi_2, v >) \le r_1 + r_2.$$

The same calculations can be applied to derive robust slew constraints. The variation in the lengths and widths of the buffers in the boundary constraints, $l_{min}x_{l_b}^{-1} \le 1$, $b \in \mathbf{B}$ and $w_{min}x_{w_b}^{-1} \le 1$, $b \in \mathbf{B}$, are normally introduced as part of the manufacturing process, so variation does not need to be considered for these constraints.

Using the above equations, the constraints of (3.2.2) considering the process variations in the buffer sizes can be reformulated using the following constraints (7) to (5):

$$\begin{aligned} (RGP) \quad \min. \quad & \text{Area}(\mathbf{x}) = \sum_{b \in \mathbf{B}} x_{w_b} x_{l_b} \\ \text{s.t.} \quad & d_i(\mathbf{x}) + r_{1_i} + r_{2_i} \le t_{\text{skew}} + d_{min}, \forall i \in \mathbf{S} \\ & \phi_{1_i}^T P\phi_{1_i} r_{1_i}^{-2} \le 1, \forall i \in \mathbf{S} \\ & \phi_{2_i}^T P\phi_{2_i} r_{2_i}^{-2} \le 1, \forall i \in \mathbf{S} \\ & slew_k(\mathbf{x}) + r_{3_k} + r_{4_k} \le slew_{\max}, \forall k \in \mathbf{B} \cup \mathbf{S} \\ & \phi_{3_i}^T P\phi_{3_i} r_{3_i}^{-2} \le 1, \forall i \in \mathbf{B} \cup \mathbf{S} \\ & \phi_{4_i}^T P\phi_{4_i} r_{4_i}^{-2} \le 1, \forall i \in \mathbf{B} \cup \mathbf{S} \\ & l_{min}x_{l_b}^{-1} \le 1, \ b \in \mathbf{B} \\ & w_{min}x_{w_b}^{-1} \le 1, \ b \in \mathbf{B}. \end{aligned}$$

This problem is still exclusively made of posynomials and monomials and can be turned to a GP problem using the technique discussed in Section 3.2.2.

4. EXPERIMENTAL RESULTS

To show the validity of the proposed algorithms, networks from the 2009 ISPD clock network synthesis (CNS) contest [1] are used as these are the only networks where all teams' clock networks have been released. Details of the benchmarks used in the experiments are given in Table 1.

From left to right, the columns tabulate the circuit names, height and width of the chip, total clock sink capacitance, number of sinks, number of buffers, total buffer area, average skew (Ave. Skew), standard deviation of skew (Std. Skew) and total power consumed by the network. The number of buffers is calculated as the equivalent number of minimum size buffers. For example, if a buffer is four times wider than a minimum buffer of the corresponding type, it counts as four buffers. The process used is 45nm and the technology files are the same as for the ISPD CNS contest. All experiments were performed on a 2.8GHz Intel Pentium 4 processor with 1GB of memory. The algorithms are implemented in C++ and GPs are solved using Mosek 6.0 [2]. All power and skew calculations are measured directly from ngspice 24 [3]. Ngspice power measurements are **SPICE accurate**, i.e. include switching, short circuit, and leakage power. Since short circuit power is included, slew also impacts the power. In all experiments, t_{slew} is set to be 100ps as set in ISPD CNS contest. t_{skew} of all the optimization problems is set to be 50ps. Note that this value is set for the mathematical delay formulation used in the GP and RGP which is not as accurate as the ngspice simulation. Therefore, the skews calculated using ngspice are expected to be different. It should be mentioned that in the experiments, the skew constraints are met after solving a single GP or RGP. The proposed technique is not compared with any previous work since there is no available work that has considered the variations in buffer sizes. However, the effects of the proposed technique are analyzed by comparing the performance of the initial network and networks obtained using GP and RGP.

In the first set of experiments, Table 2, the area obtained from solving (3.2.2) and (8), average and standard deviation of skew of the Monte Carlo simulations, and the power calculations from ngspice are compared to the results of the initial network. The area, skew and power are calculated under process variations using Monte Carlo simulations. The variations in buffer width and length are set to 25% and 20% of the nominal values, respectively [31]. When performing Monte Carlo simulations, the number of simulations is 20 times the number of buffers in the network. Since GP and RGP output solutions with continuous sizes, it is necessary to fit these sizes to the available discrete sizes for fabrication. In this paper, to obtain the discrete results, the values obtained from solving GP and RGP are rounded down to the closest available discrete buffer size, and ngspice is used to recalculate the skew to ensure feasibility. More sophisticated schemes where additional bounds are placed on buffer sizes at each iteration in the GP to obtain final integer values will be developed as part of the future work. Moreover, a tool called ASTRAN is proposed in [37] and further studied in [28] which automatically lays out cells according to the continuous sizes. However, this tool is not available and hence not used in this paper. In columns 2 to 5, the buffer area for the nominally-sized, GP, in continuous, C, and discrete, D, buffer sizes and the robust solution, RGP, in continuous, C, and discrete, D, buffer sizes are given, respectively. In columns, 6 to 9, the average skew results and in columns, 10 to 13, the standard deviation of skew are given, respectively, in the same order as columns 2 to 5. The powers are calculated using ngspice and the power improvements after using GP and RGP compared to the initial networks are shown in columns 14 to 17 of Table 2. In Columns 18 and 19, the runtime for the GP and RGP are given.

The objective of the buffer sizing, area, and total power are substantially reduced for all the test cases and in both continuous and discrete solutions. On average, the area is reduced by 72% for GP and 56% for RGP while the power is reduced by 66% for GP and 55% for RGP. In comparison with the nominally-sized GP network, the robust network has more area and power but lower skew. This means that the nominal GP solution is over-optimizing the area and power because it does not consider the variations. Looking at the runtime, both GP and RGP are scalable as the runtimes increase roughly proportional to the number of variables.

The average skews calculated using ngspice for both continuous GP and RGP decrease compared to the initial network by 15ps and 25ps, respectively. Furthermore, with the discrete sizes, GP has higher average skews than the initial network, on average, while RGP has lower by 11ps. The standard deviations of skew for both GP and RGP increase compared to the initial networks. This is because the proposed GP and RGP are able to achieve the target skew using the approximate delay model with much smaller buffers. This results in reduced area and power for the networks. These smaller buffers are however more susceptible to the variations and the standard deviation of skew increases. It can be seen from the results that the circuits with the largest area decrease have the largest increase in standard deviation of skew. For example, the most noticeable improvement in area and power happens for circuit 31 with a decrease of 77% in area and 68% power reduction in GP solution and 73% area and 62% power reduction in RGP.

The power of RGP and considering variation in the formulation can be demonstrated by comparing the standard deviations of skew. When continuous RGP is used the standard deviation of skew decreases for all test cases where 35% improvement is achieved, on average, compared to continuous GP. Using our proposed RGP results in significant power and skew reduction. This improvement comes at expense of runtime for RGP, which on average, increases by a factor of 3.7. The results in Table 2 suggest a trade-off between area/power and the skew, where reduction in area/power results in higher skew. These results also suggest that considering variations and using the proposed RGP model can result in optimal solutions where area, power and skew are all reduced. This is because the optimized solution is obtained considering the variations in buffer sizes rather than just nominal values.

The curve shown in Figure 3 is the Pareto optimal trade-off curve between area and t_{skew} for a small network with two buffers and two sinks, but the ideas apply to the larger circuits. For this circuit, a skew of about 6.15ps is not achievable as the curve grows asymptotically on the left side. Also, setting t_{skew} greater than 7.7ps makes the skew constraint inactive since 7.7ps can be achieved with less area than it takes to achieve a higher skew. A clock network designer can trace out such an optimal trade-off curve and decide upon the most appropriate design.

Finally, to show the power of the robust solution, the distribution of the skew for each Monte Carlo run for the GP and RGP networks considering spatial correlations are presented in Figure 4. For this comparison, the network for circuit 11 is sized nominally (grey) and sized considering variations (black). When GP is used, the skew is distributed between 29ps to 39ps. However, for RGP network, the skew ranges between 27ps to 34ps, where both average and standard deviation of skew decrease.

Circuit	Height (mm)	Width (mm)	Sink cap. (pF)	$|S|$	$|B|$	Area (μm^2)	Ave. Skew (ps)	Std. Skew (ps)	Power (μW)
11	11	11	4.2	121	3536	536	48	1.26	19668
12	8	13	4.1	117	3472	527	91	0.88	19685
21	13	12	3.6	117	3568	541	51	1.21	20631
22	12	5	3.4	91	2112	320	27	0.85	12125
31	17	17	9.6	273	7760	1177	228	1.09	42976
32	17	17	6.7	190	5904	895	143	0.70	33392
nb1	3	2	5.9	330	1264	192	25	1.15	5234
Ave.	-	-	-	-	-	598	88	1.02	21959

Table 1: Details of the benchmarks used for experimentation.

Table 2: Comparison of initial networks and the resultant nominally-sized networks and robust networks.

Cir.	Area (μm^2)				Ave. Skew (ps)				Std. Skew (ps)				Power **Improvement**				Runtime (s)	
	GP		RGP		GP		RGP		GP		RGP		GP		RGP			
	C	D	C	D	C	D	C	D	C	D	C	D	C	D	C	D	GP	RGP
11	175	153	303	268	35	57	30	35	1.78	3.81	1.47	1.43	63%	65%	39%	46%	49	195
12	152	132	252	222	95	108	76	86	2.16	2.60	1.09	1.22	67%	69%	49%	53%	49	182
21	170	149	244	228	42	56	34	36	0.86	1.16	0.8	0.86	62%	64%	49%	51%	53	205
22	112	98	197	175	27	49	23	27	2.27	3.48	1.31	1.23	64%	66%	40%	46%	12	126
31	262	248	318	278	161	219	130	188	3.17	9.79	2.01	3.40	68%	66%	62%	57%	221	673
32	259	219	434	379	122	184	120	135	1.61	2.04	0.52	0.85	66%	67%	47%	52%	104	342
nb1	46	36	73	62	29	452	26	32	1.87	0.21	1.76	1.86	72%	76%	59%	56%	10	113
Ave.	168	148	260	230	73	161	63	77	1.96	3.30	1.28	1.55	66%	67%	55%	52%	71	262

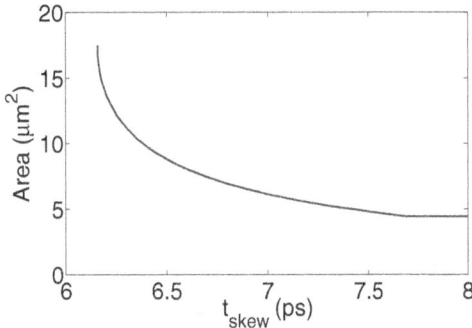

Figure 3: An illustration of the optimal area-t_{skew} trade-off for a small network.

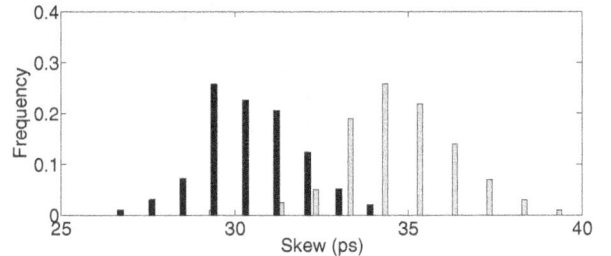

Figure 4: The Monte Carlo skew distributions for the GP network (grey) and the RGP network (black).

5. CONCLUSION AND FUTURE WORK

In this paper, it is shown that by using geometric programming, a globally optimal solution for buffer sizing problem can be obtained in an acceptable amount of runtime. In addition, a robust model is proposed that can consider the variation in the sizes of the buffers. The proposed techniques are based on state-of-the-art optimization techniques and the experimental results obtained from SPICE show improvement. By using the proposed GP formulations, designers can trace out optimal trade-off curves to decide on the most appropriate design.

For future work, a technique will be developed to discretize the buffer sizes with minimal disruption to skew. Also, the robust model can be developed as a multi-objective optimization problem with the objectives of minimizing the skew and the area simultaneously. The model can be extended to include other variations, such as wire width variations, to allow for robust designs. Finally, the proposed variation-aware techniques can be merged with other effective methods, such as using cross links, to create even more robust networks.

6. ACKNOWLEDGMENTS

This research is supported by the Natural Sciences and Engineering Research Council of Canada and Alberta Innovates Technology Futures. This research has been enabled by the use of computing resources provided by WestGrid and Compute/Calcul Canada.

7. REFERENCES

[1] ISPD 2009 clock network synthesis contest. http://www.sigda.org/ispd/contests/09/ispd09cts.html.

[2] Mosek 6.0. http://www.mosek.com.

[3] ngspice 24. http://ngspice.sourceforge.net/.

[4] C. Alpert and A. Devgan. Wire segmenting for improved buffer insertion. In *Proc. of DAC*, pages 588–593, 1998.

[5] H. Bakoglu. *Circuits, Interconnections, and Packaging for VLSI.* Addison–Wesley, 1990.

[6] H. Bakoglu, J. Walker, and J. Meindl. A symmetric clock distribution tree and optimized high speed interconnections for reduced clock skew in ULSI and WSI circuits. In *Proc. ICCAD*, pages 118–122, 1986.

[7] D. Bertsimas, D. Brown, and C. Caramanis. Theory and applications of robust optimization. *SIAM Review*, 53:464–501, 2011.

[8] S. Boyd and L. Vandenberghe. *Convex Optimization.* Cambridge University Press, 2004.

[9] H. Chang and S. Sapatnekar. Statistical timing analysis considering spatial correlations using a single pert-like traversal. In *Proc. of ICCAD*, pages 621–625, 2003.

[10] Y. Chang, C. Wang, and H. Chen. On construction low power and robust clock tree via slew budgeting. In *Proc. of ISPD*, pages 129–136, 2012.

[11] T. Chao, Y. Hsu, J. Ho, K. Boese, and A. Kahng. Zero skew clock routing with minimum wirelength. *IEEE Tran. on CAS*, 39(11):799–814, 1992.

[12] C. Chu and D. Wong. Closed form solution to simultaneous buffer insertion/sizing and wire sizing. *TODAES*, 6(6):343–371, 2001.

[13] S. Dhar and M. Franklin. Optimum buffer circuits for driving long uniform lines. *IEEE Journal of Solid-State Circuits*, 26(1):32–40, 1991.

[14] G. Flach, G. Wilke, M. Johann, and R. Reis. A Mesh-Buffer Displacement Optimization Strategy. In *Proc. Annual Symposium on VLSI*, pages 282–287, 2010.

[15] M. Guthaus, D. Sylvester, and R. Brown. Clock buffer and wire sizing using sequential programming. In *Proc. of DAC*, pages 1041–1046, 2006.

[16] M. Guthaus, D. Sylvester, and R. Brown. Process-induced skew reduction in nominal zero-skew clock trees. In *Proc. of ASPDAC*, pages 84–89, 2006.

[17] M. Jackson, A. Srinivasan, and E. Kuh. Clock routing for high-performance ics. In *Proc. DAC*, pages 573–579, 1990.

[18] A. Jagannathan, S. Hur, and J. Lillis. A fast algorithm for context-aware buffer insertion. In *Proc. DAC*, pages 368–373, 2000.

[19] L. J. M. I. H. J. Kahng, A. *VLSI Physical Design: From Graph Partitioning to Timing Closure.* Springer, 2011.

[20] N. Kurd, J. Barkarullah, R. Dizon, T. Fletcher, and P. Madland. A multigigahertz clocking scheme for the Pentium® 4 microprocessor. *IEEE Journal of SSC*, 36(11):1647–1653, 2001.

[21] M. Lai and D. Wong. Maze routing with buffer insertion and wire sizing. In *Proc. of DAC*, pages 374–378, 2000.

[22] D. Lee and I. Markov. Multilevel tree fusion for robust clock networks. In *Proc. of ICCAD*, pages 632–639, 2011.

[23] J. Lillis, C. Cheng, and T. Lin. Optimal wire sizing and buffer insertion for low power and a generalized delay model. In *Proc. of ICCAD*, pages 138–143, 1995.

[24] B. Liu, A. Kahng, X. Xu, J. Hu, and G. Venkataraman. A global minimum clock distribution network augmentation algorithm for guaranteed clock skew yield. In *Proc. ASPDAC*, pages 24–31, 2007.

[25] B. Lu, J. Hu, G. Ellis, and H. Su. Process variation aware clock tree routing. In *Proc. of ISPD*, pages 174–181, 2003.

[26] T. Mittal and C.-K. Koh. Cross link insertion for improving tolerance to variations in clock network synthesis. In *Proc. ISPD*, 2011.

[27] F. Niu, Q. Zhou, H. Yao, Y. Cai, J. Yang, and C. Sze. Obstacle-avoiding and slew-constrained buffered clock tree synthesis for skew optimization. In *Proc. of GLSVLSI*, pages 199–204, 2011.

[28] G. Posser, A. Ziesemer, D. Guimares, G. Wilke, and R. Reis. A study on layout quality of automatic generated cells. In *Proc. of ICECS*, pages 651 –654, 2010.

[29] A. Rajaram, J. Hu, and R. Mahapatra. Reducing clock skew variability via cross links. In *Proc. DAC*, pages 18–23, 2004.

[30] A. Rajaram, D. Pan, and J. Hu. Improved algorithms for link-based non-tree clock networks for skew variability reduction. In *Proc. ISPD*, pages 55–62, 2005.

[31] J. Singh, V. Nookala, Z. Luo, and S. Sapatnekar. Robust gate sizing by geometric programming. In *Proc. DAC*, pages 315–320, 2005.

[32] H. Su and S. Sapatnekar. Hybrid structured clock network construction. In *Proc. ICCAD*, pages 333 –336, 2001.

[33] M. Toyonaga, K. Kurokawa, T. Yasui, and A. Takahashi. A practical clock tree synthesis for semi-synchronous circuits. In *Proc. of ISPD*, pages 159–164, 2000.

[34] L. van Ginneken. Buffer placement in distributed RC-tree networks for minimal elmore delay. In *Proc. ISCAS*, volume 2, pages 865–868, 1990.

[35] K. Wang, Y. Ran, H. Jiang, and M. Marek-Sadowska. General skew constrained clock network sizing based on sequential linear programming. *IEEE Trans. on CAD*, 24(5):773–782, 2005.

[36] Q. Zhu. *High-Speed Clock Network Design.* Boston: Kluwer Academic Publishers., 2003.

[37] A. Ziesemer and C. Lazzar. Transistor level automatic layout generator for non-complementary cmos cells. In *Proc. of VLSI - SoC*, pages 116–121, 2007.

Local Merges for Effective Redundancy in Clock Networks

Rickard Ewetz and Cheng-Kok Koh
School of Electrical and Computer Engineering,
Purdue University,
West Lafayette, IN 47907-2035
rewetz,chengkok@purdue.edu

ABSTRACT

Process and environmental variations affect the reliability of clock networks. By synthesizing non-tree structures, the robustness of clock networks can be improved at the expense of higher capacitance. A cheap way of converting a tree structure to a non-tree structure is to insert cross links. Unfortunately, the robustness seems to improve only when the links are sufficiently short. Other non-tree structures such as meshes and multilevel fusion trees improve the robustness more effectively, but with much higher cost. In this work, we develop a new non-tree topology by merging a sub-clock tree with all other sub-clock trees that contain sequential elements that require strict synchronization. Results show that when compared with the state-of-the-art solutions, clock networks constructed with the proposed structure have similar capacitance but notable improved robustness. Moreover, the clock networks can satisfy tight skew constraints even when simulated under a more stringent variations model, with 22% lower capacitance when compared to solutions in earlier studies.

Categories and Subject Descriptors

B.7.2 [**Hardware**]: Integrated Circuits-Design Aids

General Terms

Algorithms, Design, Performance, Reliability.

Keywords

VLSI CAD; Physical Design; Clock Network; Cross Links.

1. INTRODUCTION

The synthesis of clock networks is an important design step of modern ICs. A clock network delivers a synchronizing signal to all sequential elements in a synchronized circuit to trigger them almost simultaneously. Clock skew is the difference of the arrival times between a pair of sequential elements. In this work, sequential elements that

require strict synchronization are said to be *sequentially related.* It is important to bound the skew between sequentially related elements to guarantee the correct operation of an IC. Many clock synthesis methodologies are effective in constructing clock networks that meet skew constraints under nominal conditions. However, when considering process, voltage, and temperature (PVT) variations, it becomes more challenging to reliably bound the clock skew. A "High Performance Clock Synthesis" contest [15] was held in ISPD 2010 to address such a challenge.

The presence of redundant paths in a clock network seems to be a key factor to reducing the impact of variations. While clock meshes (see [17, 14]) naturally exhibit redundancy, they have high cost (in terms of capacitance and power consumption). To better contain the cost, there have been several attempts to insert redundant paths into clock trees, which are more economical to construct in general. The insertion of cross links, each of which is a wire between two points in a clock tree, to enhance robustness was proposed in [11], and the method was further improved in [16, 13, 12, 8]. The concept of adding a cross link to create a redundant path between two points was generalized in [7], when redundant paths between multiple points are inserted using the notion of a multilevel fusion tree. In [7], an initial clock trees is constructed. Then, additional clock trees are constructed to connect sensitive locations and they are fused together with the original tree. The increased robustness is, however, achieved with relatively higher cost when compared to methods based on cross link insertion.

In this paper, we present a method of merging a subtree with multiple subtrees when these subtrees share sequentially related elements. As these subtrees are at the bottom of the clock network, the merges will span shorter distances, hence, "local". The motivation for performing local merges arises from our observation that cross links are effective when they are topologically close to sequentially related elements. Multilevel fusion trees are robust for the same reason. With the proposed method, we achieve robustness comparable to that of mesh structures, but at lower cost. Moreover, we can meet design constraints even under a more stringent variations model.

2. PROBLEM DEFINITION

We follow the same clock synthesis problem posed in the ISPD 2010 contest. In that contest, the sequential elements in a clock network are presented as clock sinks and they are to be connected to the clock source, which resides at the bottom left corner of the chip. The connections have to be realized by wires of various widths specified in a wire library and inverters of various sizes specified in an inverter library. (In this paper, we use "device" to refer to either an inverter or a buffer, i.e., a chain of two inverters). Wires may be routed over blockages, but devices cannot be placed in blockages. A constructed clock network is then evaluated by 500 Monte Carlo simulations using NGSPICE [9], with the supply voltage being affected by $\triangle V$, the voltage variations around the nominal voltage, and the width of a wire affected by $\triangle w$, the wire width variations around the nominal wire width. The skew of each pair of clock sinks separated by less than a user-specified distance is called local clock skew (LCS). Such a pair of clock sinks is said to be sequentially related. Each simulation would result in a worst local clock skew (wLCS). The 95th percentile of the wLCS's, denoted as 95% skew, is considered to be the skew of the network. The synthesized clock network must meet a specific constraint on the 95% skew. Moreover, the slew at any point in the clock network must also meet a specific constraint under variations.

The $\triangle V$ variations of the power supply that devices experienced in the Monte Carlo simulations are generated using the following model in the ISPD 2010 contest [15]: The $\triangle V$ of each device is generated independently. Because such variations are independent of each other, one can average out the effect of variations by placing a large number of devices in parallel [17, 2]. This is not realistic because PVT variations are expected to be spatially correlated. In the following, we refer to such a model as the ISPD model.

To counter the optimism in the ISPD model, a more stringent model has been proposed in [2] as follows: All devices at the same location experience the same $\triangle V$ variations, i.e., putting devices in parallel does not average out variations; it only enhances the drive strength but not the robustness. Such a model is referred to as the Single-Location Single-Voltage (SLSV) model. If we can satisfy design constraints under the SLSV model, it is likely that we can also satisfy design constraints under ISPD variations.

3. REDUNDANCY = ROBUSTNESS?

It has been demonstrated that the reliability of a clock network can be improved by introducing redundancy in the form of cross links in [8, 16, 13, 12] or in the form of multilevel fusion trees in [7]. These two structures, i.e., cross links and multilevel fusion trees, provide different levels of robustness at different costs. In this work, we seek to explore a more cost-effective way of introducing redundancy to improve robustness. We first perform a case study to answer the question of where redundancy should be introduced. Our proposed structure and methodology stem from the results of this case study.

3.1 Case study

We construct a experimental structure shown in Figure 1(a). The figure illustrates a clock tree driving a 8-stage path, with each stage consisting of a device driving a long wire. This structure is supposed to approximate a pair of sequentially related sinks that are topology wise distant; these are referred to as critical clock sink pairs in [7]. We choose to study critical pairs because their long distances in the clock tree topology makes these pairs prone to synchronization error. The 45nm technology provided in the ISPD 2010 contest [15] is used in this experiment. The inverters available are specified in Table 1. In this experiment, the devices are formed by connecting 7 parallel inv1's in series with 10 parallel inv1's. Each stage consists of a 1,000,000nm wire. Each clock sink has a capacitance of 300fF and the clock sinks are placed 400,000nm apart.

Table 1: Devices provided in the ISPD contest [15].

Device Name	input cap. (fF)	output cap. (fF)	output resist. (Ω)	inverted?
inv0	35	80	61.2	Yes
inv1	4.2	6.1	440	Yes

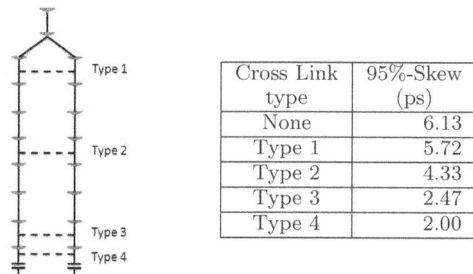

Cross Link type	95%-Skew (ps)
None	6.13
Type 1	5.72
Type 2	4.33
Type 3	2.47
Type 4	2.00

(a) Experimental structure. (b) Robustness improvement.

Figure 1: Experimental setup to determine where redundancy should be introduced.

The structure is enhanced by inserting a cross link. Four different locations for cross link insertion are considered as illustrated in Figure 1(a). The improvement in robustness of inserting the cross link at different locations is determined by performing 500 Monte Carlo NGSPICE simulations with SLSV variations on each of them. The results of the simulations can be found in Figure 1(b).

We observe that inserting any cross link improves the robustness compared to the original tree structure. Also, as the cross link is placed closer to the sinks the skew is successively reduced. Based on this we speculate that redundancy may improve the robustness more effectively when it is introduced closer to the sinks.

To grasp how the introduction of redundancy increases the cost (capacitance) of a design, the capacitance distribution of a regular clock tree is studied. A clock tree is constructed on benchmark circuit ispd10cns01 from the ISPD 2010 contest [15]. The cost at each stage is determined as the sum of sink capacitances, wire capacitances and output capacitances of driving devices. Not surprisingly, 71% and 12% of the total capacitance are respectively located at the bottom-most stage (stage 1) and at the second bottom-most stage (stage 2).

Table 2: Cost distribution over stages of a clock tree constructed on benchmark circuit ispd10cns01.

Stage	Total Cap
1	71%
2	12%
3	6%
4	4%
> 4	7%

3.2 Proposed structure

We summarize what we have learned so far to motivate our methodology. (1) Redundancy introduced in the bottom stage is more effective in improving the robustness. (2) The bulk of the capacitance of a clock network is at the bottom.

We propose a trade off between cost and robustness by introducing the redundancy in the second bottom-most stage, where the redundancy is still relatively close to the clock sinks, but the cost will be significantly lower.

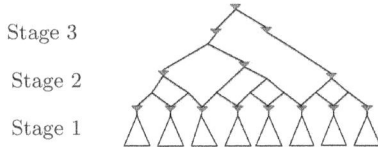

Figure 2: The proposed structure assuming that adjacent subtrees are sequentially related.

Therefore, we propose the following: First, we construct the bottom-most stage of subtrees as in all bottom-up clock tree construction algorithms. After inserting the bottom-most stage of devices to drive these subtrees, a subtree sharing sequentially related pair of clock sinks are merged. As these subtrees are typically close by because they are at the bottom of the clock network, these merges can be seen as "local" merges that span short distances. Each merge creates a new root. After this, we continue by constructing a normal tree structure on top of the set of new roots. With this proposed structure, we obtain the structure shown in Figure 2. As the redundancy is close to the clock sinks, robustness should improve compared to a tree.

4. THE PROPOSED METHODOLOGY

Our clock network synthesis (CNS) flow, shown in Figure 3, builds a clock network in a bottom up fashion. First, the bottom-most stage subtrees, each driven by a device, are constructed. Each subtree is created by performing iterative merging operations as in the DME approach [1, 4, 5]. When a subtree is sufficiently large, a device is selected to drive the subtree and a stem wire is inserted between the subtree and the device. This step inserts the first stage devices (at the bottom-most level of the buffered clock network), and is described in Section 4.1. The insertion of the second stage devices involves performing multiple local merges on subtrees that share sequentially related clock sink pairs. After that, a sparsification step is performed to reduce the cost of the multiple local merges. The details are given in Section 4.2. For the insertion of devices in the third stage and all following stages, subtrees are constructed as described in the insertion of the first stage devices. When only one subtree remains, a top down embedding step determines the locations of all nodes in the constructed topology. Blockages are handled similar to in [2]. In the sequel, we refer to the construction of the subtrees above the nth-stage devices ($n \geq 0$) and the insertion of the $n + 1$st-stage devices as the $n + 1$st-stage construction. The 0th-stage devices are the *clock sinks*, and the *sink nodes* for a nth stage subtree ($n > 1$) are the devices driving $n - 1$th-stage subtrees. In this work, we distinguish between a clock sink (a sequential element) and a sink node (a device).

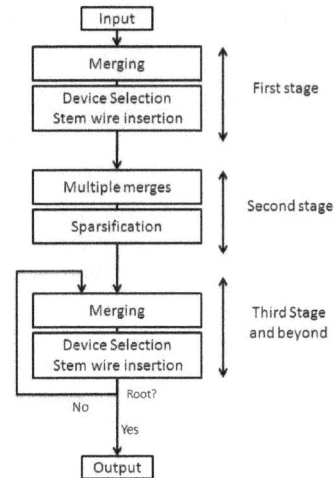

Figure 3: Proposed CNS flow.

In this flow, we use either a inverter library or a buffer library. The libraries are built from two 45 nm technology inverters named inv0 and inv1 specified in Tabel 1. The inverter library, which contains inverters made up of 14 to 35 parallel inv1's, is used when we construct clock networks that are to be evaluated under the ISPD model. The buffer library, which contains two parallel inverters connected in series, is used when the clock networks are built to meet the skew and slew constraints under the more stringent SLSV model. The first inverter in the buffer library consists of 10 parallel inv1's, and it drives an inverter made up of 17 to 32 parallel inv1's. In the following, we use d_{min} and d_{max} to refer to the smallest and largest devices, respectively, in a library.

4.1 Construction of subtrees

We adopt the general flow of bottom up tree synthesis algorithms. Subtrees are generated by iterative merging and then a driving device is selected and stem wire is inserted for each sufficiently large subtree.

4.1.1 Merging

We use the greedy DME approach outlined in [6], which uses a Nearest Neighbor Graph (NNG) to guide the iterative merging process. In this NNG, subtrees are vertices and the cost of an edge connecting two vertices corresponds to the wiring cost (capacitance) required to merge the two subtrees with zero skew. Iteratively, we alternate between merging two subtrees who are least cost neighbors in the NNG [6] and merging the subtree with the least delay with its least cost neighbor [12]. For the first stage construction, the nodes in the initial NNG are all the clock sinks. For the other stages, the driving devices of lower-stage subtrees are the sink nodes and the initial nodes in the NNG. As a pair of subtrees are merged, the corresponding nodes are deleted from the NNG and the newly formed subtree is inserted into the NNG. However, if driving the newly merged subtree with device d_{max} violates the slew constraint, denoted as S_{slew}, the two subtrees are unmerged and locked, i.e., they are not considered for further merging.

The slew of a clock sink or sink node, denoted S_{sink}, is estimated with $S_{sink} = \sqrt{S_{out}^2 + (2.2 * \delta_{elmore})^2}$, where

δ_{elmore} is the Elmore delay and S_{out} is the output slew of the device obtained from a pre-characterized Look Up Table (LUT) [10]. The LUT is constructed by performing NGSPICE simulations of an inverter/buffer with samples taken from a range of input slew values and a range of load capacitance values. The slew of the subtree, denoted $S_{subtree}$, is defined to be the maximum among the sinks in the subtree. In the synthesis, NGSPICE is used to determine a more accurate slew of a subtree if the estimated slew of a subtree is between 30ps and 90ps. However, not the entire subtree is simulated. While the subtree directly driven by the current stage-n device ($n > 1$) is simulated in full, subtrees driven by the stage-$(n-1)$ devices are modeled as lumped capacitance to reduce the simulation time. When all subtrees are locked, a device selection is performed and a stem wire is inserted for each subtree.

4.1.2 Device selection and stem wire insertion

For each subtree, we first select a device of appropriate size between $d_{\min}$ and $d_{\max}$ to drive it without violating the slew constraint S_{slew} using a simple linear search, starting from $d_{\min}$. Next, we determine, with a binary search, the maximal length of the stem wire to be inserted between the selected device and the subtree without violating the slew constraint S_{slew} using NGSPICE simulations. Recall that not the entire subtree is simulated for determining slew.

Next, we perform a round of fine tuning on all the subtrees. With the aid of NGSPICE the stem wire of all the subtrees are adjusted to obtain a similar delay δ, where δ is the delay from the root to the sinks of a subtree.

4.2 Multiple local merges and sparsification

The second stage introduces redundancy by merging a subtree with multiple subtrees that share sequentially related clock sinks. A sparsification is applied to prune some of the redundant paths that do not provide substantial robustness improvement, thereby reducing the overall cost.

4.2.1 Multiple local merges

Here, we construct a sequential relation graph (SRG). In an SRG, a vertex corresponds to a stage-1 subtree. An edge is added between two vertices if the corresponding subtrees share a pair of sequentially related clock sinks. Each subtree is then split [7, 12] into as many parts as the number of adjacent nodes the corresponding vertex has. Each of these parts is called a split-subtree. For every edge in the SRG, a zero-skew merge is performed on the two corresponding split-subtrees. The process is illustrated in Figure 4. Each double-sided arrow in Figure 4(a) implies that the pair of subtrees contain sequentially related clock sinks, and (b) is the corresponding SRG. In (c), split-subtrees are formed, and each is merged with an adjacent split-subtree.

The new root nodes in (c) are then used to create an NNG, and iterative merging is performed as in Section 4.1.1. After all subtrees have been locked, an optional sparsification process is performed, followed by device selection and stem wire insertion as in Section 4.1.2.

4.2.2 Sparsification

Sparsification is performed to ensure a unique path from the driving device of the subtree to each subtree. As multiple split-subtrees may have been split from the same subtree, there may exist multiple paths to the same subtree from the

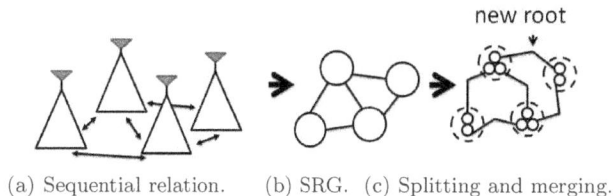

(a) Sequential relation.   (b) SRG.   (c) Splitting and merging.

Figure 4: (a) Subtrees with a double-sided arrow between them are sequentially related. (b) The sequential relation is represented by an SRG. (c) Nodes in the SRG are split and the split-subtrees are merged.

same driving device. Multiple paths from a single device, however, does not provide a good trade-off between robustness and cost (as the focus is on voltage variations). We shall show in Table 5 in Section 5.2 that we observe an improvement of only 0.08ps in skew reduction when the total cost increases by 14%.

We have a number of locked subtrees as the input to the sparsification step. All sink nodes of a locked subtree, which are all the split-subtrees in the locked subtree, are placed in an NNG. We then apply the subtree merging process in Section 4.1.1. If there are multiple split-subtrees that are identical because they are split from the same stage-1 subtree, they are of distance 0 in the NNG, and they will all be merged first. If the new subtree constructed through this process meets the slew constraint, its root node will not have multiple paths to the same sink node. However, if the new subtree cannot meet the slew constraint, we will keep the original un-sparsified version.

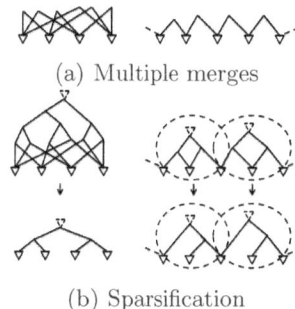

(a) Multiple merges

(b) Sparsification

Figure 5: Illustration of multiple local merges and sparsification. Left: Four sequentially related subtrees. Right: Pairwise sequentially related subtrees.

After all second-stage subtrees have been sparsified, the amounts of capacitance in the subtrees may have reduced. That presents an opportunity to perform additional merges, and if necessary, an additional sparsification process. This iterative process stops when either additional merges result in slew violation or sparsification is not successful. At this point, the second-stage devices will be inserted as described in Section 4.1.2.

Figure 5 shows an iteration of multiple local merges and sparsification. The left shows four stage-1 subtrees that are sequentially related with each other and the right five stage-1 subtrees that are sequentially related with immediate neighbors. For the former, local merges result in six new roots, and for the latter, four new roots. Applying iterative merging in Section 4.1.1 to the six new roots in the former results in a single tree, whereas, two trees are obtained for the latter. In both cases, sparsification is successfully performed, and therefore, the former is sparsified into a tree, and the latter is sparsified into two trees. The multiple paths from a driving device to a sink node have been eliminated.

5. EXPERIMENTAL RESULTS

We now present the experimental results. Our CNS flow is implemented in C++ and the experiments are run on a quad core 3.1GHz Linux machine with 7.7GB memory. Results are obtained for the ISPD 2010 Benchmarksuit [15] for both the ISPD variations model and the SLSV variations model. For brevity, we will refer to these ISPD 2010 benchmark circuits ispd10cns01–08 as BM01–08. The skew constraint is 4.99ps for BM03 and 7.5ps for all other circuits. All synthesized clock networks meet the 100ps slew constraint.

5.1 ISPD variations model

Table 3 shows the comparisons with [8, 3] on 95% skew and total capacitance under the ISPD model. Compared to [3], our proposed solutions have higher robustness in all circuits, but lose out in capacitance on BM04, 07, and 08. Overall, our proposed solutions have 1% lower total capacitance. Compared to [8], our proposed solutions have 11% lower capacitance and except for BM03, have higher robustness. The 95% skew of our solution for BM03 is a shade higher that of [8]. In fact, the maximum local clock skews (see the column labeled "Max Skew") of most of our solutions would also meet the skew constraints specified in these circuits.

To put the results in perspective, we also construct tree solutions with our tool (by disabling the second stage). From Table 3, we can observe that for BM04-08, tree solutions actually outperform all other solutions in terms of cost, which is not surprising, but also outperform [8, 3] in terms of robustness. Hence, it is evident that for smaller designs with loose design constraints, a tree will be the best solution. However, our proposed structure can still trade off cost for better robustness. On larger circuits, namely, BM01 and 02, a simple clock tree cannot meet design constraints and our proposed structure allows the cheapest and most robust solutions.

5.2 SLSV variations model

Table 4 shows the comparisons with [2, 17] on 95% skew and total capacitance under the SLSV model. In terms of total capacitance, our solutions are 22% cheaper than those of [2], and the solutions of [17] cost at least thrice as much as ours. Perhaps, our main contribution is that we can satisfy the skew constraint for BM01 and BM02 without using meshes, the only known solutions thus far, which are from [17]. Except for BM03, our solutions in general have comparable or better robustness, in terms of 95% skew. In fact, our solutions also perform well in terms of the maximum local clock skew (see the column labeled "Max Skew").

As mentioned, our solution for BM03 in Table 4 could not meet the skew constraint. Recall that the minimum buffer used in this experiment, i.e., d_{min}, is formed by using 10 parallel inv1's driving 17 parallel inv1's. We use $d_{min} = 10/17$ to denote such a configuration. We can also meet the skew constraint for BM03 by using $d_{min} = 10/26$, i.e., a larger minimum buffer, and not performing the sparsification stage. Even though this approach is costlier, the obtained solu-

Table 3: Comparisons of 95% skew and cost under the ISPD model. An '*' indicates that the solution does not meet the skew requirement.

Ckt.	Method	Max Skew (ps)	95% Skew (ps)	Cap. (fF)	Run Time (s)
BM01	[8]	-	7.32	142.64	1092
	[3]	-	6.48	137.97	472
	Our work	6.65	5.35	130.82	680
	Tree	*11.98	*8.23	105.64	618
BM02	[8]	-	7.42	265.21	4314
	[3]	-	7.38	268.29	1450
	Our work	*8.53	5.76	254.36	1621
	Tree	*12.45	*9.67	196.19	1625
BM03	[8]	-	4.49	36.61	2094
	[3]	-	4.76	34.17	383
	Our work	*5.95	4.64	31.61	150
	Tree	*7.01	*5.06	30.62	155
BM04	[8]	-	6.70	51.07	934
	[3]	-	7.14	42.77	110
	Our work	6.27	4.97	46.63	284
	Tree	*8.53	5.91	38.52	261
BM05	[8]	-	4.78	25.13	1110
	[3]	-	5.88	22.13	278
	Our work	5.46	4.02	19.49	118
	Tree	*8.17	4.79	18.30	121
BM06	[8]	-	6.41	32.68	285
	[3]	-	5.61	28.55	61
	Our work	6.52	4.59	28.03	129
	Tree	6.43	4.86	24.60	111
BM07	[8]	-	5.86	48.32	818
	[3]	-	6.62	43.91	133
	Our work	6.67	4.93	46.61	272
	Tree	*7.54	5.85	39.41	275
BM08	[8]	-	5.07	32.67	327
	[3]	-	6.50	28.41	54
	Our work	5.96	4.76	32.23	150
	Tree	6.70	4.91	26.81	135
Norm.	[8]			1.11	
	[3]			1.01	
	Our work			1.00	
	Tree			0.86	

tion is still of lower cost than those from [2, 17] (40.73fF vs. 69.15fF and 93.96, respectively), as seen in Table 5. However, we would like to point out that our current implementation does not have the ability to adaptively determine a suitable d_{min}. Such a feature was available in the work of [2], and we would consider that in a future extension of this work. Also note that for $d_{min} = 10/17$, without sparsification, the skew reduction is only 0.08ps but the total cost increases by 14%. Therefore, our implementation performs sparsification by default.

As [7] presented only multilevel fusion tree solutions to BM08, we will use only BM08 for comparison. In [7], two types of devices are used, namely, inv0 or a composite inverter similar to 8 inv1's. As both inv0 and the composite inverter are treated as a single inverter, the variations model under consideration in [7] is equivalent to the SLSV model. For inv0, the variations are assumed to follow a uniform distribution, which is what we have been using in the Monte Carlo simulations thus far. For the composite inverter, a Gaussian distribution is assumed, with $\sigma = 0.015V$. We compare in Table 6 our solution to BM08 using both distributions under the SLSV model and the least and most robust multilevel fusion solutions from [7] under the two distributions. Our solution shows an advantage in terms of both robustness and cost. In fact, the robustness of our

Table 4: Comparisons of 95% skew and cost under the SLSV model. An '*' indicates that the solution does not meet the skew requirement.

Ckt.	Method	Max Skew (ps)	95% Skew (ps)	Cap. (fF)	Run Time (s)
BM01	[17]	*9.43	7.23	1168.10	675
	[2]	*11.40	*10.29	189.06	2324
	Our work	6.65	5.69	155.14	879
	Tree	*11.85	*9.21	117.50	1139
BM02	[17]	*8.99	7.35	2099.81	2140
	[2]	*14.19	*12.30	341.08	6723
	Our work	*7.69	6.26	292.99	1978
	Tree	*14.87	*10.52	218.43	1740
BM03	[17]	4.23	3.95	93.96	21
	[2]	*5.80	4.95	69.15	1269
	Our work	*6.67	*5.61	34.57	158
	Tree	*8.65	*6.50	32.83	152
BM04	[17]	*7.64	7.25	125.33	22
	[2]	*8.23	7.17	56.59	2711
	Our work	6.76	5.57	53.08	315
	Tree	*8.79	7.04	42.38	310
BM05	[17]	*9.40	7.27	74.08	10
	[2]	5.64	5.00	26.25	1057
	Our work	5.86	4.97	22.69	142
	Tree	*10.68	6.70	24.29	180
BM06	[17]	*8.04	6.79	87.39	46
	[2]	6.40	5.95	32.57	1027
	Our work	7.32	5.87	31.78	138
	Tree	*8.87	7.04	27.03	149
BM07	[17]	6.67	5.95	128.35	27
	[2]	*8.99	7.07	56.13	2917
	Our work	7.23	6.39	51.48	317
	Tree	*11.57	*7.78	42.35	306
BM08	[17]	6.62	5.37	97.42	18
	[2]	7.39	6.53	37.40	1427
	Our work	6.54	5.88	36.22	170
	Tree	*8.78	6.95	28.85	221
Norm.	[17]			3.87	
	[2]			1.22	
	Our work			1.00	
	Tree			0.85	

Table 5: Improving robustness of BM03 by scaling up the minimum device size and not performing sparsification.

d_{min}	sparsified	95% Skew (ps)	Cap (fF)	Run Time (s)
10/17	Yes	*5.61	34.57	158
	No	*5.53	39.51	175
10/26	Yes	*5.22	35.97	145
	No	4.94	40.73	172

solution is even higher when Gaussian distribution is used. However, we would like to point out that we have an unfair advantage in that we could select appropriate device sizes in our methodology. The entries in the column labeled "Device Sizes" for our work means that $d_{min} = 10/17$ and $d_{max} = 10/32$.

6. CONCLUDING REMARKS

In this work, we analyze the effectiveness of redundant paths in reducing the effects of variations. We propose a CNS technique that introduces redundant paths close to the clock sinks by performing multiple local merges. Evaluation performed under the ISPD and SLSV models shows that with modest overhead in capacitance, our proposed solutions can improve robustness.

Table 6: Comparison with multilevel fusion trees on BM08.

Distr.	Method	Device Type	Device Sizes	95% Skew (ps)	Cap. (fF)	Run Time (s)
Uni -form	Our work	inv1	10/17-32	5.88	36.22	170
	[7]	inv0	1	22.09	37.70	340
		inv0	1	12.72	56.37	747
Gaus -sian	Our work	inv1	10/17-32	3.56	36.22	170
	[7]	inv1	8	7.16	37.27	1101
		inv1	8	4.13	51.96	3900

7. ACKNOWLEDGMENTS

This work was supported in part by the National Science Foundation under award CCF-1065318 and the Semiconductor Research Corporation under task 1292-074.

8. REFERENCES

[1] K. Boese and A. B. Kahng. Zero-skew clock routing trees with minimum wirelength. In *Proc. IEEE Int. Conf. on ASIC*, pages 1.1.1–1.1.15, 1992.

[2] S. Bujimalla and C.-K. Koh. Synthesis of low power clock trees for handling power-supply variations. In *Proc. ISPD*, pages 37–44, 2011.

[3] Y.-C. Chang, C.-K. Wang, and H.-M. Chen. On construction low power and robust clock tree via slew budgeting. In *Proc. ISPD*, pages 129–136, 2012.

[4] T.-H. Chao, J.-M. Ho, and Y.-C. Hsu. Zero skew clock net routing. In *Proc. DAC*, pages 518–523, 1992.

[5] M. Edahiro. Minimum skew and minimum path length routing in VLSI layout design. In *NEC Research and Development*, pages 569–575, 1991.

[6] M. Edahiro. A clustering-based optimization algorithm in zero-skew routings. In *Proc. DAC*, pages 612–616, 1993.

[7] D.-J. Lee and I. L. Markov. Multilevel tree fusion for robust clock networks. In *Proc. ICCAD*, pages 632–639, 2011.

[8] T. Mittal and C.-K. Koh. Cross link insertion for improving tolerance to variations in clock network synthesis. In *Proc. ISPD*, pages 29–36, 2011.

[9] NGSPICE. http://ngspice.sourceforge.net/.

[10] R. Puri, D. S. Kung, and A. D. Drumm. Fast and accurate wire delay estimation for physical synthesis of large ASICs. In *Proc. Great Lakes Symp. on VLSI*, pages 30–36, 2002.

[11] A. Rajaram, J. Hu, and R. Mahapatra. Reducing clock skew variability via cross links. In *Proc. DAC*, pages 18–23, 2004.

[12] A. Rajaram and D. Pan. Variation tolerant buffered clock network synthesis with cross links. In *Proc. ISPD*, pages 157–164, 2006.

[13] A. Rajaram, D. Z. Pan, and J. Hu. Improved algorithms for link-based non-tree clock networks for skew variability reduction. In *Proc. ISPD*, pages 55–62, 2005.

[14] X.-W. Shih, H.-C. Lee, K.-H. Ho, and Y.-W. Chang. High variation-tolerant obstacle-avoiding clock mesh synthesis with symmetrical driving trees. In *Proc. ICCAD*, pages 452–457, 2010.

[15] C. Sze. ISPD 2010 high performance clock network synthesis contest: Benchmark suite and results. pages 143–143, 2010.

[16] G. Venkataraman, N. Jayakumar, J. Hu, P. Li, S. Khatri, A. Rajaram, P. McGuinness, and C. Alpert. Practical techniques to reduce skew and its variations in buffered clock networks. In *Proc. ICCAD*, pages 592–596, 2005.

[17] L. Xiao, Z. Xiao, Z. Qian, Y. Jiang, T. Huang, H. Tian, and E. F. Y. Young. Local clock skew minimization using blockage-aware mixed tree-mesh clock network. In *Proc. ICCAD*, pages 458–462, 2010.

An Improved Benchmark Suite for the ISPD-2013 Discrete Cell Sizing Contest

Muhammet Mustafa Ozdal[1], Chirayu Amin[1], Andrey Ayupov[1], Steven Burns[1], Gustavo Wilke[2], and Cheng Zhuo[2]

[1] Strategic CAD Labs, Intel Corporation, Hillsboro, OR 97124
[2] Core CAD Technologies, Intel Corporation, Hillsboro, OR 97124
{mustafa.ozdal, chirayu.s.amin, andrey.ayupov, steven.m.burns, gustavo.r.wilke, cheng.zhuo}@intel.com

ABSTRACT

Gate sizing and threshold voltage selection is an important step in the VLSI design process to optimize power and performance of a given netlist. In this paper, we provide an overview of the *ISPD-2013 Discrete Cell Sizing Contest*. Compared to the ISPD-2012 Contest, we propose improvements in terms of the benchmark suite and the timing models utilized. In this paper, we briefly describe the contest, and provide some details about the standard cell library, benchmark suite, timing infrastructure and the evaluation metrics.

Categories and Subject Descriptors

B.7.2 [**Hardware, Integrated Circuits**]: Design Aids

General Terms

Algorithms, Design, Experimentation, Performance

Keywords

Gate sizing, Circuit optimization, Benchmarks, Physical Design

1. INTRODUCTION

Circuit optimization is an important stage in the VLSI design process. The objective is to minimize power consumption while satisfying the given performance constraints. Although this problem has been studied extensively in the literature [1, 2, 3, 4, 8, 9, 10, 11, 12], there are still various challenges that make it difficult to apply the existing academic algorithms to modern industrial designs. Some of these challenges have been outlined in [6, 7] as follows: discrete cell sizes, cell timing models, complex timing constraints, slew effects, and large design sizes.

Last year, we organized the *ISPD-2012 Discrete Cell Sizing Contest* [5], and focused on the problem of gate sizing and threshold voltage selection. The objective was to expose some of these challenges to the academic community, while keeping the problem complexity manageable. For this purpose, the standard cell library and the benchmark suite used in the contest were made public. This year, the ISPD-2013 Contest focuses on the same problem, with an improved benchmark

suite and more realistic timing models. Specifically, a simple interconnect timing model was assumed in the ISPD-2012 Contest, where each net was defined to have a lumped capacitance with zero resistance. The purpose was to allow the contestants to focus on the basic algorithmic aspects of the problem in the first year. In the ISPD-2013 Contest, we remove this simplification, and utilize the realistic interconnect timing models of an industrial timing analysis tool.

In this paper, we provide an overview of the *ISPD-2013 Discrete Cell Sizing Contest*. Similar to the ISPD-2012 Contest, we provide a standard cell library and a set of benchmarks for the contest. Each benchmark consists of the following: 1) verilog netlist, 2) interconnect parasitics, and 3) timing constraints. The contestant programs are expected to compute a discrete size and a threshold voltage level for each cell in a given netlist such that the total leakage power is minimum and the timing constraints are satisfied. An industrial tool, Synopsys PrimeTime®, will be used to evaluate the timing violations of the results.

The rest of the paper is organized as follows. In Section 2, we describe the standard cell library of the contest. In Section 3, we summarize our timing models and infrastructure. The contest benchmarks are discussed in Section 4. Finally, the evaluation metrics and the ranking criteria are described in Section 5.

2. STANDARD CELL LIBRARY

The cell library of the ISPD-2012 Contest is used for this year's contest. In this section, we provide a brief overview of this library. Further details can be found in [5].

The cell library provided is in Synopsys Liberty™ format. There are 12 different logic functions (11 combinational and 1 sequential) as listed in Figure 1. Corresponding to each combinational logic function, there are 3 threshold voltage (Vt) levels and 10 sizes for each Vt level. In other words, for each combinational cell in the given netlist, there are 30 different choices in the library, each with a different power-performance tradeoff. Note that sequential cell sizing is not allowed in this contest, and there is only a single Vt level and a single size for the sequential library cell.

The timing characteristics of the library cells are defined as lookup tables associated with the timing arcs. Specifically, the delay of a timing arc is a function of the transition time (i.e. slew) at the input pin and the effective capacitance at the output pin. This function is provided as a look-up table in the library. The transition time at the output pin is defined in a similar way. A simple current source model was used to characterize cell delays and transition times, details of which can be found in [5].

In addition, each library cell has a leakage power value; each input pin has an input capacitance, and each output pin has a maximum load constraint. A C++ class is provided to extract the relevant data from the cell library.

Prefix	Description	Function
in01	Inverter	!A
na02	2-input NAND	!(A & B)
na03	3-input NAND	!(A & B & C)
na04	4-input NAND	!(A & B & C & D)
no02	2-input NOR	!(A | B)
no03	3-input NOR	!(A | B | C)
no04	4-input NOR	!(A | B | C | D)
ao12	AND-OR-INV	!(A | (B & C))
ao22	AND-OR-INV	!((A & B) | (C & D))
oa12	OR-AND-INV	!(A & (B | C))
oa22	OR-AND-INV	!((A | B) & (C | D))
ms00	Flip Flop	sequential

Figure 1: Cell families in the ISPD-2012 and ISPD-2013 Benchmark Suites

3. TIMING INFRASTRUCTURE

The cell delay and slew functions have been provided as lookup tables in the standard cell library, as described in Section 2. A major difference from the previous contest ([5]) is that interconnects are modeled as distributed RC trees (one tree per net). For simplicity, the RC tree does not have any cross-coupling capacitances. The delay through interconnect is supposed to be non-zero except for clock nets which are assumed to be without interconnect for simplicity. Additionally, the interconnect RC will degrade transition time from driver output to receiver inputs.

As in the previous contest, all sequential cells in the benchmarks are rising-edge triggered flip flops. Sequential sizing is not allowed, and there is only a single size available in the contest library for sequential cells, for which the set-up time is 0, and there is no hold constraint. For contest purposes, the following ideal clock network assumptions hold:

- Clock port at the top level is directly connected to all sequential pins without clock buffers.

- Arrival time of the clock signal at all sequential inputs is the same (i.e. zero skew).

- Sequential delay is independent of the clock slew.

- Clock input pin capacitances are zero for all sequential cells.

- Clock net has zero parasitics (i.e. no RC).

Only a single clock domain is defined, and there are no complex timing constraints such as false paths, transparent latches, multi-cycle overrides, etc. The purpose here is to simplify the timing model so that the contestants can focus their efforts on the combinational sizing optimization. On the other hand, the cell timing and slew models are kept realistic, and this complexity needs to be taken into account during sizing optimization.

In terms of the timing engine, the contestants have the following two options: 1) Implement a timing engine to be used during optimization. 2) Use a stand-alone industrial timing engine as a black-box during optimization. If the contestants choose the second option, we have provided a timing infrastructure (with C++ and TCL APIs) for the contestants to be able to call the Synopsys PrimeTime® timing engine as a black-box from their sizing engines. To help out with delay calculations, the APIs contain functions to query effective capacitance at driver outputs and the arrival and transition times at driver outputs as well as receiver inputs. The contestants are also allowed to implement their own scripts to call Synopsys PrimeTime®.

Table 1: Statistics of the ISPD-2013 Contest Benchmarks

Name	# Inputs	# Outputs	# Comb. Cells	# Seq. Cells	# Total Cells
usb_phy	15	19	511	98	609
pci_bridge32	160	201	27316	3359	30675
des_perf	234	140	103842	8802	112644
netcard	1836	10	884427	97831	982258
cordic	34	64	40371	1230	41601
fft	1026	1026	30297	1984	32281
matrix_mult	3202	1600	152427	2898	155325
edit_dist	2562	12	125000	5661	130661

Table 2: Descriptions of the new benchmarks

cordic	fast sine and cosine calculation using CORDIC algorithm
fft	64-point fast Fourier transform using butterfly algorithm
matrix_mult	two 10x10 matrix multiplication (4 bit integer elements)
edit_dist	edit distance algorithm for strings of 128 symbols long and 4 bit alphabet

4. BENCHMARKS

For the ISPD-2013 contest, there are eight benchmarks released, as shown in Table 1. Top four circuits in the table are based on the benchmarks used in the ISPD-2012 Contest. The SPEF files with distributed RC networks were generated for them, and some netlist optimization was done. The other four benchmarks are new and generated for the contest by organizers. To generate netlists for the new benchmarks and SPEF files for the new and the last year benchmarks, Cadence ® C-to-Silicon Compiler and Encounter Digital Implementation System were used. Short descriptions of the algorithms that the new benchmarks implement are provided in Table 2.

Each benchmark consists of the following files: netlist in structural verilog format, timing constraints in Synopsys Design Constraints (SDC) format, and interconnect parasitics in IEEE SPEF format. As part of the benchmark package, we provide C++ parser classes to extract the contest-specific data from these benchmarks. Note that the parsers are not generic and are developed to parse files from the benchmark suite only.

5. EVALUATION METRICS

The contest evaluations will be done in a similar way as the ISPD-2012 Contest [5]. As mentioned before, the basic objective is to minimize the total leakage power while satisfying all the timing related constraints. The contestants will be ranked using two different evaluation metrics. The primary metric is based mainly on the solution quality as defined later in this section. The secondary metric is a combination of both solution quality and runtime. For both metrics, there are strict runtime limits for each benchmark.

The solution quality is defined in terms of both the timing violations and the total leakage power. The timing results are evaluated using an industrial timing engine, Synopsys PrimeTime®. There are 3 different violation types defined: 1) slack violations, 2) slew violations, and 3) maximum load violations. All benchmarks are set up in such a way that it is not difficult to obtain results with zero violations. For this reason, the solution quality metric is defined in lexicographic order as follows: 1) the sum of all violation values, and 2) the total leakage power. In other words, a solution with smaller violations is defined to be better than another solution with larger violations, regardless of the leakage power values. If two solutions have identical violations, then the ranking will be based on the total leakage power. Note that as long as all the timing constraints are satisfied, power will be the main metric to assess solution qualities.

As mentioned above, the contestants will be ranked using two different metrics. In the primary metric, the ranking will be done based on solution qualities. Only when two results have identical solution qualities, then the runtimes will be considered. On the other hand, in the secondary metric, both solution quality and runtime will be important. Specifically, the secondary metric is defined in lexicographic order as follows: 1) total violations, and 2) combined cost metric, which is given by the following equation:

$$combined\_cost = \frac{Power}{Power_{REF}} + \gamma \frac{Runtime}{Runtime_{REF}} \quad (1)$$

Here, the power and runtime values are normalized with respect to the reference values, which are set based on the solution with the best quality. The value of γ is set to 0.05 as in [5]. Intuitively, $\gamma = 0.5$ means that 1% degradation in solution quality can be compensated by a 20% runtime reduction with respect to the reference values.

Strict runtime limits are defined for both the primary and the secondary metrics. These limits depend on the number of cells in a given benchmark. Specifically, for the primary ranking metric, the runtime limit is defined as 3 hours plus 1 hour for every 40K cells in the netlist. For the secondary ranking metric, the contestant programs will be re-run with a special command line argument: "-fast". In the fast mode, the runtime limit of a benchmark will be $1/5$ of the corresponding limit of the primary metric. We hope that such a strict runtime limit and the combined cost metric defined in Equation 1 will encourage contestants to make use of the multiple cores available in the evaluation platforms.

6. CONCLUSIONS

The *ISPD-2013 Discrete Gate Sizing Contest* is a continuation of the contest introduced in 2012 with improvements in the benchmarks and the timing models. Our objective has been to expose some of the industrial challenges for the gate sizing problem by providing a realistic benchmark suite. We hope that this will help the research community to advance state of the art.

7. ACKNOWLEDGEMENTS

The contest organizers would like to thank Troy Wood and Robert Hoogenstryd (Synopsys), Noel Menezes, Jason Xu, Nanda Kuruganti, Rohit Vachher, Shishpal Rawat, and Robert Nguyen (Intel) for helping with the contest organization.

8. REFERENCES

[1] M. R. C. M. Berkelaar and J. A. G. Jess. Gate sizing in MOS digital circuits with linear programming. In *Proc. of DATE*, pages 217–221, 1990.

[2] C. P. Chen, C. C.-N. Chu, and D. F. Wong. Fast and exact simultaneous gate and wire sizing by Lagrangian relaxation. *IEEE Trans. on Computer-Aided Design*, 18(7):1014–1025, July 1999.

[3] D. Chinnery and K. Keutzer. Linear programming for sizing, vth and vdd assignment. In *Proc. of ISLPED*, pages 149–154, 2005.

[4] H. Chou, Y.-H. Wang, and C. C.-P. Chen. Fast and effective gate sizing with multiple-Vt assignment using generalized Lagrangian relaxation. In *Proc. of ASPDAC*, pages 381–386, 2005.

[5] M. M. Ozdal, C. Amin, A. Ayupov, S. Burns, G. Wilke, and C. Zhuo. The ISPD-2012 discrete cell sizing contest and benchmark suite. In *ACM International Symposium on Physical Design*, pages 161–164, 2012.

[6] M. M. Ozdal, S. Burns, and J. Hu. Algorithms for gate sizing and device parameter selection for high-performance designs. *IEEE Transactions on Computer-Aided Design of Integrated Circuits and Systems (TCAD)*, 31:1558–1571, 2012.

[7] M. M. Ozdal, S. Burns, and J. Hu. Gate sizing and device technology selection algorithms for high-performance industrial designs. In *Proc. of ICCAD*, Nov. 2011.

[8] M. Rahman, H. Tennakoon, and C. Sechen. Power reduction via near-optimal library-based cell-size selection. In *Proc. of DATE*, 2011.

[9] H. Ren and S. Dutt. A network-flow based cell sizing algorithm. In *Workshop Notes, Int'l Workshop on Logic Synthesis*, 2008.

[10] S. Roy, W. Chen, C. C.-P. Chen, and Y. H. Hu. Numerically convex forms and their application in gate sizing. *IEEE Trans. on Computer-Aided Design*, 26(9):1637–1647, Sept. 2007.

[11] H. Tennakoon and C. Sechen. Gate sizing using Lagrangian relaxation combined with a fast gradient-based pre-processing step. In *Proc. of ICCAD*, pages 395–402, 2002.

[12] J. Wang, D. Das, and H. Zhou. Gate sizing by Lagrangian relaxation revisited. *IEEE Trans. on Computer-Aided Design*, 28(7):1071–1084, July 2009.

TAU 2013 Variation Aware Timing Analysis Contest

Debjit Sinha[1], Luís Guerra e Silva[2], Jia Wang[3], Shesha Raghunathan[4],
Dileep Netrabile[5], and Ahmed Shebaita[6]

[1,5]IBM Systems and Technology Group, [1]Hopewell Junction/[5]Essex Junction, USA
[2]INESC-ID / IST - TU Lisbon, Portugal
[3]Illinois Institute of Technology, Chicago, USA
[4]IBM Systems and Technology Group, Bangalore, India
[6]Synopsys, Sunnyvale, USA
tau.contest@gmail.com

ABSTRACT

Timing analysis is a key component of any integrated circuit (IC) chip design-closure flow, and is employed at various stages of the flow including pre/post-route timing optimization and timing signoff. While accurate timing analysis is important, the run-time of the analysis is equally critical with growing chip design sizes and complexity (for example, increasing number of clocks domains, voltage islands, etc.). In addition, the increasing significance of variability in the chip manufacturing process as well as environmental variability necessitates use of variation aware techniques (e.g., statistical, multi-corner) for chip timing analysis which significantly impacts the analysis run-time.

The aim of the TAU 2013 variation aware timing contest is to seek novel ideas for fast variation aware timing analysis, by means of the following: (a) increase awareness of variation aware timing analysis and provide insight into some challenging aspects of the analysis, (b) encourage novel parallelization techniques (including multi-threading) for timing analysis, and (c) facilitate creation of a publicly available variation aware timing analysis framework and benchmarks to further advance research in this area.

Categories and Subject Descriptors

B.8.2 [**Hardware**]: Performance and Reliability—*Performance Analysis and Design Aids*

General Terms

Algorithms, Design, Experimentation, Performance

Keywords

Timing analysis; Variability

1. INTRODUCTION

Accurate and fast chip timing analysis is essential at multiple stages in the chip design-closure flow including pre/post-route timing optimization and timing signoff. Growing chip design sizes and complexities (for example, increasing number of clock domains, voltage islands, etc.), as well as more complex and accurate timing models (for example, current source models) lead to longer timing analysis run-times thereby degrading designer productivity.

The increased significance of variability in the deep submicron chip manufacturing process necessitates variation aware timing analysis. Manufacturing sources of variability include device front-end variability (e.g., variations in channel length, oxide thickness, dopant concentration, etc.) and back-end-of-line variability (metal variability). In addition, environmental sources of variation like voltage and temperature strongly impact circuit timing. Sources of variation that impact circuit timing are termed *parameters*.

Variation aware timing analysis can be performed in different ways. One approach is to perform multiple deterministic timing runs at various corners in the parameter space, wherein a corner indicates a condition for each parameter (e.g., at some corner C_i, parameter voltage is at 1.05 volts that denotes a 3 sigma corner for parameter voltage, temperature is at 50 degrees Celsius that denotes a 2 sigma corner for parameter temperature, parameter metal-layer 4 is at the thick condition that denotes a -3 sigma corner for parameter metal-layer 4, etc.). This approach requires a large number of runs to cover the parameter space. In some cases, more than two hundred corners need to be analyzed [7]. As a result, multithreading or parallelization techniques are often employed for multi-corner timing. An alternate approach is to perform a single statistical timing analysis, wherein each timing quantity (e.g., delay, slew, arrival time) is modeled as a random variable with a known distribution to represent uncertainty due to variability. While the primary advantage of this approach is the ability to cover the parametric variation space in a single/fewer run(s) and pessimism reduction by means of root sum square (RSS) of parametric variations, analytic linear modeling assumptions may lead to some inaccuracies in the final result. Monte Carlo based statistical timing analysis overcomes the accuracy problem, but is too run-time expensive for modern large designs. A broad overview of this topic is presented in [3].

This paper describes the TAU 2013 variation aware timing contest. The aim of the contest is to seek novel ideas for fast variation aware timing analysis, by means of the following: (a) increase awareness of variation aware timing analysis and provide insight into some challenging aspects of the analysis, (b) encourage novel parallelization techniques (including multi-threading) for timing analysis, and (c) facilitate creation of a publicly available variation aware timing analysis framework and benchmarks to further advance research in this area. Additional details regarding the contest can be obtained from [2].

2. TIMING ANALYSIS

Timing analysis involves computing timing information as signal transitions propagate from the inputs to the outputs of a digital circuit, usually described by a netlist of circuit elements. This is achieved by estimating signal propagation through the circuit elements, from the netlist inputs to outputs. Each element introduces a delay on signal transition propagation. Signal transitions are characterized by their transition times or *slew*. Circuit elements affect the signal transitions at their inputs by modifying their slew at the outputs. The general model is illustrated in Figure 1, where delay is designated by d, input slew by s_i and output slew by s_o.

Early and late mode *arrival times* (designated by at) quantify the earliest or latest time instant that a signal transition can reach the corresponding circuit node, when traveling from a circuit input. Arrival times are thus computed by adding edge delays across a path and computing the min or max (for early or late mode) of such delays when they converge at a given circuit node. For example, assuming at_A^{early} and at_B^{early} to be the early arrival times at pins A and B, respectively, of the circuit element represented in Figure 1, the early arrival time at the output pin Y is given by

$$at_Y^{early} = \min(\ at_A^{early} + d_{AY},\ at_B^{early} + d_{BY}\ ). \quad (1)$$

The late arrival time at the output pin Y is given by

$$at_Y^{late} = \max(\ at_A^{late} + d_{AY},\ at_B^{late} + d_{BY}\ ). \quad (2)$$

Required arrival times (designated by rat) are constraints or limits imposed on the arrival times, at particular nodes of the circuit. Such limits are usually necessary to ensure proper circuit operation. Assuming either early or late mode, when a required arrival time is defined for a particular circuit node, the following conditions must hold.

$$at^{early} \geq rat^{early} \quad (3)$$

$$at^{late} \leq rat^{late} \quad (4)$$

Slacks (designated by $slack$) are the difference between arrival times and required arrival times, and measure how well the constraints of (3) and (4) are met.

$$slack^{early} = at^{early} - rat^{early} \quad (5)$$

$$slack^{late} = rat^{late} - at^{late} \quad (6)$$

Slacks are positive when the required arrival time constraints are met, and negative otherwise.

Slew propagation is another essential task of timing analysis, since cell and interconnect delays are a function of the input slew. The contest assumes worst-slew propagation, wherein

the smallest or the largest slew is propagated for early or late mode, respectively.

$$s_{oY}^{early} = \min(\ s_{oAY}^{early}(\ s_{iA}^{early}),\ s_{oBY}^{early}(\ s_{iB}^{early})\ ) \quad (7)$$

$$s_{oY}^{late} = \max(\ s_{oAY}^{late}(\ s_{iA}^{late}),\ s_{oBY}^{late}(\ s_{iB}^{late})\ ) \quad (8)$$

Slew propagation need not be related to the arc causing delay propagation: for the example circuit element of Figure 1, the delay can be propagated from input A and the slew from input B. It is to be noted that industrial timers use more sophisticated slew propagation techniques than used in this contest.

2.1 Parametric timing analysis

A linear parametric timing model is used for timing analysis in this contest. Similar to the work in [13], timing quantities are represented by a linear function of the parameters. In addition, all parameters are assumed to be independent. Each parameter is modeled as a Gaussian random variable. Any timing quantity (e.g., delay) is thus a weighted sum of Gaussians.

Variation aware parametric timing analysis involves propagating timing information across the parameter space through the circuit. This could be done using multiple techniques including multiple corner analysis, Monte Carlo analysis, or analytical statistical timing analysis. The contest does not require adoption of any particular technique; and novel techniques are encouraged.

For analytical statistical timing analysis using timing quantities as a first order sum of random variables, addition and subtraction operations are performed easily. All terms are linearly added, while the random terms are root-sum-squared [13]. Maximum and minimum operations, on the other hand are complicated, and involve approximations. Clark's method [5] to obtain the moments of the maximum of two Gaussian distributions may be applied and a Gaussian with these two moments may be constructed to denote the approximate result of the maximum operation. Relevant mathematical details have been presented in [4, 5, 13]. For multiple Gaussians, the final result is obtained using iterative *pair-wise* maximum operations. It should be noted that approximation errors in the final result depend on the order of the pair-wise operations, and techniques may be adopted to reduce error in the ordering process [12]. The same idea applies to statistical minimum operations as well. While performing maximum/minimum operations using Monte Carlo simulations may be most accurate, they are run-time expensive and should be used cautiously. The contest does not advocate any particular approach for these operations.

2.2 Sources of variability and sensitivity

Each timing quantity may be sensitive to the following global inter-chip sources of variability.

- environmental: voltage (V), temperature (T)

- front end of line process: channel length (L), device width (W), voltage threshold (H)

- back end of line: metal (M)

For simplicity, only a single parameter M is assumed for all metal layers in this contest. These parameters denote systematic chip-to-chip (or inter-chip) sources of variation, while

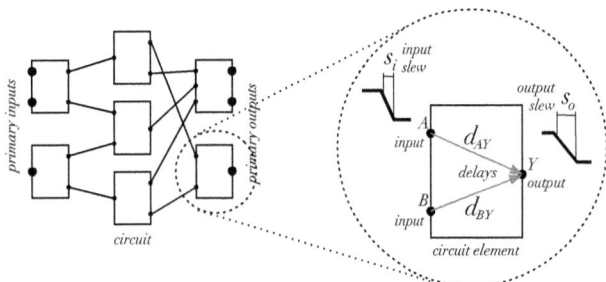

Figure 1: Circuit model

systematic intra-chip variability from all parameters are ignored. In addition to the above global inter-chip parameters, each timing quantity may contain an independently random source of variability (R) that denotes random inter-chip as well as random intra-chip variation. Any timing quantity may thus be denoted in the following notation:

$$\mu + a_v\Delta V + a_t\Delta T + a_l\Delta L + a_w\Delta W + a_h\Delta H + a_m\Delta M + a_r\Delta R, \quad (9)$$

where μ denotes the nominal value of the quantity (value in absence of variability), and each parameter (ΔV, ΔT, ..., ΔR) denotes a unit normal Gaussian $N(0,1)$ and may vary from -3 to $+3$ sigmas.

Parameter sensitivities are denoted as time units per sigma values (e.g., $a_v = 5$ pico seconds per sigma), and are obtained either as *asserted* values, or via *finite differencing*. In the former case, the sensitivity of a timing quantity to a parameter is available directly as an input (e.g., voltage sensitivity for cells are available as asserted values in the cell library). Finite differencing in context of any parameter X implies that the value of the timing quantity Q is available (or can be computed) for at least two (sigma) corners of X. The first order sensitivity is then computed as the ratio of the change in Q between the two corners of X and the difference in the two corners of X. Assuming two sigma corners of X as $+3$ and -3 sigma, the finite differenced sensitivity a_X is computed as

$$a_X = \frac{Q_{|X=+3\sigma} - Q_{|X=-3\sigma}}{3 - (-3)}.$$

If Q is truly a linear function of X, the choice of sigma corners does not matter during finite differencing. For models where Q is not a linear function of X, finite differenced sensitivity calculation is ideally sigma corner dependent but allows the creation of an approximate linear model.

3. MODELS

For each benchmark circuit in the contest, two files are available: a netlist file and a library file. The netlist file contains circuit information, topology, and other circuit related data. The netlist is composed of a set of interconnected elements, namely cell instances and interconnecting circuitry. The library file contains timing information regarding the available cell elements as well as variability information. The models to be encountered during timing analysis, are of two types: interconnect and cell circuit elements.

3.1 Interconnect

The basic instance of interconnect (wire) is a *net*, which is assumed to have an input pin, designated by *port*, and one or more output pins, designated by *taps*, as illustrated in Figure 2 (left). For each net, the netlist of its parasitic RC tree is provided in the netlist file. An example of a parasitic RC tree is presented in Figure 2 (right). Parasitic RC trees only contain grounded capacitors and resistors between nodes in the tree (there are no grounded resistors or floating capacitors).

The computation of port-to-tap delays can be accurately performed through electrical simulation. However, for sake of simplicity, a simpler delay model, namely the Elmore delay model [6] is considered. In this model, the delay is approximated by the value of the first moment of the impulse response. For RC tree networks, [9] provides a simple topological method for computing this value by an appropriate

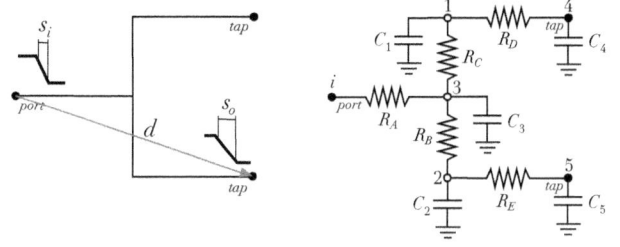

Figure 2: Interconnect characterization.

traversal of the netlist of the parasitic RC tree. For the example net illustrated in Figure 2 (right),

$$d_5 = R_A(C_1 + C_3 + C_4) + (R_A + R_B)\,C_2 + (R_A + R_B + R_E)\,C_5.$$

This value provides the nominal or mean wire delay between the port and the tap. For variation aware parametric delay computation, finite differencing is performed to compute the sensitivity of delay to the metal parameter M. Corner specific metal resistance m_R^σ and metal capacitance m_C^σ scalar values are provided (in the cell library) to obtain the updated resistance and capacitance values of the interconnect network when the metal parameter is set to a given corner ($\Delta M = \sigma$). Each interconnect resistance and capacitance is scaled by the provided scalar, and another deterministic delay computation is performed to compute the delay when ΔM is at the σ corner. It should be noted that the component of tap capacitance that comes from the cell pin capacitance should not be scaled.

For the example above, assuming that the tap capacitance C_5 comes partly from the cell pin capacitance $C_{p,5}$ connected at node 5, the remaining capacitance ($C_5 - C_{p,5}$) is part of the interconnect network, using similar notation for tap capacitance C_4, the delay at this corner is given by

$$\begin{aligned}
d_{5|\Delta M=\sigma} &= m_R^\sigma R_A(m_C^\sigma[C_1 + C_3 + C_4 - C_{p,4}] + C_{p,4}) + \\
&\quad (m_R^\sigma[R_A + R_B + R_E])\,(C_{p,5} + m_C^\sigma(C_5 - C_{p,5})) \\
&\quad + (m_R^\sigma[R_A + R_B])\,m_C^\sigma C_2. \quad (10)
\end{aligned}$$

The above computation is performed for the provided sigma corner of the metal parameter (typically $+3$ sigma). The sensitivity to parameter ΔM is now computed using finite differencing between this corner and the nominal corner as follows.

$$\alpha_{m,5}^D = \frac{d_{5|\Delta M=\sigma} - d_{5|\Delta M=0}}{\sigma - 0} = \frac{d_{5|\Delta M=\sigma} - d_5}{\sigma}. \quad (11)$$

The parametric delay model of the interconnect from the port to tap node 5 thus involves two deterministic delay calculations and is finally approximated to a linear model as

$$d_5 + \alpha_{m,5}^D\Delta M.$$

In this contest, wire delays consequently do not contain any sensitivity to other parameters (including random variation). The above parametric wire delay model as a function of the normal Gaussian parameter ΔM is assumed as the golden model for the contest.

The value of the nominal output slew on any given tap node o can be approximately computed by a two-step procedure. First, one computes the nominal output slew of the impulse response on o, which was observed in [6,8] to be well approximated by the following expression,

$$\hat{s}_o \approx \sqrt{2\beta_o - d_o^2}, \quad (12)$$

173

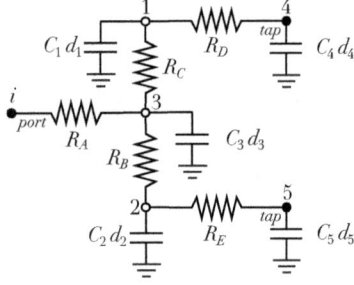

Figure 3: Modified RC tree for computing the second moment of impulse response.

where β_o is the second moment of the impulse response at node o, and d_o is the corresponding Elmore delay for node o. The value of β_o can be computed through the efficient path-tracing algorithm for moment computation proposed in [11], which is a generalization of the algorithm proposed in [9] and described earlier.

For computing β_o, we start by replacing all capacitance values C_k by $C_k d_k$, where d_k is the Elmore delay. Figure 3 illustrates the modified parasitic RC tree for the example of Figure 2. Next, we follow the same procedure as before for computing β_e as

$$\beta_e = \sum_k R_{ke} C_k d_k. \tag{13}$$

Therefore, for the example parasitic RC tree illustrated in Figure 3, we obtain,

$$\begin{aligned} \beta_5 = & \ R_A \left(C_1 d_1 + C_3 d_3 + C_4 d_4 \right) + (R_A + R_B) C_2 d_2 + \\ & (R_A + R_B + R_E) C_5 d_5. \end{aligned} \tag{14}$$

After computing $\hat{s}_o$ from (12), we proceed to compute the mean slew of the response to the input ramp, s_o, for which a good approximation is proposed in [10], given by the simple expression,

$$s_o \approx \sqrt{s_i^2 + \hat{s}_o^2}, \tag{15}$$

where s_i is the nominal or mean input slew.

Parametric output slew calculation involves a bit of complex finite differencing. Under metal variability, $\hat{s}_o$ is a function of parameter ΔM only since both β_o and d_o are dependent on metal scalars. For a metal sigma corner σ,

$$\hat{s}_{o|\Delta M=\sigma} \approx \sqrt{2\beta_{o|\Delta M=\sigma} - d_{o|\Delta M=\sigma}^2}, \tag{16}$$

where, $d_{o|\Delta M=\sigma}$ is the Elmore delay value to the tap at node o at the given metal corner[1] and $\beta_{o|\Delta M=\sigma}$ may be computed by scaling the interconnect resistance and capacitance values similar to (10). The sensitivity to the metal parameter may now be computed via finite differencing the value of $\hat{s}_o$ between the two sigma corners for ΔM, as

$$\alpha_m^{\hat{s}_o} = \frac{\hat{s}_{o|\Delta M=\sigma} - \hat{s}_{o|\Delta M=0}}{\sigma - 0} = \frac{\hat{s}_{o|\Delta M=\sigma} - \hat{s}_o}{\sigma} \tag{17}$$

It follows that an additional deterministic computation for β_o at the metal corner σ is required to compute the parametric output slew $\hat{S}_o$ to an impulse response. We thus have

$$\hat{S}_o = \hat{s}_o + \alpha_m^{\hat{S}_o} \Delta M. \tag{18}$$

[1]already calculated during parametric wire delay calculation in (11)

This model is treated as the golden model for the contest. Given parametric models for S_i (should be available from upstream calculations) and $\hat{S}_o$, it is not straightforward to compute a linear parametric output slew S_o given the non-linearity of (15). Novel approaches to modeling the parametric output slew (possibly as a linear model) are encouraged. For evaluation using a Monte Carlo simulator, samples for S_i and $\hat{S}_o$ would be applied to (15) to obtain a sample for the output slew.

A naive finite differencing based approach to model S_o in a linear parametric form is as follows (it is not required to follow this approach in the contest). It is assumed that S_i is available in the following linear parametric form

$$S_i = s_i + s_m \Delta M + \ldots + s_x \Delta X + \ldots,$$

where, X denotes any non metal parameter (e.g., V, T, ...). For any non-metal parameter (including random) variation ΔX, the sensitivity of output slew to ΔX is computed by perturbing that parameter between its extreme sigma corners and observing the change in the projected value of S_i when all other parameters are at their nominal conditions (0 sigma). Note that $\hat{S}_o$ is not a function of ΔX. Using finite differencing, the sensitivity of the parametric output slew to ΔX is thus given by

$$\begin{aligned} \alpha_x^S &= \frac{s_{o|\Delta X=+3} - s_{o|\Delta X=-3}}{+3 - (-3)} \\ &= \frac{\sqrt{s_{i|\Delta X=+3}^2 + \hat{s}_o^2} - \sqrt{s_{i|\Delta X=-3}^2 + \hat{s}_o^2}}{6} \\ &= \frac{\sqrt{(s_i + 3s_x)^2 + \hat{s}_o^2} - \sqrt{(s_i - 3s_x)^2 + \hat{s}_o^2}}{6}. \end{aligned} \tag{19}$$

For computing the sensitivity to the metal parameter, both S_i and $\hat{S}_o$ need to be perturbed during finite differencing, since both are functions of ΔM. Mathematically,

$$\begin{aligned} \alpha_m^S &= \frac{s_{o|\Delta M=+3} - s_{o|\Delta M=0}}{+3 - 0} \\ &= \frac{\sqrt{s_{i|\Delta M=+3}^2 + \hat{s}_{o|\Delta M=+3}^2} - \sqrt{s_{i|\Delta M=0}^2 + \hat{s}_{o|\Delta M=0}^2}}{3} \\ &= \frac{\sqrt{(s_i + 3s_m)^2 + (\hat{s}_o + 3\alpha_m^{\hat{S}_o})^2} - \sqrt{s_i^2 + \hat{s}_o^2}}{3} \\ &= \frac{\sqrt{(s_i + 3s_m)^2 + (\hat{s}_o + 3\alpha_m^{\hat{S}_o})^2} - s_o^2}{3}. \end{aligned} \tag{20}$$

The output slew is thus expressed in the following parametric linear form.

$$S_o = s_o + \alpha_m^S \Delta M + \ldots + \alpha_x^S \Delta X + \ldots,$$

where, X denotes any non metal parameter (e.g., V, T, ...).

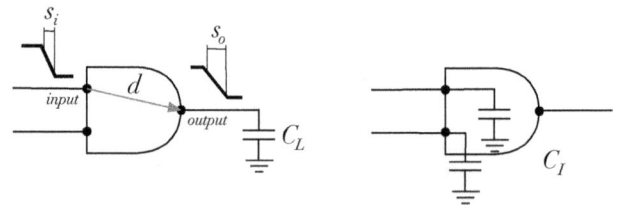

Figure 4: Combinational cell characterization.

174

3.2 Combinational cells

In the contest, it is assumed that cell delay, D, and output slew, S_o, can be approximated, for a given combinational cell input/output pin pair, by the following formulas.

$$D = a(1 + k_{d,v}\,\Delta V + k_{d,t}\,\Delta T + k_{d,l}\,\Delta L + k_{d,w}\,\Delta W \\ + k_{d,h}\,\Delta H + k_{d,r}\,\Delta R) + b\,C_L + c\,S_i \quad (21)$$

$$S_o = x(1 + k_{s,v}\,\Delta V + k_{s,t}\,\Delta T + k_{s,l}\,\Delta L + k_{s,w}\,\Delta W \\ + k_{s,h}\,\Delta H + k_{s,r}\,\Delta R) + y\,C_L + z\,S_i \quad (22)$$

where a, b, c, x, y, z and each k_i term are cell-dependent constants; C_L and S_i are the parametric output load and parametric input slew, respectively. The cell dependent constants are provided for each cell and for rise/fall transition, in the cell library file. Note that the k_i terms represent sensitivity to front end parameters as a fractional value of a or x. As an example, the sensitivity of delay to parameter ΔV would be given by $(a * k_{d,v} + c * s_v)$, where s_v denotes the sensitivity of input slew S_i to ΔV. However, the delay sensitivity to random variation ΔR is given by $\sqrt{(a * k_{d,r})^2 + (c * s_r)^2}$.

C_L denotes the equivalent downstream capacitance seen from the output pin of the cell. Several sophisticated models have been proposed for computing C_L. For simplicity, the application of such models is considered to be out of the scope of the present contest, and a simple model is adopted. C_L is assumed to be the sum of all the capacitances in the parasitic RC tree including cell pin capacitances at the taps of the interconnect network.

$$C_L = \sum_k C_k \quad (23)$$

For the example net illustrated in Figure 2 (right), at the nominal metal corner, we trivially have

$$C_{L|\Delta M=0} = C_1 + C_2 + C_3 + C_4 + C_5. \quad (24)$$

Since the interconnect capacitances are dependent on ΔM, C_L is also a function of ΔM. In the above example, at a given sigma corner σ for ΔM (from the cell library),

$$C_{L|\Delta M=\sigma} = m_C^\sigma [C_1 + C_2 + C_3 + (C_4 - C_{p,4}) \\ + (C_5 - C_{p,5})] + C_{p,4} + C_{p,5}, \quad (25)$$

where, $C_{p,4}$ and $C_{p,5}$ denote the cell pin capacitance contributions in the tap capacitances C_4 and C_5, respectively, and m_C^σ is the interconnect capacitance scalar value obtained from the cell library to denote impact of metal variation at corner σ. Cell input pin capacitances are provided as fixed values for each pin and for each rise/fall transition, in the cell library file.

The effective load's sensitivity to ΔM is computed via finite differencing as

$$l_m = \frac{C_{L|\Delta M=\sigma} - C_{L|\Delta M=0}}{\sigma - 0} = \frac{C_{L|\Delta M=\sigma} - C_{L|\Delta M=0}}{\sigma}. \quad (26)$$

The effective load C_L is then expressed in a linear parametric form as

$$C_{L|\Delta M=0} + l_m \Delta M,$$

and plugged into the cell delay and slew models in (21) and (22), thus making cell delay and slew a function of ΔM.

3.3 Flip-flops

Sequential circuits consist of combinational blocks interleaved by registers, usually implemented with flip-flops. Typically they are composed of several stages, where a register captures data from the outputs of a combinational block and injects it into the inputs of the combinational block in the next stage. Register operation is synchronized by clock signals generated by one or multiple clock sources. Clock signals that reach distinct flip-flops (sinks in the clock tree) are delayed from the clock source by a given *clock latency*, that we will designated by l.

A flip-flop (D flip-flop, more specifically) is a storage element that captures a given logic value at its input data pin D, when a given clock edge is detected at its clock pin CK, and subsequently presents the captured value and its complement at the output pins Q and $\overline{Q}$. The flip-flop also enables asynchronous preset (set) and clear (reset) of the output pins, through the S and R input pins. Proper operation of a flip-flop requires the logic value of the input data pin to be stable for a specific period of time *before* the earliest capturing clock edge. This period of time is called *setup time*, and we will represent it by t_{setup}. Additionally, the logic value of the input data pin must also be stable for a specific period of time *after* the latest capturing clock edge. This period of time is called *hold time*, and we will represent it by t_{hold}. Setup/hold times are one of the standard performance figures provided in cell specification libraries for storage elements. Other figures include delay from clock to output, $d_{CK \to Q}/d_{CK \to \overline{Q}}$ and asynchronous preset and clear delays, d_{preset} and d_{clear}. All flip-flop standard timing figures are illustrated in Figure 5.

Setup and hold constraints are modeled as functions of the input slews at both the clock pin, CK, and the data input pin, D, respectively, as follows.

$$t_{setup} = g + h\,S_i^{CK} + j\,S_i^D \quad (27)$$

$$t_{hold} = m + n\,S_i^{CK} + p\,S_i^D \quad (28)$$

In the above equations, the input slews are parametric, and consequently, the setup and hold times are parametric as well. Let us consider the usual case of signal propagation between two flip-flops, as illustrated in Figure 6. Assuming that the clock edge is generated in the clock source at time 0, then it will reach the injecting flip-flop at time l_i, making the data available at the input of the combinational block $d_{CK \to Q}$ time later. If the propagation delay in the combinational block is d_{comb}, then the data will be available at the input of the capturing flip-flop at time $l_i + d_{CK \to Q} + d_{comb}$. Assuming the clock period to be a deterministic constant T, then the next clock edge will reach the capturing flip-flop at time $T + l_o$. For correct operation, the data must be be available at the input of the capturing flip-flop t_{setup} before the next clock edge reaches the capturing flip-flop. Therefore, at the data input pin, D, we have the following.

$$at_D^{late} = l_i^{late} + d_{CK \to Q} + d_{comb}^{late} \quad (29)$$

$$rat_{setup} = rat_D^{late} = T + l_o^{late} - t_{setup} \quad (30)$$

A similar condition can be derived for ensuring that the hold time is respected. The data input of the capturing flip-flop must remain stable for at least t_{hold} after the clock edge reaches the corresponding CK pin. Consequently, at the data input pin, D, we have the following.

$$at_D^{early} = l_i^{early} + d_{CK \to Q} + d_{comb}^{early} \quad (31)$$

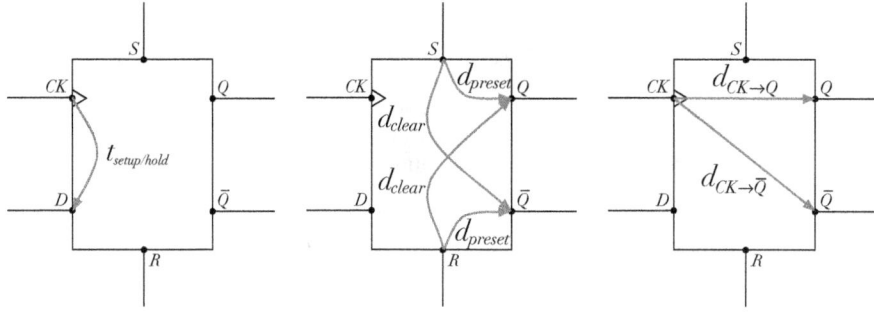

Figure 5: Flip-flop characterization.

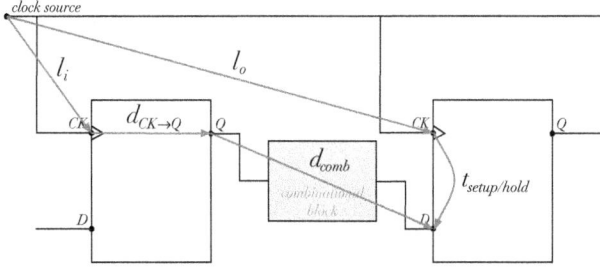

Figure 6: Propagation between two flip-flops.

$$rat_{hold} = rat_D^{early} = l_o^{early} + t_{hold} \qquad (32)$$

The previous arrival times and required arrival times induce setup and hold slacks, which can be computed from (5) and (6). Preset and clear values are presented for completeness, and are ignored for the timing analysis task proposed in this contest.

4. EVALUATION

4.1 Projection of timing quantities

Various projection techniques are possible for any parametric timing quantity T expressed as

$$\mu + a_v\Delta V + a_t\Delta T + a_l\Delta L + a_w\Delta W + a_h\Delta H + a_m\Delta M + a_r\Delta R.$$

Three projection techniques are required in this contest for accuracy evaluation and are as follows (the first two are primarily for debug purposes).

1. MEAN_ONLY: In this projection approach, only the nominal or mean value μ of T is desired.

2. SIGMA_ONLY: In this projection approach, the standard deviation σ_T of T is desired. σ_T is computed as $\sqrt{a_v^2 + a_t^2 + a_l^2 + a_w^2 + a_h^2 + a_m^2 + a_r^2}$.

3. WORST_CASE: In this projection approach, all device parameter sensitivities are first root sum squared to a *device* sensitivity a_d. Next, a_d, the random sensitivity a_r, and the metal sensitivity a_m are each projected to a $+3$ or -3 sigma corner to reflect a worst case projection of T. Mathematically, $a_d = \sqrt{a_v^2 + a_t^2 + a_l^2 + a_w^2 + a_h^2}$. For all late mode timing quantities like late mode arrival time or slew, the returned value is $(\mu + 3|a_d| + 3|a_m| + 3|a_r|)$, while for all early mode quantities like early mode arrival time or slew and all slacks (both early and late mode), the returned value is $(\mu - 3|a_d| - 3|a_m| - 3|a_r|)$.

Projected arrival times and slacks for various nodes in the circuit are used for accuracy evaluation. Details of expected tool output and a set of benchmarks are available on the contest web-page [2].

4.2 Evaluation metric

The utilization of parallelization techniques (especially multi-threading) is strongly encouraged for the contest. Entries are judged based on a weighted combination of accuracy of results, tool run-time, and tool peak-memory consumption. The results from a Monte Carlo simulator are used as the golden reference for accuracy. Several inaccuracy range buckets (e.g., $0-5\%$, $5-10\%$, $10-20\%$) are assumed wherein results within any bucket are awarded the same score, and results in larger inaccuracy buckets are awarded lower scores. Results outside the largest inaccuracy bucket are awarded 0 points irrespective of their run-time and memory consumption. Inaccuracy measurements are based on both average inaccuracy and peak inaccuracy across circuit nodes.

At least half of the final score is based on run-time and peak memory. Elapsed time will be used instead of CPU time, thus enabling contestants to take advantage of parallelization techniques. Numerous runs over larger testcases will be executed, so as to enable accurate elapsed time measurements. Since netlist and library reading/parsing, as well as output generation times could be a non-negligible fraction of the overall run time, contestants are encouraged to apply parallelization/multi-threaded techniques for these steps as well.

5. FILE FORMATS

5.1 Netlist

The netlist file, which contains the description of the circuit topology, is formatted as follows.

```
input <node>
output <node>
instance <cell name> <pin name>:<node> ...
wire <port node> <tap node> ... <tap node>
    res <node> <node> <resistance>
    ...
    cap <node> <capacitance>
    ...
slew <node> <slew fall> <slew rise>
clock <node> <period>
at <node> <at fall early> <at fall late> \
        <at rise early> <at rise late>
rat <node> <mode> <rat fall> <rat rise>
```

Keywords:

- input, primary input node;

- output, primary output node;

- instance, cell instance;

- wire, interconnect net;

- res, cap, resistor and capacitor of a parasitic RC tree (can appear in any order);

- slew, input slew at the primary inputs;

- clock, clock input information (node and period);

- at, arrival time constraint (only used for primary input nodes);

- rat, required arrival time constraint.

Variable fields:

- <node>, <port node> and <tap node> are node names, of up to 64 characters in length, which can contain alphanumeric characters, the underscore or the dash (the first character must be a letter);

- <cell name> is the name of the library cell (exactly as it will appear in the cell library file), of up to 32 characters in length, which can contain only alphanumeric characters (the first character must be a letter);

- <pin name> is the name of a pin of the cell (exactly as it will appear in the cell library file), of up to 32 characters in length, which can contain only alphanumeric characters;

- <resistance> is the value of the resistance in Ohms, represented in scientific notation;

- <capacitance> is the value of the capacitance in Farads, represented in scientific notation;

- <slew fall> and <slew rise> are the fall and rise slews for the corresponding primary input, in seconds, represented in scientific notation. Early and late slews at the inputs are assumed to be identical;

- <period> is the clock period in seconds, represented in scientific notation;

- <at fall early>, <at fall late>, <at rise early> and <at rise late> are real numbers, represented in scientific notation, which represent arrival time constraints for fall/rise transitions in early/late mode, at the primary inputs, in seconds;

- <mode>, is the mode of operation and can be either early or late;

- <rat fall> and <rat rise> are real numbers, represented in scientific notation, which represent required arrival time constraints for rise/fall transitions and early/late mode, in seconds.

If no input slew (arrival time) is defined for any given primary input, it should be assumed to be 1e-12 (0), for both fall and rise transitions. The design will have one clock input pin (one clock domain) at most. The netlist file does not contain any parametric values, and a deterministic timing quantity (e.g., arrival time) may be considered a parametric quantity with 0 sensitivity to all parameters.

5.2 Cell Library

The cell library file, that contains the timing information of each cell, is formatted as follows.

```
metal <sigma corner> <resistance scale factor> \
        <capacitance scale factor>
...
cell <cell name>
  pin <pin name> input <fall cap> <rise cap>
  pin <pin name> output
  pin <pin name> clock
...
  timing <i/p pin name> <o/p pin name> <sense> \
        <fall slw> <rise slw> <fall dly> <rise dly>
  setup <clk pin name> <i/p pin name> <edge type> \
        <fall constraint> <rise constraint>
  hold <clk pin name> <i/p pin name> <edge type> \
        <fall constraint> <rise constraint>
```

Keywords:

- metal, metal parameter scalars at given sigma corner;

- cell, start of cell definition;

- pin, start of pin definition;

- input, output and clock, pin type;

- timing, delay;

- setup, setup time;

- hold, hold time;

Variable fields:

- <sigma corner> is the sigma corner value (σ) of metal parameter ΔM for which resistance and capacitance scale factors are provided.

- <resistance scale factor> is the value (m_R^σ) by which the nominal interconnect resistance provided in the netlist should be scaled at the given metal sigma corner;

- <capacitance scale factor> is the value (m_C^σ) by which the nominal interconnect capacitance provided in the netlist should be scaled at the given metal sigma corner;

- <cell name> is the name of the cell, of up to 32 characters in length, which can contain only alphanumeric characters (the first character must be a letter);

- <pin name> is the name of a pin of the cell, of up to 32 characters in length, which can contain only alphanumeric characters (the first character must be a letter);

- <fall cap> and <rise cap> are values of the pin's input capacitances in Farads, for rise/fall transitions, represented in scientific notation;

- <i/p pin name>, <o/p pin name> and <clk pin name> are the names of the input, output and clock pins of a given delay or constraint specification, of up to 32 characters in length, which can contain only alphanumeric characters (the first character must be a letter);

- <sense>, can be one of:

- `positive_unate`, transition direction is maintained from input to output (rise→rise, fall→fall);
- `negative_unate`, transition direction is reversed from input to output (rise→fall, fall→rise);
- `non_unate`, transition direction cannot be inferred from a single input (take the worst, among rise/fall);

- `<fall slw>`, `<rise slw>`, are each given by 9 real numbers separated by white spaces, which correspond to the parameters x, y, z, $k_{s,v}$, $k_{s,t}$, $k_{s,l}$, $k_{s,w}$, $k_{s,h}$, and $k_{s,r}$ of Eqn. (22) (fall/rise refers to the transition direction in the output pin);

- `<fall dly>` and `<rise dly>`, are each given by 9 real numbers separated by white spaces, which correspond to the parameters a, b, c, $k_{d,v}$, $k_{d,t}$, $k_{d,l}$, $k_{d,w}$, $k_{d,h}$, and $k_{d,r}$ of Eqn. (21) (fall/rise refers to the transition direction in the output pin);

- `<edge type>`, can be one of:
  - `falling`, constraint applies to the falling clock edge;
  - `rising`, constraint applies to the rising clock edge;

- `<fall constraint>` and `<rise constraint>`, are each given by 3 real numbers separated by white spaces, which correspond to the parameters g, h and j of (27), or m, n and p of (28), if we are dealing with setup constraints or hold constraints, respectively.

6. CONCLUSIONS

Variation aware timing analysis is expected to be employed at various stages of any modern deep sub-micron chip design-closure flow. Reducing the run-time of this analysis is imperative to designer productivity. The aim of the TAU 2013 variation aware timing contest is to seek novel ideas for fast variation aware timing analysis, by means of the following: (a) increase awareness of variation aware timing analysis and provide insight into some challenging aspects of the analysis, (b) encourage novel parallelization techniques (including multi-threading) for timing analysis, and (c) facilitate creation of a publicly available variation aware timing analysis framework and benchmarks to further advance research in this area including common path pessimism reduction (CPPR), incremental timing, and spatial variations.

Simplified timing models for gates and wires are used in the contest specifically to steer contest focus towards development of novel run-time reduction techniques (e.g., parallelization or multi-threaded techniques) and towards development of novel techniques to handle impact of variability on timing.

Acknowledgments

The netlist and library formats used in the contest are extensions of those used in the 2011 "International Workshop on Power and Timing Modeling, Optimization and Simulation (PATMOS)" timing analysis contest [1]. The authors acknowledge the help and support of the PATMOS 2011 contest committee for providing sample benchmarks and library.

The authors also acknowledge Prof. Yao-Wen Chang and his research group from National Taiwan University, Taiwan for sharing the source code of their deterministic timing analysis tool (winners of [1]) with the participants of the TAU 2013 variation aware timing contest to avoid a complete re-write of the netlist and library file parsers.

7. REFERENCES

[1] *2011 PATMOS timing analysis contest.* http://patmos-tac.inesc-id.pt.

[2] *2013 TAU variation aware timing analysis contest.* https://sites.google.com/site/taucontest2013/.

[3] D Blaauw, K Chopra, A Srivastava, and L Scheffer. Statistical Timing Analysis: From Basic Principles to State-of-the-art. In *IEEE Transactions on Computer-Aided Design, 27(4) April 2008*, pages 589–607.

[4] H. Chang and S. S. Sapatnekar. Statistical timing analysis considering spatial correlations using a single PERT-like traversal. In *International Conference on Computer-Aided Design*, pages 621–625, 2003.

[5] C. E. Clark. The greatest of a finite set of random variables. In *Operations Research, Vol. 9, No. 2 (Mar - Apr)*, pages 145–162, 1961.

[6] W. C. Elmore. The Transient Response of Damped Linear Networks with Particular Regard to Wide-Band Amplifiers. *Journal of Applied Physics*, 19(1):55–63, January 1948.

[7] R. Goering. *Why Multi-Mode, Multi-Corner (MMMC) ECO Closure Requires a New Signoff Approach.* http://www.cadence.com/Community/blogs/ii/archive/2012/08/06/why-multi-mode-multi-corner-mmmc-eco-closure-requires-a-new-signoff-approach.aspx.

[8] R. Gupta, B. Tutuianu, and L. T. Pileggi. The Elmore Delay as a Bound for RC Trees with Generalized Input Signals. *IEEE Transactions on Computer-Aided Design of Integrated Circuits and Systems*, 16(1):95–104, January 1997.

[9] P. Penfield Jr. and J. Rubinstein. Signal Delay in RC Tree Networks. In *Design Automation Conference*, pages 613–617, 1981.

[10] C. V. Kashyap, C. J. Alpert, F. Liu, and A. Devgan. Closed-Form Expressions for Extending Step Delay and Slew Metrics to Ramp Inputs for RC Trees. *IEEE Transactions on Computer-Aided Design of Integrated Circuits and Systems*, 23(4):509–516, April 2004.

[11] C. L. Ratzlaff and L. T. Pillage. RICE: Rapid Interconnect Circuit Evaluation Using AWE. *IEEE Transactions on Computer-Aided Design of Integrated Circuits and Systems*, 13(6):763–776, June 1994.

[12] D. Sinha, H. Zhou, and N. V. Shenoy. Advances in computation of the maximum of a set of Gaussian random variables. In *IEEE Transactions on Computer-Aided Design, 26(8) August 2007*, pages 1522–1533.

[13] C. Visweswariah, K. Ravindran, K. Kalafala, S. G. Walker, and S. Narayan. First-order incremental block-based statistical timing analysis. In *Design Automation Conference*, pages 331–336, 2004.

Opportunities and Challenges for High Performance Microprocessor Designs and Design Automation

Ruchir Puri

IBM T. J. Watson Research Center

ruchir@us.ibm.com

ABSTRACT

With end of an era of classical technology scaling and exponential frequency increases, high end microprocessor designs and design automation methodologies are at an inflection point. With power and current demands reaching breaking points, and significant challenges in application software stack, we are also reaching diminishing returns from simply adding more cores. In design methodologies for high end microprocessors, although chip physical design efficiency has seen tremendous improvements, strong indications are emerging for maturing of those gains as well. In order to continue the cost-performance scaling in systems in light of these maturing trends, we must innovate up the design stack, moving focus from technology and physical design implementation to new IP and methodologies at logic, architecture, and at the boundary of hardware and software, solving key bottlenecks through application acceleration. This new era of innovation, which moves the focus up the design stack presents new challenges and opportunities to the design and design automation communities. This talk will motivate these trends and focus on challenges for high performance microprocessor design and design automation in the years to come.

Categories and Subject Descriptors

B.7.2 [**Integrated Circuits**]: Design Aids – *layout, placement and routing, simulation, verification.*

General Terms: Algorithms, Design, Verification.

Keywords: VLSI, Microprocessor, Design, Design Automation.

Bio

Ruchir Puri is an IBM Fellow at Thomas J Watson Research Center, Yorktown Hts, NY where his efforts have focused on high performance design and methodology solutions for all of IBM's enterprise server and system chip designs. Most recently, he lead the design methodology innovations for IBM's latest Power7 and zEnterprise microprocessors and is currently leading design methodology research efforts on future processors. Ruchir has received numerous IBM awards including the highest technical honor – IBM Fellow, which was awarded for his transformational role in microprocessor design methodology. In addition, he has received "Best of IBM" awards in both 2011 and 2012 and IBM Corporate Award from IBM's CEO, and several IBM Outstanding Technical Achievement awards.

Dr. Puri is a Fellow of the IEEE, a member of IBM Academy of Technology and IBM Master Inventor, an ACM Distinguished Speaker and IEEE Distinguished Lecturer. He is recipient of SRC outstanding mentor award and has been an adjunct professor at Dept. of Electrical Engineering, Columbia University, NY and was also honored with John Von-Neumann Chair at Institute of Discrete Mathematics at Bonn University, Germany.

To Do or Not To Do Hierarchical Timing?

Florentin Dartu
Synopsys Inc.
fdartu@synopsys.com

Qiuyang Wu
Synopsys Inc.
qwu@synopsys.com

ABSTRACT

The latest design specs have arrived and it is now clear that timing runs will not fit in our machines. What to do! Is hierarchical STA the perfect solution? Will it work exactly like flat? How will it interact with the hierarchical design implementation? This talk will present the advantages and limitations of using hierarchical STA and how it impacts the design process.

Categories and Subject Descriptors

B.7.2 [**Integrated Circuits**]: Design Aids – *simulation, verification.*

General Terms

Algorithms, Design, Verification.

Keywords

VLSI, Design Automation, Timing Analysis.

Variability Aware Hierarchical Implementation of Big Chips

Vidyamani Parkhe
Mentor Graphics, Fremont CA
vidyamani_parkhe@mentor.com

ABSTRACT

Over the last decade, transistor dimensions have shrunk rapidly by a factor of ten, while chip sizes have increased many folds and so have the scenarios in which they are expected to function. Silo technology development of yester years has all but disappeared as clear boundaries between various stages of the flow evaporate-- need for resynthesis can be felt as late as post routing and necessity for understanding signal integrity and manufacturing as early as pre-placement. This leads designers to explore newer techniques to meet their power and performance targets and EDA vendors to retool and reinvent algorithms for sub teen nanometer technologies.

In classical implementation tools which invested heavily in placement and routing technologies while optimization steps resolved to simple buffering and sizing techniques, RTL synthesis and technology mapping had to do the heavy lifting in terms of achieving quality of results. In contrast, optimization techniques in current P&R tools are poised to take center stage: ranging from traditional buffering, shielding, sizing and remapping to more physically aware methodologies which involve topology planning, wire spreading, timing aware location reassignment and VT assignment. Hand held and mobile devices running power hungry apps, necessitate the need to trade off performance and area metrics for lure of longer usage and stand-by times. Designers of chips and SoCs at the heart of these gadgets squeeze out every last milli-Watt by innovative and aggressive usage of clocking, multi-voltage and switchable regions and islands.

Growing design sizes and shrinking project cycles makes is imperative for designers to employ replicated partitions that co-exists and interact amongst themselves and other modules on chip. Complex timing constraints and critical timing paths necessitate the need to analyze complete chips in memory, aided by timing abstractions to battle a multitude of timing scenarios co-existing and at times competing for valuable real estate and routing resources. Faux pas and simplified assumptions made at floor planning, pin assignment, synthesis, initial placement, clock tree building stages need to be revised. Tight schedules and phased tape-outs demand methodologies like metal-only ECOs and spare cell optimization to replace manual or script based iterative approaches to squeeze out every last picosecond of performance.

In this talk we will look at various such methodologies designer and EDA vendors use to conquer these ever exploding challenges.

Categories & Subject Descriptors
B.7.2 [Integrated Circuits]: Placement and Routing

Author Keywords
Timing Optimization, Place & Route, Algorithms, Full Chip, Complex Timing Scenarios and Abstractions

Bio
The speaker is an Architect at Mentor Graphics and Technical Lead of Olympus-SoC Optimization team with 12 years of experience in EDA industry. His background includes timing optimization, synthesis, static timing analysis, parallel and randomized algorithms. His primary responsibility involves architecture and development of robust algorithms for performance and power optimization and research of innovative techniques in the area of timing optimization and physical synthesis. His passion is clean and scalable software architecture, devising simple solutions for complex problems and writing C++11 code. He received his Bachelors degree in Electrical Engineering from University of Pune and M.S. in Computer Science from University of Florida, Gainesville.

Challenges in Managing Timing and Wiring Contracts during Hierarchical Floorplanning and Design Closure

Shyam Ramji
IBM Systems and Technology Group
Hopewell Junction, NY
ramji@us.ibm.com

ABSTRACT

The growing complexity and size of designs have driven chip implementation teams to adopt hierarchical design methodologies that divide-and-conquer the design closure task. Wherein, the large chip is partitioned into physical blocks with boundary constraints for physical synthesis which are then integrated at the top-level to achieve overall design closure. Each block could be further partitioned until the block sizes are manageable from a tool turnaround time perspective. A key aspect to efficient hierarchical design involves generation of realistic block-level timing budgets and physical constraints (contracts) to enable parallel implementation of each block. In a typical hierarchical design flow, these block-level contracts are created based on early estimations of timing and wiring from the floorplanning phase but evolve as the design progresses, often requiring several iterations of design integration to converge. High-performance designs require efficient contract management while allowing seamless design closure optimizations across levels of hierarchy. This talk will focus on some of the challenges in managing contracts with emphasis on the implications for physical synthesis tools used in hierarchical design closure.

Categories and Subject Descriptors

B.7.1 [**Integrated Circuits**]: Types and Design Styles - VLSI (very large scale integration), Microprocessors and microcomputers

General Terms

Design, Management, Performance

Keywords

Hierarchical Design, Physical Synthesis

ISPD'13, March 24–27, 2013, Stateline, Nevada, USA.
ACM 978-1-4503-1954-6/13/03.

Author Index

www.ingramcontent.com/pod-product-compliance
Lightning Source LLC
Chambersburg PA
CBHW081527220326
41598CB00036B/6354